Geometric Transformations for 3D Modeling

Michael E. Mortenson

Industrial Press Inc.

New York

Library of Congress Cataloging-in-Publication Data

Mortenson, Michael E., 1939-
  Geometric transformations for 3D modeling / Michael Mortenson. -- 2nd ed.
    p. cm.
  Originally published: Geometric transformations. 1995.
  ISBN 978-0-8311-3338-2
  1. Transformations (Mathematics) I. Mortenson, Michael E., 1939-
Geometric transformations. II. Title. III. Title: Geometric transformations
for three dimensional modeling.
  QA601.M85 2007
  006.6'93015161--dc22

2007006588

Industrial Press Inc.
989 Avenue of the Americas
New York, NY 10018

Copyright © 2007 by Industrial Press Inc., New York,
Printed in the United States of America. All rights reserved.
This book, or parts thereof may not be reproduced, stored in a retrieval
system, or transmitted in any form without the permission of the publisher.

1   2   3   4   5   6   7   8   9   10

TO

JAM

Contents

1. Geometry   1

1.1 What Is Geometry?   1
What kind of answer?, fractals, Kock snowflake, limits of intuition and visualization

1.2 History   5
Origins in Babylon and Egypt, Rhind papyrus, Euclid's *Elements*, geometry as a logical system, Archimedes, René Descartes, Riemann, Klein, Lobachevsky, Euler, non-Euclidean geometry

1.3 Geometric Objects   10
Geometric objects and nongeometric objects, continuous sets, Mandelbrot sets, figure and form, lines, planes, curves, surfaces, polygons, polyhedra, vectors, tensors

1.4 Space   11
Euclidean space, abstract spaces, coordinate systems in the plane and their notation, rectangular and oblique coordinate systems, curvilinear coordinate systems, Riemann spaces

1.5 Geometry Is ...   14
Coordinate-free geometry, ways of thinking about geometry: visually synthetically, analytically, abstractly, metrically

1.6 *E Pluribus Unum* – Transformation and Invariance   20
Classification of geometries, geometry of incongruent figures, geometry of similar figures, affine geometry, projective geometry, topology, invariance of distance with transformation of coordinates

2. Theory of Transformations   26

2.1 Functions, Mappings, and Transformations   26
Domain and range, a function machine, *into* and *onto* mappings, mapping $\Re$ into and onto itself, one-to-one functions, point-transformation machines, transformation of a set onto itself, the product of transformations, composition of functions, associative property, and more

2.2 Linear Transformations — 44
Systems of linear equations, the Einstein convention, matrices and linear transformations, singular and nonsingular transformations

2.3 Geometric Invariants — 51
The group of homogeneous linear substitutions for coordinates of a point, fixed points; line-preserving, distance-preserving, angle-preserving, and orientation-preserving transformations

2.4 Isometries — 54
Simple isometries, reflections in the plane, reflections in a line, reflection is orientation-reversing, inversion in a point and halfturns in the plane, the product of two halfturns, the product of three halfturns, the product of two reflections in the plane, the product of three or more reflections in the plane, glide reflection, isometries in space, the group of isometries, rotation followed by translation

2.5 Similarities — 75
Similar triangles, similar but oppositely oriented triangles, a product of a dilation and isometry, dilations in the plane, the product of two dilations with the same center, the product of two dilations about different centers, dilations in space, the group of similarities

2.6 Affinities — 82
Affine deformation of a cube, affine geometry in the plane, shear, area-preserving shear transformation, strain, equiareal transformations, distance in affine geometry, the affine group, parallel projection

2.7 Projectivities — 96
The equations of projective transformations, central projection of a plane figure, central projection of a figure in space

2.8 Topological Transformations — 99
Topological invariants, Möbius strip, Klein bottle, sidedness, simple arcs, knottedness, nonlinear transformations

3. Vector Spaces — 101

3.1 Introduction to Linear Vector Spaces — 101
Linear vector spaces, linear dependence, linear dependence of three vectors in the plane, vectors qua vectors, parallelogram law, vector components, vector magnitude, direction cosines, scalar product, vector product, vector geometry

3.2 Basis Vectors — 114
A three-dimensional basis, an oblique coordinate system, change of basis, basis vectors and coordinate systems, affine transformation of a Cartesian coordinate system, scaling parameters for an affine transformation, parallel projections, objectivity of a vector, reciprocal basis vectors, orthogonal basis vectors and matrices

3.3 Eigenvalues and Eigenvectors — 128
Homogeneous affine transformation of a point, eigenvectors (a geometric interpretation) the characteristic equation, similarity transformations, symmetric transformations, diagonalization of matrices

3.4 Tensors — 136
Contravariant and covariant vectors, index notation, Einstein summation convention, the Kronecker delta, contravariant and covariant vectors in an orthonormal coordinate system, tensor notation, the metric tensor (two views)

4. Rigid-Body Motion — 145
4.1 Translation — 145
Vectors and translation, translation of a line, coordinate system translation, a succession of translations, unequal translations over two points, invariants under translation, the translation group

4.2 Rotation — 152
A proper rotation, a direct isometry, rotation about the origin, rotation conventions, geometry of a plane rotation about the origin, rotation and the polar coordinate system, successive rotations about the origin, rotation of a line segment about the origin, rotating a line, rotation about an arbitrary point, successive rotations about two different points, rotation of the coordinate system, rotations about the principal axes in space, angle of rotation about the principal axes, successive rotations about the principal axes, rotation of the coordinate system in space, Euler angle rotations, roll, pitch, and yaw, rotation about an arbitrary axis in space, equivalent rotations in space, eigenvectors and equivalent rotations, rotation groups, rotations and quaternions,

4.3 Composite Motion — 185
Homogeneous coordinates, the homogeneous transformation matrix, the screw transformation, sweeps, sweeps with twists

4.4 Kinematics — 190
Degrees of freedom, joint motions, links, articulated mechanism, local coordinate systems, multiple joint positions for a given end position

5. Reflection and Symmetry — 193

5.1 Central Inversion — 195
Inversion fixing the origin, inversion fixing an arbitrary point in space, the product of two inversions, the product of three inversions, equivalent points of inversion, center of symmetry

5.2 Reflections in the Plane — 204
Reflection fixing a line parallel to a principal axis, reflection fixing an arbitrary line through the origin, reflection fixing the principal diagonals in the x, y plane, reflection fixing an arbitrary line, reflection in two parallel lines, reflection in three parallel lines, reflection in intersecting lines, oblique reflection, lines of symmetry

5.3 Reflections in Space — 212
Reflection fixing a line in space, reflection in a principal axis, reflection fixing a plane parallel to a principal plane, reflection in a plane parallel to a principal plane, reflection fixing an arbitrary plane, reflection in a plane through the origin, reflection in an arbitrary plane, vector solutions to reflections, product of reflection in parallel planes, reflection in intersecting or perpendicular planes

5.4 Summary of Reflection Matrices — 219
Reflection in the y axis, in the x axis, central inversion, reflection in the yz plane, in the xz plane, in the xy plane, halfturn about the x axis, halfturn about the y axis, halfturn about the z axis

5.5 Symmetry Basics — 221
Six possible symmetries of an equilateral triangle, centers, lines, and planes of symmetry, analysis of symmetry

5.6 Symmetry Groups — 226
Symmetry group for a frieze pattern, symmetry group for a parquet pattern, the cyclic groups, the general cyclic group, the dihedral groups, symmetry of a square, the general dihedral group $\mathbf{D_n}$, finite symmetry groups on space, cyclic symmetry in space, equivalence of a reflection and a halfturn

5.7 Ornamental Groups — 235
The seven frieze groups, the F-notation, classifying frieze patterns, 17 wallpaper groups, a lattice defined by two nonparallel translations, the crystallographic restriction, plane lattices, unit cells in a lattice

5.8 Polygonal Symmetry and Tiling — 247
Examples of tilings, regular tilings, Schläfli symbol, congruent tilings, duals of regular tilings, three regular tilings theory, semiregular tilings, a variety of vertices in tilings

Contents

5.9 Polyhedral Symmetry 253
Rotational symmetry, vertex and face-center correspondences, polyhedral duals, the cube and octahedron duals, the 4.3.2, 5.3.2, and 3.2 symmetry groups, the rotational symmetry of a cube, the tetrahedron and 3.2 symmetry, truncations of the tetrahedron

6. More Linear Transformations 257

6.1 Isotropic Dilation 257
Uniform expansion or contraction about a fixed point, isotropic dilation fixing the origin, uniform expansion, centerless uniform expansion, isotropic dilation fixing an arbitrary point, the product of two dilations with the same center, the product of two dilations with different centers

6.2 Anisotropic Dilation 262
Anisotropic dilation fixing the origin, anisotropic dilation in the plane, general orthogonal dilations, anisotropic dilation fixing an arbitrary point, anisotropic dilation fixing a line in the plane, anisotropic dilation fixing a line in space, the product of anisotropic dilations

6.3 Shear 268
A slippery deck of cards, shear in the plane fixing the x axis, rotation as a product of three shears, shear in space, equiareal transformations

6.4 Projective Geometry 272
Brunelleschi, Klein, algebraic foundations, Desargues, invariant properties, projection theorems of trigonometry, vector projection, projective collineations, equations of projective geometry, the projective plane, Euclidean three-space mapped onto the projective plane, homogeneous coordinates and projective transformations, homogeneous coordinates of points, parallel lines, ideal points and homogeneous coordinates, special points on the projective plane, and more

6.5 Parallel Projection 287
Parallel projection onto a plane, parallel projection onto a line, parallel projection of a line, orthogonal projection of a line, nonorthogonal projection, parallel projection of a plane, parallel projection of points in space, oblique projection of a point onto a plane

6.6 Central Projection 294
Central projection and perspective drawing, central projection of parallel lines, central projection of intersecting lines, central projection of parallel planes, central projection of intersecting planes, central projection of a line, central projection of points in space, vector solution for a general perspective projection, telescopic projection, thin-lens geometry

6.7 Map Projections 301
Area-preserving mappings, geodesic mappings, conformal mappings, continuous mapping, stereographic projection, projection of a sphere onto a developable surface, projection onto a tangent plane, projection onto a cone, conical projection with one standard parallel, conic projection with two standard parallels, cylindrical projection

6.8 Display Projection 306
A variety of coordinate systems, the picture-plane coordinate system, orthographic projection, oblique projection, perspective projection

7. Nonlinear Transformations 310

7.1 Linear and Nonlinear Equations 310
Comparing linear and nonlinear equations, reciprocal radii transformation

7.2 Inversion in a Circle 312
Inversion in a circle by geometric construction, inversion in a circle of infinite radius, circular inversion of a straight line not containing the origin, the inverse of a circle

7.3 Curvilinear Coordinate Systems 316
Parametric representation of curves, parametric representation of a circle, parametric representation of surfaces, a surface as the locus of a moving deforming curve, parametric representation of a solid, parametric variable transformation

7.4 Deformations 324
A bivariate deformation, a trivariate deformation, sweep transformation and deformation, tapering, global tapering along the z axis, twisting, global twisting about the z axis

Answers to Selected Exercises 331
Index 347

PREFACE

Geometric Transformations for 3D Modeling is written for students, teachers, and professionals who require a more advanced treatment of the mathematics of geometric transformations than is found in current textbooks on computer graphics, 3D modeling, and engineering. It is the second edition of a textbook whose first edition title is *Geometric Transformations*. The new title more accurately describes the book's underlying focus and applicability to 3D modeling and other computer graphics applications. It is a comprehensive introduction to the theory and mathematics of transformations, written from a standpoint accessible to students, teachers, and professionals studying or practicing in engineering, computer science, mathematics, or physics. It will appeal particularly to those working in 3D modeling, geometric modeling, computer graphics applications, and animation.

Most texts in these fields touch only briefly on the nature and use of geometric transformations, some with just a slim chapter on rigid-body transformations. This text explores and develops the subject in much greater breadth and depth, offering the reader a better understanding of transformation theory, the role of invariants, the uses of various notation systems, and the relationships between transformations. It describes how geometric objects may change position, orientation, or even shape when subjected to mathematical operations we call geometric transformations, while properties characterizing their geometric identity and integrity remain unchanged. This work may serve as a textbook for upper division and graduate students and as a supplementary resource for students, teachers at all levels, researchers, and other professionals.

A course in geometric transformations presupposes a minimum foundation of mathematical knowledge and a certain maturity of mind. The usual analytic geometry and first-year calculus courses taught to students in engineering and the sciences are prerequisites. A working knowledge of basic matrix and vector algebra is also necessary. Eigenvalues and eigenvectors – special characteristics of matrices – are less familiar to the average reader, so the text introduces them with ample motivation, as is also the case with some elementary tensor concepts. Although the text presents eigenvalues, eigenvectors, and tensors with a certain mathematical rigor, it tries to appeal to the reader's intuition to make these concepts easier to understand. The student, teacher, or professional unacquainted with these techniques will be pleasantly surprised at how quickly he or she grasps their meaning and potential for application.

One of the most challenging problems in writing this book was where and how to present the rather formidable amount of foundation mathematics necessary so that the book can stand alone as comprehensive and reasonably self-contained. Introducing this foundation mathematics only as needed would seriously fragment its discussion, not to mention interrupt the discussion of other topics. To avoid this condition, most of the more abstract foundation material appears in the first three chapters, particularly the second and third chapters. This reduces the clutter of theoretical derivation and development in the remainder of the text and introduces the operational and more application-oriented tools and concepts as their need arises. Formal proofs of theorems and assertions have been omitted, because these are easy to find in appropriate specialized literature.

Here is a summary of the contents: Chapter 1 is a set of introductory essays about geometry, its history and content. Its purpose is to create an attitude that we might call "post-Euclidean," one of openness and expectation for the new views of geometry to follow. Chapters 2 and 3 present somewhat more rigorous foundation material: the theory of transformations and vector spaces, respectively. Chapter 2 reviews functions, mappings, and transformations; and then it introduces linear transformations, geometric invariants, isometries, similarities, affinities, projectivities, and topological transformations. It functions as a transition between the synthetic and analytic modes of expressing transformations, whereas the chapters that follow it are primarily analytical. Chapter 3 introduces vector spaces, basis vectors, eigenvalues, eigenvectors, and tensors. Although there is some pedagogically useful repetition in the analytical representations, particularly in Chapters 2, 5, and 6, the settings are quite different. Chapter 4 focuses on the rigid-body transformations and introduces homogeneous coordinates and transformation matrices. Chapter 5 covers reflections and symmetry, including central inversion, reflection in the plane and in space, symmetry groups, ornamental groups, tilings, and polyhedral symmetry. Chapter 6 completes the study of linear transformations, introducing isotropic and anisotropic dilations, shear, and a variety of projective transformations, including a brief introduction to projective geometry, parallel, central, and map projections, as well as computer graphic display transformations. Chapter 7 departs from the familiar linear transformations, where at least straight lines always manage to remain straight, and enters the fascinating world of nonlinear transformations.

This second edition integrates into the primary text the material that appeared in "boxes" in the first edition. The figures are revised and improved, many are done in grayscale with gradient shading to enhance three-dimensional effects and important features. With the exception of the first, each chapter is sprinkled with exercises, and answers to some are given at the end of the book.

Mike Mortenson
May 2007

Acknowledgments

Every written work on a technical subject begins with a lot of research. When I made this statement in the first edition, I didn't realize how true that is of second editions, too. I thank the readers and instructors using the first edition who took time to suggest improvements and corrections. Thanks, once again, to John Carleo, Editorial Director, for his encouragement and patience, and Janet Romano, for the beautiful cover design, both of Industrial Press, and both of whom have guided through to publication this and several other works of mine. And finally, fond thanks to my wife Janet, still a relentless advocate for simplicity and clarity.

1 GEOMETRY

Here we look at geometry from three different points of view. First, we examine various attempts to define it and try to develop an appropriate attitude toward the subject. Next, we review its history and explore its scope and diversity, discovering that there are many geometries. Finally, we discover how a single powerful principle codifies, unites and simplifies our understanding of geometry. We see change and permanence—transformation and invariance—emerge as agents of this principle.

1.1 What is Geometry?

Arrange three straight sticks end to end so that they form a triangle. If you randomly choose the sticks from an assortment of varying lengths, you won't always succeed in your construction. You will soon observe that if the length of any one of the three sticks exceeds the total length of the remaining two, then it is impossible to form a triangle. You have discovered a simple yet fundamental principle of the geometry of triangles. Next, leap to a study of Einstein's theory of general relativity. In it, you find that gravitational forces are attributable to curvature in the four-dimensional geometry of spacetime. How are two such seemingly distinct and diverse phenomena encompassed by a single discipline? Indeed, what is geometry?

Euclid and his students would have had no trouble answering this question, but does geometry mean the same thing today as it did more than two thousand years ago? Does it mean the same thing today as it did only fifty years ago? It may come as a surprise to many, but geometry and the geometer's concept of it changes and grows as we extend and refine it, developing new principles, new applications and new computational techniques. Certainly, geometry is not static. It is an evolving, growing body of knowledge, with a unique beauty and an awesome logical structure. Fractals are a prime example.

Fractals

Beginning in the nineteenth century some mathematicians began to create a new class of curves that they soon came to describe as "monsters." The behavior of these curves seemed unorthodox and pathological. Among these mathematicians we find Giuseppe Peano (1858 - 1932) of Italy, David Hilbert (1862 - 1943) of Germany,

Helge von Koch (1870 - 1924) of Sweden, Waclaw Sierpinski (1882 -1969) of Poland, Gaston Julia (1893 - 1978), and Karl Menger (1902-1985).

Koch's construction is particularly interesting (Figure 1.1): Begin with an equilateral triangle. Trisect each side and replace the middle third of each with two sides of a new equilateral triangle of appropriate size. Imagine this process continued indefinitely. You have constructed a so-called Koch snowflake. After n cycles, the perimeter of the snowflake is $3s(4/3)^n$, where s is the length of a side of the original triangle. Remarkably, for $n \to \infty$ this "curve" has no tangent at any point. What is even more astonishing is this property: Given any two points on the curve, the arc length between the two points is infinite!

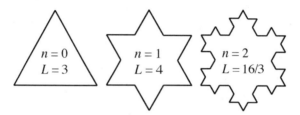

Figure 1.1 Kock "snowflake" construction.

Benoit Mandelbrot's work has immeasurably increased our understanding and clarified our way of thinking about such curves. It is Mandelbrot who has given these self-similar curves the name *fractals* and who produced marvelous insights into their properties and behavior. He shows that for curves of infinite length, spanning a finite distance on a plane surface, their dimension D is greater than one and less than or equal to two, depending on properties of the generator. For the Koch snowflake $D = 1.2618...$, approximately. The magnitude D is a measure of a fractal curve's roughness. Using computer graphics, we are able to explore these objects more fully than ever-dreamed possible by earlier mathematicians.

Return to the Koch snowflake (Figure 1.2). The equilateral triangle we started with is the initiator and each of its sides is a four-segment generator. The first cycle created a curve of twelve straight-line segments. In the next cycle, we replace each of these twelve segments with a properly rescaled, but similar, four-segment generator, and so on, indefinitely.

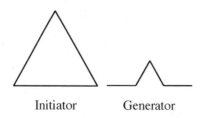

Initiator   Generator

Figure 1.2. Kock snowflake initiator and generator.

Here is another fractal-producing combination of initiator and generator (Figure 1.3). It has a fractal dimension $D = 1.5$:

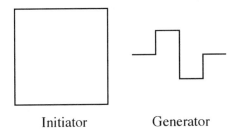

Initiator   Generator

Figure 1.3 Fractal with $D = 1.5$.

There is much more. Research the literature for Peano curves, Cantor dusts, Sierpinski gaskets, Julia sets and the Menger sponge, for starters.

What Kind of Answer?

We should not expect a simple short answer to the question "What is geometry?". We should not expect a succinct, enlightening pronouncement, nor a lightning flash of insight into the nature of geometry. "Geometry is what geometers do," we have heard. While no doubt true, and we will see this later in this sequel, there is much more we would like to know. We may intuitively know that geometry has something to do with the forms and structures of the world around us, their measurement and interrelationships, in both concrete and abstract terms. Yet, somehow we find that this, too, is not much more meaningful than saying that *Whistler's Mother* is a study in grays and white. Asking the question is important, but the attitude we assume in seeking the answer is also important. Even though we will find that the answer won't come suddenly and complete, it will form slowly and subtly with each hour of study and reflection. Like rain nourishing an acorn, one drop at a time, until after many seasons a complex and beautiful tree will have grown. We do, however, begin, like the acorn, with much of the answer already within us.

We seem to be born with a certain capacity to form geometric concepts, much like our capacity for speech or language. Upon this, we acquire and build our knowledge of geometry through learning and experience. We come to recognize form and structure. From what begins as something close to perceptual chaos, we soon differentiate individual objects of a more-or-less permanent existence. We grasp, crawl and squirm through an object-rich environment, filling each waking hour of our earliest years with new wonders. These early haptic and kinesthetic experiences help us build and elaborate mental geometric models of the world around us. We abstract a notion of permanence of form through the movement of our bodies coupled with our visual and tactile sensibilities. As we mature, our efforts to negotiate and manipulate these obstacles and objects demand less and less focus of our immediate attention and budding intellectual powers. We gradually begin to abstract these experiences and acquire a system of beliefs and expectations of a more or less well-ordered, predictable physical world. Early instruction in geometry relies on these acquired and intuitive skills, particularly those associated with seeing and visualization. We gradually approach real limits to visualization, however, which we should understand and use to advantage.

Limits of Intuition and Visualization

Are there limits to our ability to visualize and intuit the elements of geometric knowledge? The short answer appears to be "yes," but with interesting consequences. Consider the following situations:

1. We easily visualize and distinguish between a triangle and a square, but it is impossible to distinguish between a regular polygon with 1000 sides and one with 1001.

2. We easily visualize and count the number of diagonals in a square or pentagon, and probably even a hexagon, but we simply cannot do it for a regular polygon of 1000 sides.

3. Although we readily visualize a right angle, we certainly cannot distinguish it from an angle of $89° 59'$.

4. Approximate or badly drawn figures are the norm in geometry, yet such pictorial deficiencies do not stop us from forming correct hypotheses about their intended, implied, or assumed properties.

In the situations in statements 1 and 2, we would not even trust the result of counting the diagonals of a 1000-sided regular polygon in a drawing. However, by study and experiment with the graphic representation of the few-sided regular polygons, we quickly discover an analytical relationship between the number of vertices and consequent number of diagonals.

Statements 3 and 4 tell us that we can successfully use crudely drawn figures to support a demonstration of the truth of a geometric hypothesis. From these few examples, and many more we could cite, we conclude that it is not visualization, per se, that is the source of the correctness of our work (since inexact images still seem to promote correct inferences), but it is instead the process of abstraction and logic that we attach to these images. The compelling fact is not the visualization but the logic implicit in the images.

We see here a cognitive process that takes us from the concrete to the abstract, from the visual-intuitive to the logical-analytical. Visualization and intuition are not so much limits beyond which we must abandon all further consideration of problems we can no longer visualize, but they are instead very direct and concrete ways to organize, interpret and express logical structures and relationships—eminently useful, if not indispensable in the opening game of geometry.

GEOMETRY

With few exceptions, and those only in the more abstract branches of geometry, we encounter things we are tempted to call geometric objects. These might be things like points, lines, curves and polygons or other somewhat different things like vectors and tensors. The list could go on considerably further. When we think much about it, we conceive of these geometric objects as being contained in a space of some kind, perhaps only the abstract plane represented by a page in an elementary geometry textbook.

On your way through this work, you will discover that geometry is not at all a world of static geometric objects fixed forever in some featureless space. You will find that the essence of many geometric objects lies in their plasticity and that any current manifestation is simply one among many. You will find, too, that space, far from being a passive and featureless stage upon which geometric objects perform, is an active and often richly characterized participant.

We first survey briefly the history of geometry. It will inform and motivate us as it reveals the results of generations of creative efforts sustained over thousands of years to produce today's geometry. Next, we will come to grips with geometric objects and the nature of space.

1.2 History

Geometry is one of the three major branches of modern mathematics, the other two are analysis and algebra, including the theory of numbers. Geometry is by far the oldest of the three. In fact, as a distinct discipline attracting and sustaining the creative energies of the human intellect, it is even older than any of the major religions of the world.

Perhaps we will never know the complete history of early geometry, but certain broad outlines are clear. Geometry most certainly has its origins in the engineering and agricultural practices of ancient Babylon and Egypt, possibly as long ago as 7000 years. The farmers and builders of antiquity measured lands and estimated crop yields; they built granaries of sufficient size and number to store the harvest; and of course they planned and built the monumental pyramids, some now close to 5000 years old. These works required them to apply many rather complex geometric principles. However empirical and practical the origin of these principles from even more remote ages, by the time of the Babylonian and Egyptian civilizations they had attained some degree of sophistication and abstraction to accommodate the variety of uses, sites, and resources to which they were applied.

The so-called Rhind papyrus, first deciphered in 1877, dates from about 1700 BCE. The scribe Ahmes copied it from an earlier work of about 3400 BCE. In it we find a variety of geometric formulas, including rules for the area of a triangle and trapezoid and an expression for the area of a circle that assumes $\pi=3.1604\ldots$. Imagine—this is over 3000 years before Euclid.

The geometry with which you are already most familiar is, of course, Euclidean geometry. Euclid was the first to organize into a logical system much of what he and his contemporaries then knew about geometry. His work is preserved for all time in his *Elements,* written around 300 BCE. The *Elements* is the most widely known and read scientific treatise of all antiquity. To this day, it has been published in more editions and languages than any other secular book. It so far surpassed all earlier works that little trace of them remains.

The *Elements* is far from being an exhaustive encyclopedia of the accumulated knowledge of geometry of its time. Nor does it appear to have been Euclid's intention to make it one. For example, he omitted the theories of conics and higher-degree curves, which for sometime he and his contemporaries had been extending and developing. The subject matter Euclid did choose, however, was sufficiently developed and coherent to allow him to emphasize the logical connections and to present this geometry as a closed system, with no emphasis on practical application. This was a clear demonstration that geometry long had enjoyed both an abstract and secular vitality.

Geometry as a Logical System

1. If we consider geometries as logical systems, then we assume certain properties and try to deduce other properties that these imply.

2. In any geometry, some of the elements we assert and accept without formal definition; all other elements we must define in terms of these undefined elements.

3. In any geometry, some of the relationships among the elements we must accept without formal proof.

4. Postulates (axioms), are assumptions about relationships among elements.

5. Theorems are other relationships that we prove or deduce.

6. Definitions allow us to associate names with elements and relationships that we express in terms of the undefined elements, postulates, previously defined elements, and previously proved relationships.

7. Postulates and definitions we combine according to the rules of logic to obtain statements of the properties of the geometry.

8. Postulates must satisfy certain conditions before being generally accepted:

    a. They must be easily understood.

    b. They must be few in number.

    c. They must require only a few accepted undefined terms.

d. No two should be contrary.

e. No two theorems derived from the postulates should be contrary.

f. They must be independent (no one set is a consequence of the others).

We may speculate that Euclid intended the *Elements* to be not only an introduction to geometry and other mathematical knowledge, but also to be useful as preparation for philosophical studies, emphasizing as it does axiomatics and a disciplined way of thinking. Like Thales before him, who was the first we recognize to have used the axiomatic method in geometric proofs, and the incomparable mathematician and engineer, Archimedes, who followed, Euclid sought truth through the creative powers of the human intellect to reason and synthesize.

For the next two thousand years, until the seventeenth century and the great work of René Descartes (1596-1650), mathematicians studied and developed geometry mainly in the nonalgebraic, synthetic method of Euclid and were concerned largely with the metric properties of figures—properties involving the measurement of length, angle, area, and volume. In the seventeenth century, Descartes' *Geometry* (1637) changed all that. From then on, at least two new developments arose within the framework of Euclidean geometry: The first was the application of algebraic methods and coordinate systems to geometry, leading to the development of analytic geometry. Subsequent study of geometric figures as the locus of points corresponding to algebraic equations revolutionized our view of both geometry and algebra. The second was a growing interest in the nonmetric properties of figures, leading to a profusion of new systems of geometry: affine geometry, conformal geometry, projective geometry, and topology, to name just four.

Prior to Descartes, mathematicians interpreted and solved equations geometrically. For example, they treated expressions such as $y^2 + xy$ as areas and added them as such. This meant that an expression like $y^2 + y$ was meaningless, since its individual terms were of a differing dimension. It was Descartes' great insight that led him to treat all such terms simply as magnitudes, so that these expressions could be graphed and their properties studied with techniques we now call analytic geometry. Although today we treat such expressions as pure numbers, Descartes treated them as a hybrid somewhere between a mathematical and a physical or mechanical object. He probably didn't make nor appreciate our fine distinction between these two concepts. Descartes intended to lead a life of tranquility and repose, but he probably never achieved this happy state, as a study of his life and work would reveal. In the end, he was obliged to answer the call of Queen Christine of Sweden, to become her tutor in mathematics and philosophy. Either the cold, inhospitable climate of the north, or Christine's vigorous life style, or both, were his downfall. He died in Stockholm in 1650, at the age of 54.

Thus, beginning in the seventeenth century, mathematicians began to depart from the classical approach to geometry. We see them classifying theorems into groups whose members have certain properties in common. Using geometric constructions and algebraic formulas, they transformed figures into what seemed at first to be gross distortions of the originals. They soon discovered that certain properties of the transformed figure remained unchanged, the properties we now know as invariants under the given transformation. The notion of invariance predates Euclid; for in the physical geometry of antiquity it was taken for granted that measuring devices were rigid bodies. They remained fixed, invariant in size and shape as they were moved about during the course of their use. However, by the mid-nineteenth century various mutations of Euclidean geometry had emerged.

These efforts produced two seminal works: Georg Friedrich Bernhard Riemann's *On the Hypotheses which Lie at the Foundation of Geometry* (1854) and Felix Klein's *Erlanger Program* (1872). Riemann revolutionized differential geometry and pointed the way toward the geometrization of physics. Klein took inspiration from Riemann's work and from developments in group theory to create a powerful method for classifying the rapidly multiplying geometries. In his *Erlanger Program* he asserts that every geometry is the theory of invariants of a particular transformation group. Therefore, it seems appropriate that we next consider the work of both Riemann and Klein.

When you see or hear the expression *non-Euclidean geometry* used in mathematics, you can be reasonably certain that it refers to the geometries of Lobachevsky (1793-1856) and Riemann (1826-1866). Riemann was well acquainted with the earlier work of Leonhard Euler (1703-1783), the prolific Swiss mathematician. In fact Riemann was so influenced by him that some historians report that Riemann's writings often have a distinctly Eulerian touch. Curiously, Lobachevshy's work seems little affected by Euler, perhaps to his detriment. However, Nikolas Lobachevsky's work did break the grip of Euclid's parallel postulate, which had held geometers in thrall for two thousand years. It opened the way to other new geometries and ultimately to new ways of thinking about the world. Riemann's 1854 treatise on the foundations of geometry, given as his inaugural lecture at the University of Göttingen, is one of the greatest works in mathematics. It went well beyond Lobachevsky's work and planted the seed that later blossomed into general relativity, geometrodynamics, and ultimately the "geometrization of everything." Riemann gave us the concept of a manifold with which he created a new concept of space. As if this were not enough, his work included a new definition for distance and the notion of curvature in a manifold (space). Had he lived beyond his forty years, who can say how much more he would have given the world of mathematics and physics? Specifics of his contributions aside, what deeply motivates us here is the new outlook, the open and fresh attitude toward geometry that his work inspired.

Riemann, himself, believed that the worth of his new geometry went beyond the world of pure mathematics and that it would prove to be of use in the other sciences, particularly physics. In the conclusion to his 1854 memoir he says,

"Either therefore the reality which underlies space must form a discrete manifold, or we must seek the ground of its metric relations outside it in binding forces which act upon it.

"The answer to these questions can only be got by starting from the conception of phenomena which has hitherto been justified by experience, and which Newton assumed as a foundation, and by making in this conception the successive changes required by facts which it cannot explain."

He continues, noting that his work and that of others of similar attitude

"… can be useful in preventing this work from becoming hampered by too narrow views, and progress of knowledge of the interdependence of things from being checked by traditional prejudices.

"This leads us to the domain of another science, that of physics, into which the object of this work does not allow us to go today."

In an attempt to obtain order and coherence out of an ever-increasing mass of new geometries, Felix Klein (1849-1925) in 1872 defined a geometry as

"the study of those properties that are invariant when the elements (points) of a given set are subjected to the transformations of a given transformation group."

Klein reasoned that each of the many new species of geometry could be regarded as a special instance of an underlying whole. He used the theory of groups to achieve this. By successive orderly generalizations of Euclidean geometry, he obtained conformal, affine, projective and other geometries. He introduced and elaborated these ideas in his *Erlanger Program*, named for the University of Erlangen where he was a professor of mathematics. Klein's great work unified geometry and revealed a hierarchy of geometries, where each geometry is characterized by a particular group of transformations producing a set of invariant properties. The more general and inclusive the geometry is, the smaller the set of invariant properties. Also, the more general the geometry is, the more kinds of transformations are allowed. Beyond this, Klein's work also revealed some unanticipated connections between different geometries. For example, some geometries turned out to have identical groups and are therefore the same geometry. However, this codification, while extremely useful even today, as this text will amply demonstrate, is no longer the all-encompassing system that it was for over fifty years. Many new geometries have been invented that do not fit Klein's scheme. The new geometries that followed general relativity rendered Klein's synthesis inadequate.

If we think of space as an empty, featureless arena without measurable properties of its own and in which objects are moved about and their invariant properties compared, then Klein's synthesis applies. It is the invariance under certain prescribed transformations, this malleability and identity that allow us to distinguish between geometries. However, there are geometries, Riemannian and others, that do not yield to this scheme. We find that there are spaces that we cannot look at as passive arenas in which independent geometric objects play. We find that certain spaces have a structure and dynamics of their own.

The *Erlanger Program*, as mentioned above, does not encompass everything we call geometry today. It accommodates the geometry of special relativity, for example, but not that of general relativity. However, even in the new geometries this idea of invariance remains important, even though the associated sets of transformations are not, in general, groups. Felix Klein's life and work make fascinating reading, as is of course true of many mathematicians. With Klein, however, we see someone closer to us than Riemann (or other nineteenth century mathematicians) in outlook and method and whose life spanned both sides of the intellectual revolution in mathematics and physics of the late nineteenth and early twentieth centuries. In this text most of the topics fit nicely within Klein's *Erlanger Program*, and we take full advantage of this. Those that do not are easy to recognize (the position of the *Erlanger Program* does not mislead, but is merely silent), and they are concerned primarily with nonlinear transformations.

1.3 Geometric Objects

Geometry has commonly associated with it an entourage of identifiable "things" which convey objective characteristics. These things we call geometric objects. What is a geometric object? Let's not fall into the bottomless pit awaiting those who demand or attempt overly precise definitions. Instead, we'll look at some familiar, simple examples of geometric objects and see what they have in common, and whether they suggest an obvious classification scheme.

Our first example of a geometric object is a set of points. If the set is continuous, then it defines what we usually call a geometric figure. Examples are lines, curves, planes, polygons and polyhedra. The Mandelbrot set certainly qualifies as a geometric object, and we commonly think of it as a figure of sorts, although the exact nature of its continuity is not altogether clear. When we use the terms *figure* or *form*, we will usually mean a geometric object of this kind. If the set of points is not continuous, we refer to them merely as a set of points. Note that in any geometry a more rigorous definition of a figure depends on the elements of that geometry. The elements, or rather the undefined elements of a geometry, in other circumstances may be figures (for example, points, lines or planes). Here we are dangerously close to a vicious circle.

There are other kinds of geometric objects. The tangent vector and metric tensor are two such examples. The tangent vector associates a magnitude (sometimes treated as a length) with a direction, and the metric tensor yields a distance or distance function. An electromagnetic field, that is, the mathematical representation of the field (or gravitational field, or thermal field, and so on) is yet another kind of geometric object. All affective phenomena and their sources have some kind of geometric description or interpretation and are often best understood when interpreted under a geometric metaphor. We describe the shape of a structure or mechanical part, the magnitude and distribution of forces acting on it, and the interior state of stresses and strains with some form of geometric metaphor and interpretation. We often express phenomena such as the later of these in very abstract mathematical terms, although we also readily interpret them geometrically. We often apply a variety of geometric metaphors in an explanatory way, making use of the common geometric notions of distance, direction, curvature, content, betweenness, orientation and many others.

Perhaps the most significant difference between one geometric object and another that we are first aware of is their relative scope. At one end of a spectrum, we find the localized individual geometric objects such as a point or tangent vector; at the other end, we find extended geometric objects such as a vector field. Somewhere in between are things like lines and planes, conic curves and quadric surfaces, and the bounded figures such as spheres, cubes, and circles. We can define all of these objects independently of any coordinate system, and we notice that their underlying character survives certain kinds of linear transformations. We may treat space itself as a geometric object. In fact, in the new geometries all other geometric objects are often treated as subsets of a space.

Finally, we must ask, "What distinguishes geometric objects from nongeometric objects?" Although we can stretch the geometric metaphor almost without limit, at some point it becomes more cumbersome than useful. In number theory, for example, the study of prime numbers seems to do very well indeed without recourse to some form of geometric metaphor or interpretation. However, even here we must hedge a bit, since a proof of Fermat's Last Theorem (once thought to be a problem in algebra and number theory) employs elliptic curves and automorphic representations…clearly, a geometric interpretation. Given the right conditions, it is a singularly most productive tool available to mathematics, science and technology.

Have we found a definitive answer to our initial question? No, but we have some important clues. There are those objects we can refer to as having figure or form and others, such as fields, we would be hard put to characterize in this way. However, repeatedly in this sequel we'll see that the integrity of such diverse objects under specific kinds of linear transformations is the key we are seeking.

1.4 Space

Space is not what it used to be. The space of Euclid and Newton was simply the empty vessel in which the sun, planets, men, and beasts moved about. It was empty of properties, it was unmoved by causes and effects, and it had no influence. The *Elements* had no need to consider it.

In the seventeenth century, things began to change. Descartes" work demonstrated the power and versatility of coordinate geometry. This was the first and most important step in the mathematization of space. Mathematicians of that age did not yet achieve it, but most of the ingredients were in place to distinguish between physical space and mathematical space. The invention of the non-Euclidean geometries helped accomplish this distinction. Today, if we consider the several geometries to be mathematically distinct but *a priori* and generically equivalent, then we must ask: Which geometry describes physical reality, as there is no *a priori* need to single out Euclidean geometry? One approach to this problem soon became obvious: Mathematics describes the possible spaces, and physics determines which one of them corresponds to physical space. According to this approach, the question is subject to experimental resolution. The space of physical experience and intuition gives rise to the concepts of distance and measurement, and presents the notion and possibility of spatial relationships.

Two kinds of mathematical spaces are of interest to us: coordinate-free spaces and those with imposed coordinate systems. A coordinate-free space is easy to define. For example, in Euclidean three-space the locus of points equidistant from two arbitrary points is a plane (two-space). Note that there are several hidden assumptions here. First, points are undefined elements. Next, "equidistant" presumes an undefined distance-measuring operation. Finally, the whole process assumes a three-dimensional space in which the construction of the plane takes place. (As important as these assumptions are to mathematicians and philosophers we will not consider them further here.) We again use the idea of a locus of points to define geometric figures in this plane and use synthetic, axiomatic geometry to deduce their properties and relationships. Vectors and tensors are versatile and powerful tools of geometric modeling, differential geometry, mathematical physics and many other disciplines because of their coordinate-free nature.

You may be accustomed to using the term "space" as an undefined term, but now is the time to add some mathematical rigor to this notion. The nature of reality is such that the terms "Euclidean two-space" or "Euclidean three-space" have a meaning understood by all geometers: These terms presume a one-to-one correspondence between the points of such spaces and the sets of all real-number pairs or real-number triples, respectively. In fact, you will find, for example, that it is common to say that the set of all real-number triples *is* the three-space. Mathematicians often use the notation $\mathbf{R}$ or $\mathbf{R}^1$, $\mathbf{R}^2$, $\mathbf{R}^3$, ..., $\mathbf{R}^n$ to describe these spaces; some older or specialized texts may use E^1, E^2, E^3, ... E^n, or other symbols. We call a set of values $\{x^1, x^2, ..., x^n\}$ the coordinates of a point in n-space. The individual variables, x^i, are the coordinates, and the totality of points corresponding to all their values within specified ranges defines a space of n dimensions. There are several things worth noting here. First, we have identified the coordinates with superscript labels. We will discuss the reasons for this convention later, in the context of vectors and tensors. Although at first you may confuse these labels with powers, it is a well-established convention, and experience and application will soon allay any anxieties you may have about this. Second, each coordinate has a specified range, depending upon the type of coordinate system we construct. For example, a three-dimensional cylindrical system has coordinates r, ϕ, z with ranges $0 \leq r < +\infty$, $0 \leq \phi < 2\pi$, $-\infty < z < +\infty$.

Third, since we are accustomed to using the term "space" to mean two- or three-dimensional Euclidean space, we use the terms "manifold" or "hyper space" to indicate a more general, possibly higher dimensional space. (There is a more rigorous definition of the term "manifold" involving continuity conditions and differentiability, which we won't discuss here.) When we work with ordinary two- and three-dimensional Euclidean spaces, we will use the familiar x, y, and z coordinate notation to develop new concepts and then occasionally restate results using the more general notation scheme. See the following review of some elementary coordinate systems in the plane and their notation.

Coordinate Systems in the Plane and Their Notation

The plane rectangular Cartesian coordinate system is, of course, the one we most frequently encounter. The notation x, y labels the principal axes, or equivalently x^1, x^2.

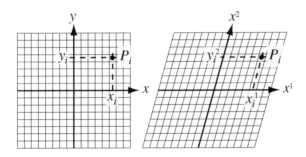

Figure 1.4 Rectangular and oblique coordinate systems.

The angle ω between the positive principal axes is the coordinate angle. If $\omega = 90°$, then the coordinate system is rectangular; otherwise it is oblique, as shown on the right (Figure 1.4). We label the principal axes as in the rectangular system or with some other notation if there is a chance of confusion.

A curvilinear coordinate (Figure 1.5) system is a generalization of the linear system, whereas the polar coordinate system is a specialization of the curvilinear.

All of these methods for installing coordinates on the plane have obvious analogs in three- and higher-dimension spaces. If we understand the mathematical nature of the space implied and denoted by an imposed coordinate system, then the specifics of any particular notation scheme will become less important and less confusing.

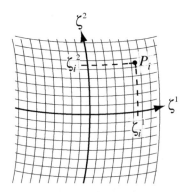

Figure 1.5 Curvilinear coordinate system.

A coordinate system can be less than perfect in many ways. Look at the convergence of the meridians of longitude at the poles of a globe. This convergence imposes two singularities. These are strictly artifacts of the coordinate system and are not characteristic of the underlying space defined by the surface of the globe. Is there an alternative? Is there a single coordinate system that will cover the globe without introducing singularities? There is a theorem that says no, that two is the minimum number of coordinate patches needed to cover a sphere without a singularity. Try this on a torus or other closed doubly curved surface.

The concept of distance is implicit in Euclidean and Riemannian spaces, so we call them metric spaces. For these metric spaces, Riemann showed that a generalization of the Pythagorean theorem applies. Using this generalization, Riemann constructed a differential expression for defining length or distance. It is what we now call the metric tensor. Therefore, we say that a continuous n-dimensional manifold is a Riemannian or metric space if there is in it a metric tensor. We may then explore details of the geometry of a space by analyzing its metric tensor. Don't be lulled into a false sense of complacency, for you should know that mathematicians can construct logically consistent non-Riemannian, or nonmetric, spaces in which the notion of distance is not only absent but unnecessary.

1.5 Geometry Is ...

Let's return briefly to the question: What is geometry? Keep in mind that we are not after a precise definition. Even if it were possible to achieve one, it is doubtful that it would serve any operationally useful purpose. Remember, our motivation in pursuing this question remains that of acquiring a certain attitude and point of view. It is a way to help us step out of the classical Euclidean world of geometry that most of us are still taught in school, and into one of transformation and invariance, one with not only a more generalized Euclidean view but also one with the beginnings of non-Euclidean notions.

GEOMETRY

History tells us that much of early geometry was concerned with certain measurable, quantitative properties of objects and the arrangement of these objects in the physical space in which they exist. It also tells us of another geometry of ancient origin: synthetic, deductive geometry, concerned with abstract logical relationships among geometric objects. These two systems of geometry we closely associate with Euclidean geometry. Synthetic geometry gives us ways to study points, lines, planes and curves, not by their coordinate-dependent equations, but by their intrinsic geometric properties and relationships. Analytic geometry allows us to translate these geometric conditions into algebraic conditions. We usually find that geometric ideas guide the algebraic development, but that the algebra often reveals new and unsuspected geometric information.

Coordinate-Free Geometry of Conic Curves and Their Coordinate-Dependent Interpretation

Coordinate-Free Geometry

An ellipse is the set of all points P for which the sum of the distances from two fixed points A and B (the foci) is a constant, K, so that $AP + BP = K$ (Figure 1.6). If $AB = 0$, then a circle is produced with a diameter $K = 2R$.

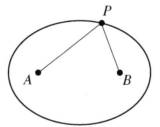

Figure 1.6 An ellipse.

A parabola is the set of all points P that are equally distant from a fixed point A (the focus) and a fixed line l (the directrix), so that $AP - d = 0$ (Figure 1.7).

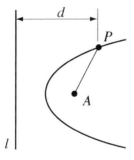

Figure 1.7 A parabola.

A hyperbola is the set of points P for which the difference of the distances from two fixed points A and B is a constant, K, so that $AP - BP = K$ (Figure 1.8).

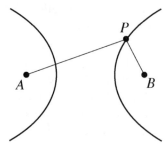

Figure 1.8 A hyperbola.

Coordinate-Dependent Interpretation

Install a rectangular coordinate system so that the x axis contains points A and B and the origin O is at their midpoint (Figure 1.9). Then, from the coordinate-free definition, find

$$\frac{x^2}{a^2} + \frac{y^2}{b^2} = 1, \ A = (-c, 0), \ B = (c, 0), \ K = 2a, \text{ and } c = \sqrt{a^2 - b^2}$$

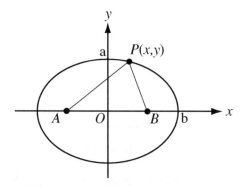

Figure 1.9 Ellipse: a coordinate-dependent definition

Install a rectangular coordinate system with the focus at $a/2$ and the directrix through $-a/2$ and perpendicular to the x axis (Figure 1.10). From the coordinate-free definition find $y^2 - 2ax = 0$

GEOMETRY

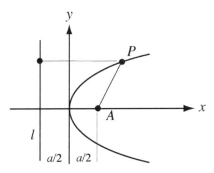

Figure 1.10 Parabola: a coordinate-dependent definition.

Install a rectangular coordinate system so that the x axis contains points A and B and the y axis is the perpendicular bisector of AB (Figure 1.11). From the coordinate-free definition find

$$\frac{x^2}{a^2}-\frac{y^2}{b^2}=1, \ A=(-c,0), \ B=(c,0), \ K=2a, \ \text{and} \ c=\sqrt{a^2+b^2}$$

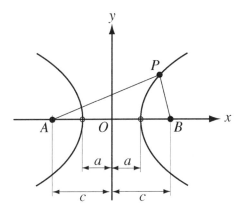

Figure 1.11 Hyperbola: a coordinate-dependent definition.

The geometry we will study in the chapters that follow concerns the properties of geometric objects that remain unchanged, or invariant, when we do something to the objects themselves. We may move the objects about, changing their positions; we may stretch or bend them, changing their shape; we may even alter the underlying space in which they are embedded, defining entirely new geometries and consequent properties and relationships. A geometry, Klein observed, is defined by a group of transformations, and investigates everything that is invariant under the transformations of this group.

We identify four important characteristics of geometry. The geometric objects, such as points, lines planes, tangent vectors, tensors and fields, give geometry its objective characteristic. The measurable properties of objects and spaces, such as length, angle, ratio, curvature and content, give it its quantitative characteristic. The modes of comparison, such as equality, less than, and greater than, give it its comparative characteristic. Finally, the spatial relationships, such as inside, outside, and between, give geometry its relative characteristic. Objective, quantitative, comparative, and relative—all these characteristics are present in any endeavor claiming to require geometric processes. If one or another is absent, it is this way by axiomatic intention, and this produces a more restricted geometry.

Here are some of the ways geometers come to grips with their subject:

1. Visually—a pictorial treatment of geometry (diagrams, not algebra)

2. Synthetically—coordinate-free geometry (deduction, not induction)

3. Analytically—coordinate geometry (algebra and graphs)

4. Abstractly—coordinate-free vector geometry (vectors qua vectors, not decomposed into components)

5. Metrically—geometry expressed through vector components and coordinate systems (indispensable in programming geometry for computer applications)

We use all of these approaches in this text, since no one of them offers a complete picture of geometry as an attitude, philosophy and technique for representing and understanding both the natural world and the products of our own invention.

Ways of Thinking about Geometry

Visually (diagrams, not algebra)

Given the radii and points of intersection A and B of two circles, find their centers. Even a crude sketch demonstrates that there are four possible solutions to this problem (Figure 1.12).

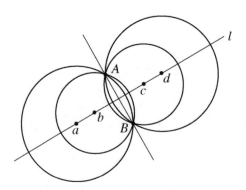

Figure 1.12 Visual interpretation of geometry.

GEOMETRY

Find the perpendicular bisector *l* of line segment *AB*. The centers must lie on this line. With compass at *A* or *B*, mark off the intersections of the circles' radii with *l*, producing the solution points *a, b, c,* and *d*.

Synthetically (deduction, not induction)

Given that *AC* is the diameter of the circle whose center is at *O* and that point *B* lies on its circumference, prove that $\angle ABC = \pi/2$ (Figure 1.13).

1. Construct *OB*: Δ's *AOB* and *BOC* are isosceles.
2. $\angle OAB = \angle OBA$ and $\angle OBC = \angle OCB$
3. $\angle OAB + \angle OBA + \angle OBC + \angle OCB = \pi$
4. $2(\angle OBA + \angle OBC) = 2\angle ABC = \pi$
5. $\therefore \angle ABC = \pi/2$

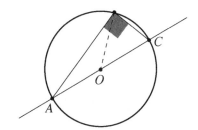

Figure 1.13 Synthetic geometry.

Analytically (algebra, calculus, and graphs)

Given two curves: $y = f_1(x)$ and $y = f_2(x)$ find their points of intersection (Figure 1.14). Solve the equation $f_1(x) - f_2(x) = 0$.

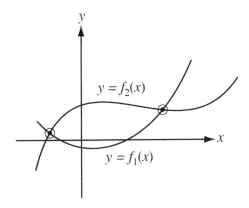

Figure 1.14 Analytical geometry.

Abstractly (vectors qua vectors)

Given a vector equation $\mathbf{a} + u\mathbf{b} + w\mathbf{c} = \mathbf{d} + t\mathbf{e}$, solve for the scalar unknowns u, w and t (Figure 1.15). To isolate t apply $\mathbf{b} \times \mathbf{c}$ as follows:

$$(\mathbf{b} \times \mathbf{c}) \bullet (\mathbf{a} + u\mathbf{b} + w\mathbf{c}) = (\mathbf{b} \times \mathbf{c}) \bullet (\mathbf{d} + t\mathbf{e})$$

yielding $(\mathbf{b} \times \mathbf{c}) \bullet \mathbf{a} = (\mathbf{b} \times \mathbf{c}) \bullet \mathbf{d} + t(\mathbf{b} \times \mathbf{c}) \bullet \mathbf{e}$.

Now solve for t: $t = \dfrac{(\mathbf{b} \times \mathbf{c}) \bullet \mathbf{a} - (\mathbf{b} \times \mathbf{c}) \bullet \mathbf{d}}{(\mathbf{b} \times \mathbf{c}) \bullet \mathbf{e}}$, and similarly for u and w.

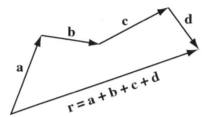

Figure 1.15 Vectors

Metrically (vector components and a coordinate system)

Find the vector $\mathbf{t}$ perpendicular to vectors $\mathbf{r}$ and $\mathbf{s}$.

$$\mathbf{t} = \mathbf{r} \times \mathbf{s} = \begin{bmatrix} \mathbf{i} & \mathbf{j} & \mathbf{k} \\ r_x & r_y & r_z \\ s_x & s_y & s_z \end{bmatrix}$$

Where does it all end? Well, actually it doesn't. Modern geometry is like that, but for now we have enough to proceed. We have reviewed some history, and we have looked beyond the elementary descriptions of geometric objects and spaces. All this is to cultivate an attitude toward geometry that will prove useful later on. We may not fully escape the familiar and secure Euclidean point-of-view, but we have taken a big step on the way. We have raised one issue we must now return to and explore in more detail: the apparent multiplicity of geometries.

1.6 *E Pluribus Unum* — Transformation and Invariance

Synthetic geometry, analytic geometry, affine geometry, projective geometry, topology, differential geometry, nonmetric geometry…these and others, plus many refinements and variations, comprise the discipline of geometry. It is not just a vaguely related collection of subdisciplines. Many strong historical, evolutionary and operational links exist. Let's look at a major subgroup of geometries, one on which we will focus our attention in the following chapters.

GEOMETRY

A Hierarchical Classification of Geometries in the Plane

Geometry of Incongruent Figures

This geometry applies only to figures of identical shape and size (Figure 1.16). However, the figures may be in arbitrary positions and orientations.

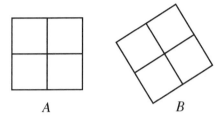

Figure 1.16 Geometry of congruent figures.

Geometry of Similar Figures

This encompasses Euclidean geometry: the study of properties unchanged by rigid motions (translation, rotation and reflection) and uniform change of scale (dilation) (Figure 1.17).

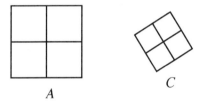

Figure 1.17 Geometry of similar figures.

Affine Geometry

We compare distances only on the same line or on parallel lines. Affine geometry becomes Euclidean geometry when we define perpendicularity (Figure 1.18).

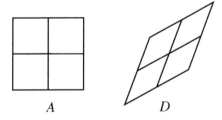

Figure 1.18 Affine geometry.

Projective Geometry

Projective geometry encompasses all the above transformations plus projections. A smaller set of properties remains invariant under projection transformations than under affine (e.g., the degree of a curve is a projective property) (Figure 1.19).

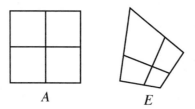

Figure 1.19 Projective geometry.

Topology

Topology encompasses the projective, affine and classical Euclidean geometries (Figure 1.20). An even smaller set of properties is invariant under topological transformations.

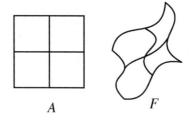

Figure 1.20 Topology.

We transform figure A into B using simple rigid body motions (translation and rotation) and state that A is congruent to B, and vice versa. We transform A into C using the rigid body motions (including possible reflections) plus a uniform scaling, making A and C similar figures. We produce D when we subject A to general affine transformations, in which parallel lines transform into parallel lines (A circle transforms into an ellipse under this regime), and we assert that A and D are affinely equivalent. The projective transformation of A into E preserves straight lines, and, as you might imagine, circles do not transform into circles. A and E are equivalent under projective transformations. Finally, figures A and F are topologically equivalent, preserving closed curves, order, and connectivity. All of these transformations have direct natural extensions into three-dimensional space and higher spaces.

The geometry of congruent and similar figures is nothing more than familiar high school Euclidean geometry. It is the study of properties of figures whose size and shape do not change as we move them about in space. Distances and angles are preserved under these conditions. We say these are invariant properties. If we allow the size of a figure to grow or shrink uniformly, then angles between lines do not change (they are invariant), and we can then study the properties of similar figures. (Trigonometry is based upon the properties of similar triangles and the ratios of their sides.) We must conclude that the geometry of similar figures is less restrictive, and more general, than the geometry of congruent figures.

If we eliminate the restrictions on both the size and shape of a figure, but do this in such a way that the only changes allowed are those that permit parallel lines to remain parallel, then we create the even more general affine geometry. Under all affine transformations, parallel lines must correspond to parallel lines and parallel planes must correspond to parallel planes. Parallelism is thus an important invariant of affine geometry. Circles and ellipses are equivalent in affine geometry. Note, the geometries of congruent and similar figures are subsets of affine geometry.

Next, we relax the restrictions preserving parallel lines but require that straight lines remain straight lines for any changes we might impose on a figure. Projective geometry is the result. Linearity and triangularity are both projective properties: straight lines and triangles always transform into other lines and triangles by projection. Circles, ellipses, parabolas and hyperbolas readily transform into one another by projective transformations.

Topology is the most general geometry we will encounter here. When we do topology, we study those properties of a figure that are invariant under the arbitrary stretching, bending or twisting of that figure. For example, think of a plane figure drawn on a rubber sheet. We may distort the sheet, and consequently the figure, in any way except by tearing or cutting holes (no fusing or gluing is allowed, either). Connectivity and order (of points on a line or curve, say) are among the few properties preserved. All the other geometries discussed here are subsets of topology. Formally, it is more usual to begin with topology and then to deduce the others by imposing on it a series of cumulative restrictions.

The term transformation has several meanings in mathematics. It may simply mean any change to an equation or expression to simplify some operation, such as computing an integral, graphing a function, or finding maximums and minimums. Another meaning, one more pertinent to our discussion here, is that expressing a functional relationship between geometric objects. The variables in a functional relationship, such as "y is a function of x", may be almost any kind of mathematical object. If the object is geometric, then we call the functional relationship a transformation or a mapping. Instead of the independent and dependent variables, x and y, we speak of the original and the image. Although geometric transformations may involve any kind of geometric object, we will most often use point transformations.

Geometers assert that the geometric properties of any figure must be expressed in formulas that are not changed (that are invariant) when they change the coordinate system; and, conversely, any formula that is invariant under a transformation of coordinates must express a geometric property. The formula for the distance between two points in a plane is a very simple demonstration of this.

Invariance of Distance with Transformation of Coordinates

In the xy plane we use the Pythagorean theorem to find the distance between points A and B (Figure 1.21); thus

$$\overline{AB} = +\sqrt{(x_B - x_A)^2 + (y_B - y_A)^2}$$

and in the x', y' coordinate system we have

$$\overline{A'B'} = +\sqrt{(x'_B - x'_A)^2 + (y'_B - y'_A)^2}$$

But $x' = x - x_T$ and $y' = y - y_T$, so that

$$\overline{A'B'} = +\sqrt{(x_B - x_A)^2 + (y_B - y_A)^2}$$

Therefore, $\overline{A'B'} = \overline{AB}$

Thus, in the plane we see that the distance between two points is independent of coordinate systems related to one another by simple translation transformations.

One of Einstein's great contributions to our understanding of the laws of physics was simply that those laws must be invariant under transformations reconciling observations made in coordinate systems that are in uniform motion relative to one another.

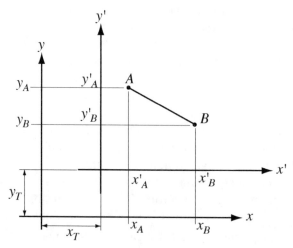

Figure 1.21 Invariance of distance.

The Lorentz Transformation

In physics, the classic Galilean transformations predict that the velocity of light c is different when measured in two coordinate systems in uniform relative motion with respect to one another, v. These transformations produce $c' = c \pm v$, assuming that the direction of the light rays is parallel to the direction of motion. A long series of experiments in the late nineteenth century culminating in those of Albert Michelson and Edward Morley (1887), failed to demonstrate this. In fact, the Michelson-Morley experiments showed that the velocity of light is always the same in all directions and is independent of the relative uniform motions of the source, observer, and the propagating medium. Furthermore, the Galilean transformations do not preserve the form of Maxwell's equations, which represent electromagnetic phenomena. Since Galileo and Newton, physicists have subscribed to the postulate of equivalence, which asserts that all physical phenomena should appear the same in all systems moving uniformly relative to each other. This postulate requires that the laws of physics must be expressed in a mathematically equivalent form for all uniformly moving systems. This means they must be "covariant" under a Galilean transformation. Maxwell's equations did not satisfy this condition, either theoretically or experimentally. Einstein reaffirmed the primacy of the postulate of equivalence and reasoned that the Galilean transformations must be incorrect.

The transformations Einstein proposed to use were named for the Dutch physicist and 1902 Nobel Prize winner H. A. Lorentz (1853-1928), whose work led directly to their formulation. Einstein showed that the laws of physics, including of course Maxwell's equations, are indeed invariant under the Lorentz transformations. If one system, say the primed one, moves with uniform velocity v relative to the unprimed one in a direction parallel to the z axis, then the Lorentz transformation equations between the coordinates and times measured in the two systems are

$$x' = x$$
$$y' = y$$
$$z' = \frac{z - vt}{\sqrt{1 - (v/c)^2}}$$
$$t' = \frac{t - vz/c^2}{\sqrt{1 - (v/c)^2}}$$

As we should expect, the inverse transformation has the identical from with the velocity now given as $-v$.

This brief presentation of the Lorentz transformation equations barely does the subject justice. Their analysis and meaning both for physics and philosophy are very deep, and of course far beyond the scope of the present work.

The literature of geometry's history is extensive and accessible. Pursue it. It will enrich your studies. Now we can begin to do transformation geometry.

2 THEORY OF TRANSFORMATIONS

The title of this chapter seems to suggest a plunge into the depths of abstraction, populated with great hypotheses, axioms, lemmas, theorems, proofs and other mathematical arcana. This is not the case, and the formalities here, such as they are, are friendly. The theory of transformations does not have hard boundaries, recognized and certified by those so disposed to do these things. The material presented here, however, does lay an adequate and important foundation for what follows. Here you will meet functions, mappings and transformations ... key players in this drama. Linear transformations and geometric invariants immediately follow and round out the preliminaries. The final sections introduce isometries, similarities, affinities, projectivities, and topological transformations. Here the text describes fundamental concepts, usually informally and without recourse to a more rigorous axiomatic approach. For the isometries, et al, the mode is at first synthetic and diagrammatic, to foster an intuitive understanding. The mode then shifts to the discussion of classical Cartesian or analytic equations describing each of these transformations.

2.1 Functions, Mappings, and Transformations

The terms function, mapping, and transformation are nearly synonymous. We will inspect them and make some distinctions in the context of *domain* and *range*. Next, we will investigate the product of transformations, or functions of functions, including their associative commutative, identity, and inverse properties. This will prepare us to study the concept of a transformation group and group properties.

Domain and Range

A function requires three things to characterize its nature: two non-empty sets, A and B, and a rule associating each element of A with a corresponding element in B. The correspondence indicated by the arrows shown in Figure 2.1 is an example of a function from A into B, showing that some rule f defines a correspondence between elements of sets A and B.

Theory of Transformations

The letter *f* denotes a particular function, and we say that *f* maps *a* to *r*. This means that *r* is the unique element of *B* that corresponds to *a* according to the rule given by the function *f*. It is also common usage to say that *r* is the image of *a* under *f* or that *r* is the value of *f* given *a*. In more concise mathematical notation we write

$$r = f(a)$$

and read this as "*r* equals *f* of *a*". The other mappings shown in the figure are expressed as $s = f(b)$ and $t = f(c)$.

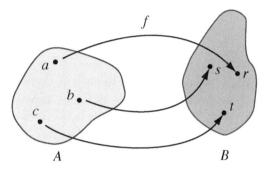

Figure 2.1 Domain and range.

Set *A* is the domain of the function *f*, and the set of all values of *f* is the range of the function. We denote these as $Dom(f)$ and $Ran(f)$, respectively. The domain of a function is the set of inputs. The range is the set of outputs. So we write

$$Ran(f) = \{r : r = f(a) \text{ for some } a \text{ in } Dom(f)\}$$

Think of a function as a machine with an input slot or hopper and some sort of internal workings that generate an output (Figure 2.1). Do you see any limits to this analogy?

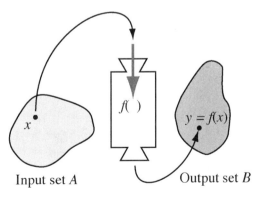

Figure 2.2 A function machine.

The input set, or domain, might be the set of real numbers, and the internal workings of the machine might be a simple mathematical expression, such as $3x^2$, or perhaps a complicated algorithm. The machine's output values constitute the range. So, simply insert a value from the set of values contained in the domain, and the machine cranks out the corresponding value in the range set. (Notice that in this example, the function produces the same output value for some pairs of input values)

Let $f(x) = 3x^2$, then:
$$\begin{cases} f(-1) = 3 \\ f(0) = 0 \\ f(1) = 3 \\ \vdots \\ f(a) = 3a^2 \\ f(x_1) = 3x_1^2 \end{cases}$$

The terms "into" and "onto" have very special meanings when used with functions. If f is a function from A into B, then $A = Dom(f)$ and $Ran(f)$ is a subset of B. However, if $Ran(f) = B$, then f is a function from A onto B, where the term "onto" implies that every element of B is an image of a corresponding element of A. Let's look at some examples. If A is the set of all integers $(1, 2, \ldots, n)$ and B is the set of all even integers $(2, 4, \ldots, 2n)$, then $f(n) = 2n$ is a mapping from A onto B, because $Ran(f) = B$. For the same sets A and B, $f(n) = 4n$ is a mapping from A into B, because $Ran(f)$ in this case is clearly a subset of B. The function $f(n) = 2n$ generates all the elements of B, while the function $f(n) = 4n$ generates only integers divisible by four (forming a subset of B). Figure 2.3 illustrates the distinction between *into* and *onto* mappings.

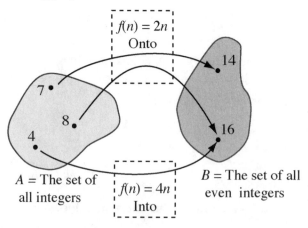

Figure 2.3 The distinction between *into* and *onto* mappings.

THEORY OF TRANSFORMATIONS

Nothing prevents sets A and B from being identical, that is $A = B$. This lets us investigate mapping a set onto or into itself, an important notion in geometric transformations. The set of real numbers $\Re$, is a good example. The domain of the function $f(x) = x^2$ is the set of all real numbers, but its range includes only positive numbers. Therefore, it is an example of mapping a set into itself. If $f(x) = x^3$, then it is a mapping of a set onto itself. Figure 2.4 gives you a visual interpretation of this mapping, showing a mapping of $\Re$ into itself and a mapping of $\Re$ onto itself. You can easily verify this by trying a few real numbers, positive and negative, as input to these functions. What do you conclude about the function: $f(x) = \sin x$?

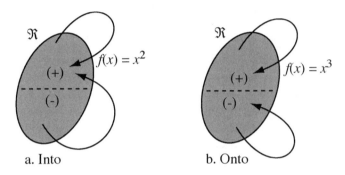

a. Into   b. Onto

Figure 2.4 A mapping of $\Re$ into and onto itself.

A word about the f notation: There is nothing special about the letter f. We may use it and other letters to distinguish between functions operating simultaneously or in the same application. For example, $f(x) = x^2 - x + 1$, $g(x) = 4x + 3$, and $h(x) = -x$. Different letters may also indicate different variables, as with $f(x)$, $g(y)$ and $h(z)$.

A brief discussion of mappings of the real line helps to elaborate these notions. Figure 2.4 illustrates the notion of mapping a set both into and onto itself. We choose to work with $\Re$, the set of real numbers. Mathematicians, particularly geometers, like to think of these numbers as points on a line, the so-called real line or number line extending indefinitely in both the positive and negative directions. This analogy provides us with an excellent way to further elaborate onto and one-to-one transformations.

Now, let's specialize our notion of a function still further. If and only if the elements of two sets can be paired so that every member of either set has a unique mate in the other set, then we say that the two sets are in a one-to-one correspondence. If f is a function such that for x_1 and x_2 in $Dom(f)$, then $x_1 \neq x_2$ implies that $f(x_1) \neq f(x_2)$, and we call f a one-to-one function (Figure 2.5). An equivalent statement is: If f is one-to-one, then $f(x_1) = f(x_2) \Rightarrow x_1 = x_2$.

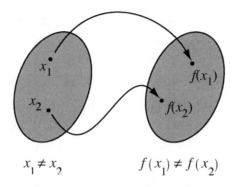

$x_1 \neq x_2 \qquad f(x_1) \neq f(x_2)$

Figure 2.5 A one-to-one function.

Let's begin with the function $f(x) = x+1$. This maps every real number exactly one unit to the right. For example, $x = -2.7$ transforms to $x' = -1.7$, $x = -.2$ to $x' = .8$, and $x = 1$ to $x' = 2$ (Figure 2.6). Every point in $\Re$ has a unique image point, and every point in $\Re$ is the unique image of some other. There are no gaps or overlaps produced by this transformation.

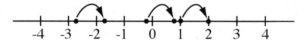

Figure 2.6 Mapping $f(x) = x+1$

Now try $f(x) = -2x$ (Figure 2.7). Do you see why it is both onto and one-to-one?

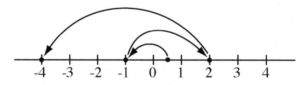

Figure 2.7 Mapping $f(x) = -2x$.

The terms "transformation" and "mapping" are suggestive of geometric processes. The term "function" as it is used in mathematics and, of course, as we have used it here, is inclusive of the terms "transformation" and "mapping" and has a broader application. The variables in a functional relation may be any kind of mathematical objects. When we refer to mappings and geometric transformations (or simply refer to transformations), we imply that the variables are geometric objects. Point transformations are the focus of our studies, and we will restrict our notion of mappings or transformations by requiring them to be functions that are both one-to-one and onto. Our points will inhabit rectangular Cartesian space unless otherwise noted.

THEORY OF TRANSFORMATIONS

Point transformations are specialized versions of function machines. They may have one or more input slots, depending on the dimension of the space in which they are operating. Their input is constrained if they map one figure onto another, or unconstrained if the entire space is mapped.

We change the notation from $f(\ )$ to $M[\ ]$. (This change is only temporary for the sake of clarity in this example.) The $f(\ )$ functions are there but embedded in another machine, $M[\ ]$. This new machine directs each coordinate to its appropriate transformation (Figure 2.8): $f(x)$, $g(y)$, or $h(z)$. Hint: $M[\ ]$ might indicate an operation involving a transformation matrix and point vectors.

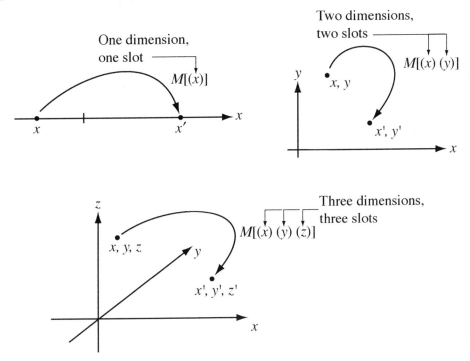

Figure 2.8 Point-transformation machines.

If we find that the domain and range of a point transformation M is not specified, then we may assume that $Ran(M) = Dom(M)$. In other words, we are then mapping a space one-to-one onto itself.

Study Figure 2.9. It is a simple example of a point transformation, the mapping of a circle onto itself. Here the image of each point on the circle we define to be the diametrically opposite point. We describe this transformation somewhat more rigorously as the reflection or inversion of a circle in its center. In this particular example, no point corresponds to its image. If a transformation carries each point through a rotation of 2π about the center, then the image of each point corresponds to its original. What happens if we reflect each point on the circle through a fixed diameter? In this case, only the images of the end points of the diameter coincide with their originals. (There is no restriction on transformations with respect to coincidence of originals and images.) If a transformation maps one geometric figure onto itself or onto another geometric figure, it maps the points that form the original figure onto itself or onto the set of points forming the second figure.

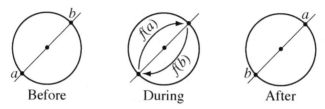

Figure 2.9 The transformation of a set onto itself.

This raises some intriguing philosophical questions, because there are two ways to interpret the phenomena of point transformations: Does the transformation in, effect, change the coordinate system, or does it move the points? And, what does "move" mean? Let's consider the latter first. Although we will often speak of moving an object (i.e., the set of points defining the object), no actual movement is made nor do we imply it. There is, of course, no path or trajectory given for a point to follow. The notion of movement is nothing more than a convenient fiction (more about motions in Chapter 4). Now, back to the first question, where we will generally take Felix Klein's point of view that the points move.

In his *Elementary Mathematics* Felix Klein observed that geometric transformations are nothing more than a generalization of the simple notion of function, and that point transformations constitute the simplest class of geometric transformations. Our concept of a function has broadened considerably, but Klein, of course, correctly recognized the role of functions as the fountainhead of geometric transformations.

Point transformations allow the point to persist—unchanged, invariant—in its role as a space element, bringing every point into correspondence with another point. Klein reminds us that the analytic expression of a point transformation is what analysis calls the introduction of new variables x', y', and z', with

$$x' = f(x, y, z)$$
$$y' = g(x, y, z)$$
$$z' = h(x, y, z)$$

There are two ways we can interpret this set of equations, passively or actively. If we make a passive interpretation, then the equations represent a change in the system of coordinates. We assign new coordinates x', y', z' to the point with the given coordinates x, y, z. If we make an active interpretation, then we fix the coordinate system and space changes. "To every point x, y, z the point x', y', z' is made

THEORY OF TRANSFORMATIONS

to correspond, so that there is, in fact, a transformation of the points in space. It is with this conception that we shall be concerned in what follows," as Klein put it. The active view is a sort of motion, but it is a motion without a path. We pluck the points from their original positions here and place them there. There is no in between.

The Product of Transformations

If f is a function mapping the elements x of set A on set B so that $f(x) = x'$, and if g is a function mapping the elements x' of set B on set C so that $g(x') = x''$, then the element $f(x)$ in B corresponds under g to the element $g(f(x))$ in C. In effect this means that we can find directly for each element of A the corresponding element of C (Figure 2.10). We call this new function the *composition* of f and g, and denote it by $g \circ f$ or $g(f)$, summarizing this algebraically as follows: $g(f(x)) = g \circ f(x)$, $\forall \{x : Dom(f)\}$.

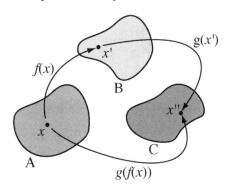

Figure 2.10 Composition of functions.

If $x \in Dom(f)$ and f is an element of B, then $f \in Dom(g)$, and we conclude that $g \circ f$ is the image of f under g. The single and double prime superscripts on x indicate the result of a transformation, so that x' is the image of x, and x'' is the image of x'. (The prime signs do not indicate differentiation.)

Because geometric transformations are functions, albeit limited to the "onto" and "one-to-one" variety, they too behave according to this composition rule. Accordingly, Figure 2.11 represents a point transformation of P to P' under α followed by the transformation of P' to P'' under β. We call this the product of transformations α and β and denote it $\beta \circ \alpha = \gamma$.

Note that we will frequently use lowercase Greek letters to denote geometric transformations or mappings. Also note the order of the transformations on the left side of the equation; we apply α first and then β. We read "$\beta \circ \alpha$" as "Alpha followed by beta," and we write transformations from right to left in the order of their execution or application. This convention also applies to matrix algebra.

Here is a concrete example of a geometric transformation product operation in the plane: Remember that an analogous process applies to a space of any dimension. If α is the mapping $x' = 2x+1$, $y' = -y$ and β is the mapping $x'' = x'-4$, $y'' = 3y'+2$, then a simple algebraic substitution shows that γ is the mapping $x'' = 2x-3$, $y'' = -3y+2$, (Figure 2.11). It is evident that the mapping γ is, indeed, the equivalent of the product of the mappings α and β. We conclude that the product of two successive transformations is itself a transformation. In fact, we might apply the same transformation repeatedly. For example, $\alpha \circ \alpha$ we denote as α^2, or $\alpha \circ \alpha \circ \alpha$ as α^3. The generalization to α^n is obvious.

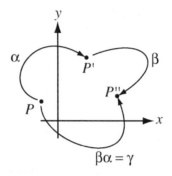

Figure 2.11 Product of transformations in the plane.

It is natural to ask if products of transformations have the same properties as real numbers with respect to associativity and commutativity. We find that the product of transformations always obeys the associative law. From Figure 2.12, for example, we see that $\alpha(P_1) = P_2$, $\beta(P_2) = P_3$, and $\gamma(P_3) = P_4$. Because $\beta \circ \alpha$ sends P_1 to P_3 and γ sends P_3 to P_4, then $\gamma \circ (\beta \circ \alpha)$ sends P_1 to P_4. Similarly, $(\gamma \circ \beta) \circ \alpha$ also sends P_1 to P_4. We conclude that $\gamma \circ (\beta \circ \alpha)$ and $(\gamma \circ \beta) \circ \alpha$ represent the same product of transformations as $\gamma \circ \beta \circ \alpha$.

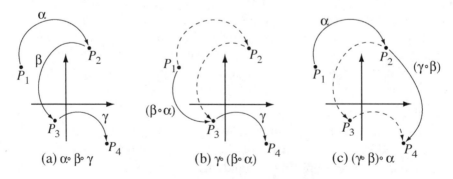

Figure 2.12 Associative property.

THEORY OF TRANSFORMATIONS

The associative law applies to any finite number of transformations. We may insert parentheses at will when forming products of transformations, if the order of the transformations is not changed. If the product $\delta \circ \gamma \circ \beta \circ \alpha$ is formed, then we may compose it in a number of ways: $(\delta \circ (\gamma \circ \beta)) \circ \alpha$ or $(\delta \circ \gamma) \circ (\beta \circ \alpha)$, for example. On the other hand, the product of transformations is not generally commutative. Confirming examples are easy to find, especially when rotations are present. For example, reflect the point C_1 on the circumference of a circle O through a diameter l and then rotate the resulting point C_2 through 90° clockwise about O to produce C_3. Call the reflection α and the rotation β; then $\beta(C_2) \circ \alpha(C_1) = C_3$. The commutation of this product yields $\alpha(C_2') \circ \beta(C_1) = C_3'$. As you see in Figure 2.13a and 2.13b, $C_3' \neq C_3$. This particular set of transformations is not commutative. However, look at Figure 2.13c. Here, $\omega \circ \phi = \phi \circ \omega$, and we see that ϕ and ω are commutative.

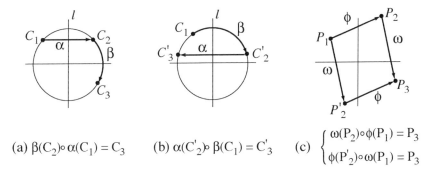

(a) $\beta(C_2) \circ \alpha(C_1) = C_3$ (b) $\alpha(C_2') \circ \beta(C_1) = C_3'$ (c) $\begin{cases} \omega(P_2) \circ \phi(P_1) = P_3 \\ \phi(P_2') \circ \omega(P_1) = P_3 \end{cases}$

Figure 2.13 Commutative property.

We cannot end our discussion of products of transformations without introducing a special transformation we call the identity transformation. It, and only it, we denote by the Greek letter iota, ι. The identity transformation has much the same function and properties as the number "one" when used in real number multiplication. Thus, we may write $\iota(P) = P$, meaning that ι transforms a point back onto itself (Figure 2.14a).

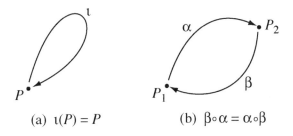

(a) $\iota(P) = P$ (b) $\beta \circ \alpha = \alpha \circ \beta$

Figure 2.14 Identity property.

If a transformation β reverses the result of a prior transformation α (Figure 2.14b), then $\beta\alpha = \alpha\beta = \iota$. (We may omit the operator symbol "∘" when our intent is obvious.) Studying the figure we see that $\alpha(P_1) = P_2$ and $\beta(P_2) = P_1$ so that $\beta(\alpha(P_1)) = P_1$ or $\beta\alpha(P_1) = P_1$, demonstrating the assertion that $\beta\alpha = \iota$. We say that β is the inverse of α, or $\beta = \alpha^{-1}$. Clearly $\alpha\alpha^{-1} = \iota$.

If α is a geometric transformation, then α^{-1} is its inverse; we say "alpha inverse." It is important to recognize that in general α^{-1} is not the same as $1/\alpha$. For example, if α maps x onto $x' = 2x - 3$, then α^{-1} must map x' onto $x = (x' + 3)/2$. The inverse transformation α^{-1} takes the output x' back to the input x. Furthermore, the domain of a transformation matches the range of its inverse. The inputs to α^{-1} are the outputs from α. The inputs to α are the outputs from α^{-1}. Finally, if $\beta = \alpha^{-1}$, then $\alpha = \beta^{-1}$ and $\beta\alpha = \alpha\beta$. In words, if β is the inverse of α, then α is the inverse of β. The relationship is completely symmetric.

It may be of interest to recall from calculus the definition of the inverse function: if $y = f(x)$, then $x = f^{-1}(y)$; if $x = f^{-1}(y)$, then $y = f(x)$. From this we derive the identity function $f^{-1}(f(x))$. This leads to the important observation that $(dy/dx)(dx/dy) = 1$. For example, let $y = f(x) = 2x - 3$, then $f^{-1}(y) = x = (y + 3)/2$ and $dy/dx = 2$, $dx/dy = 1/2$, so that $(dy/dx)(dx/dy) = (2)(1/2) = 1$.

If $\alpha\alpha = \iota$, then we call α an involution. A rotation of 180° is a trivial example of an involution because, applied twice in succession, it brings a point back to its origin. Every reflection is also an involution. (These last two examples do anticipate later discussions, but it may be helpful to think about them now.)

A transformation has the idempotent property if $\alpha^2 = \alpha$. This property arises in projective transformations, but it is easy to see that if α is a rotation of 360°, then $\alpha\alpha$ (i.e., α^2) is indeed the equivalent of α.

The Group Property

A group G is a special mathematical structure consisting of a set of elements S and an operator (∘) that combines elements two at a time. It is the mathematical structure relating the elements that distinguishes a group from a set. A group must have the following four properties:

1. **Closure**: This means that the combination of any two elements of the set under the action of the operator produces an element that is also a member of the set. Let the operation defined on a set S associate with every ordered pair of elements A, B of S a new element C that also belongs to S. C is the product of A, and B and we write: $C = A \circ B$.

Theory of Transformations

2. **Identity**: The set S must have an identity element, I. The set S contains an element I satisfying the relation $A \circ I = I \circ A = A$ for all elements of S. We also call I the unit or neutral element of the group.

3. **Inverse**: Each element must have a corresponding inverse. For every element A of S there is an inverse element A^{-1} that is also a member of S and that satisfies $A \circ A^{-1} = A^{-1} \circ A = I$.

4. **Associative**: An ordered sequence of the binary operations must be associative; that is, $(A \circ B) \circ C = A \circ (B \circ C)$.

5. **Commutative** (Optional): A group that has this fifth property we call a commutative group, or sometimes Abelian (after the Norwegian mathematician Neils Henrik Abel (1802-1829)). Many groups do not have it, but if they do then $A \circ B = B \circ A$, where A and B are elements of S.

Here are some nongeometric examples of groups. You can probably imagine and construct many others.

1. The integers.
Group operation: Addition
Identity element: 0, $n + 0 = n$
Inverse element of any element n is $-n$
Example (Figure 2.15):

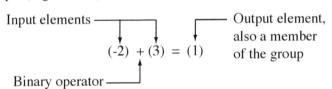

Figure 2.15 The group of integers.

2. All nonzero rational numbers
Group operation: Multiplication
Identity element: 1, $n \times 1 = n$
Inverse element of any element n is $1/n$
Example (Figure 2.16):

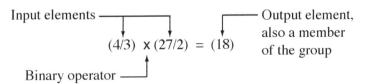

Figure 2.16 Nonzero rational numbers.

3. Any even number or any odd number
Group operation: Addition
Identity element: 0

Inverse element of any element: $E^{-1} = E$ and $O^{-1} = O$.

The following rules of addition apply: $E + E = E$, $E + O = O$, and $O + O = E$
(Figure 2.17).

This group contains a finite number of elements, and we can describe it with a group table. Note: we disregard the identity and inverse elements.

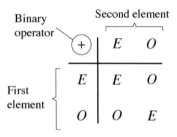

Figure 2.17 Even and odd number group.

There are also operators that are not associative, and here is a good example. Do not automatically assume that an operation is associative. Here is an example of one that is not: Let a proposed group consist of all the points of the plane and an operator whose input is any pair of points and whose output is their midpoint. So, given three points A, B, C (Figure 2.18), first find $C \circ B \circ A = P$ and then find $(C \circ B) \circ A = Q$. It is obvious from the figure below that $P \neq Q$ and that the operation is not associative. Nor is it a legitimate transformation. Do you see why? We can also prove this assertion using vector algebra.

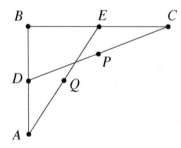

Figure 2.18 A non-associative operator.

For $C \circ B \circ A = P$ (remember, the operations proceed from right to left) first find the midpoint of A and B—generated by $B \circ A$ and labeled as point D. Next, find the midpoint of C and D using $C \circ D$ or $C \circ (B \circ A)$. This produces point P.

For $(C \circ B) \circ A$, find $(C \circ B) = E$ and then $E \circ A = (C \circ B) \circ A = Q$.

THEORY OF TRANSFORMATIONS

What does all this have to do with geometric transformations? Quite a lot, as it turns out. We frequently encounter sets of transformations of a geometric figure onto itself. If the inverse of each member of a set is also a member of the set, and if the product of any two members of the set is a member of the set, then that particular set of transformations is a group of transformations. We saw that transformation products are associative, and if the set also contains an inverse α^{-1} of each transformation α in the set, then the identity transformation ι must be a member of the set. All four properties are present, and the set does indeed fulfill all the qualifications required to form a group. Inverting or forming products of transformations belonging to a group never produces transformations that are outside of the group. Thus, a group of transformations is closed with respect to these operations.

Rotations of a circle about its center provide us with many opportunities to construct transformation groups. A small group in this category is the set comprised of rotations through 0° and π. Each is its own inverse. The product of each with itself is a rotation through 0°, and the product of the two is a rotation through π. The set of rotations 0° and $\pi/2$ is not a group because it does not contain the inverse of $\pi/2$ nor the resultant of $\pi/2$ with itself.

There are several ways to characterize groups, and here are some criteria we often use: If a group has exactly n elements, then it is finite and has order n. Otherwise, the group is infinite. If there is a smallest positive integer n so that $\alpha^n = \iota$, then α has order n; otherwise, the order is infinite. If every element of a group containing α is a power of α, then we say that the group is cyclic and α is the generator. Sometimes we denote such a group as $\langle \alpha \rangle$. The powers of a transformation always commute; for example, $\alpha^n \alpha^m = \alpha^{n+m} = \alpha^{m+n} = \alpha^m \alpha^n$, so we conclude that a cyclic group is always abelian.

We can construct a table of products for a finite transformation group. Mathematicians call it a Cayley table to honor the mathematician Arthur Cayley (1821-1895). Table 2.1 is the Cayley table for the cyclic transformation group of order four whose generator is $\rho = \pi/2$.

Table 2.1 The Cayley table for a cyclic transformation group of order 4.

	ι	ρ	ρ^2	ρ^3
ι	ι	ρ	ρ^2	ρ^3
ρ	ρ	ρ^2	ρ^3	ι
ρ^2	ρ^2	ρ^3	ι	ρ
ρ^3	ρ^3	ι	ρ	ρ^2

Remember, here $\iota = 0$, $\rho = \pi/2$, $\rho^2 = \pi$, and $\rho^3 = 3\pi/2$. The table makes it easy to check the closure and inverse properties. For our example, we see that every product permutation of transformations generates a member of the original set. And every element has an inverse: ρ and ρ^3 are mutually inverse, ρ^2 is self-inverse, and so is ι. Here are some more examples of geometric transformation groups.

1. This is a cyclic transformation group of order eight whose generator is the rotation $\rho = \pi/4$ and whose elements are $0, \rho, \rho^2, \rho^3, \rho^4, \rho^5, \rho^6$, and ρ^7. The identity element ι is 0, and the inverse of any element ρ^n is ρ^{8-n}. Its Cayley table of permutations is shown in Table 2.2. This group rotates a circle about its center, transforming it one-to-one onto itself.

Table 2.2 The Cayley table for a cyclic transformation group of order 8.

	ι	ρ	ρ^2	ρ^3	ρ^4	ρ^5	ρ^6	ρ^7
ι	ι	ρ	ρ^2	ρ^3	ρ^4	ρ^5	ρ^6	ρ^7
ρ	ρ	ρ^2	ρ^3	ρ^4	ρ^5	ρ^6	ρ^7	ι
ρ^2	ρ^2	ρ^3	ρ^4	ρ^5	ρ^6	ρ^7	ι	ρ
ρ^3	ρ^3	ρ^4	ρ^5	ρ^6	ρ^7	ι	ρ	ρ^2
ρ^4	ρ^4	ρ^5	ρ^6	ρ^7	ι	ρ	ρ^2	ρ^3
ρ^5	ρ^5	ρ^6	ρ^7	ι	ρ	ρ^2	ρ^3	ρ^4
ρ^6	ρ^6	ρ^7	ι	ρ	ρ^2	ρ^3	ρ^4	ρ^5
ρ^7	ρ^7	ι	ρ	ρ^2	ρ^3	ρ^4	ρ^5	ρ^6

2. This is an abelian transformation group of order four that maps the plane one-to-one and onto itself, whose elements are $\sigma_x(x,y) = (x,-y)$, $\sigma_y(x,y) = (-x,y)$, $\sigma_o(x,y) = (-x,-y)$, and $\iota(x,y) = (x,y)$. Each element is its own inverse. The elements σ_x, σ_y, and σ_o are involutions. (It so happens that involutions form transformation groups of order 2.) The Cayley table of permutations is shown in Table 2.3.

3. Next is an abelian transformation group of order four whose elements consist of the identity transformation ι and the reflections in each angle bisector l, m, n of an equilateral triangle (Figure 2.19 and Table 2.4).

Theory of Transformations

Table 2.3 The Cayley table for a group of order 4.

	ι	σ_x	σ_y	σ_o
ι	ι	σ_x	σ_y	σ_o
σ_x	σ_x	ι	σ_o	σ_y
σ_y	σ_y	σ_o	ι	σ_x
σ_o	σ_o	σ_y	σ_x	ι

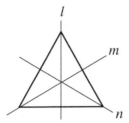

Figure 2.19 Reflections of an equilateral triangle.

Table 2.4 The Cayley table for reflections of an equilateral triangle.

	ι	σ_l	σ_m	σ_n
ι	ι	σ_l	σ_m	σ_n
σ_l	σ_l	ι	σ_n	σ_m
σ_m	σ_m	σ_n	ι	σ_l
σ_n	σ_n	σ_m	σ_l	ι

4. Finally, here is an abelian transformation group of order four whose elements consist of the identity transformation ι and the reflections of a circle in each of two mutually perpendicular diameters a and b and its center O (Figure 2.20 and Table 2.5).

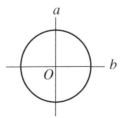

Figure 2.20 Identity transformation and reflections of a circle.

Let's review and summarize: A set of transformations forms a group under multiplication if the product of any two transformations of the set is also a uniquely determined transformation of the set (closure); the transformations of the set are associative under multiplication (i.e., $(\gamma\beta)\alpha = \gamma(\beta\alpha)$); the set contains an identity transformation ι such that $\alpha\iota = \iota\alpha = \alpha$ for every transformation α of the set; and every transformation α of the set has an inverse α^{-1} under multiplication, where α^{-1} is a transformation of the set and where $\alpha\alpha^{-1} = \alpha^{-1}\alpha = \iota$. A group may or may not be finite and cyclic or commutative (abelian). The presence or absence of the group property tells us whether or not we are dealing with a geometry and, if so, something about that geometry. Any set of transformations that distinguishes one geometry from another always forms a group.

The brevity of this section does not mean that the concept of the group is relatively unimportant. Quite the contrary, for we shall return to it repeatedly to elaborate many of the transformations we'll explore.

Table 2.5 Cayley table for reflections of a circle.

	ι	σ_a	σ_b	σ_o
ι	ι	σ_a	σ_b	σ_o
σ_a	σ_a	ι	σ_o	σ_b
σ_b	σ_b	σ_o	ι	σ_a
σ_o	σ_o	σ_b	σ_a	ι

Section 2.1 Exercises

1. If an algebraic expression defines a function without reference to its domain, then we assume that the domain is the set of all real numbers for which the expression is applicable. Find the domain and range of the following functions.

   a. $-2x$
   b. $3x + 2$
   c. x^2
   d. $-x^3 + 1$
   e. $-x^2 + 2$
   f. $\sin x$
   g. $(x+1)^{1/2}$
   h. $(1 - x^2)^{-2}$

2. Is the rule or function that sends a point (x, y) to (x^2, y) a geometric transformation? Explain your answer.

3. Is the rule or function that sends a point (x, y) to (x, y^3) a geometric transformation? Explain your answer.

4. Let σ_l and σ_m be reflections of a circle in a pair of perpendicular diameters l and m. Show that $\sigma_m \sigma_l$ is equivalent to a reflection of the circle in its center.

5. Find the inverse of each of the following functions.
   a. $x' = f(x) = -x$
   b. $x' = f(x) = 2x + 1$
   c. $x' = f(x) = 3x - 5$
   d. $x' = f(x) = -x + 1$
   e. $x' = f(x) = x^3$

6. Decide whether or not each of the following sets and operators form a group. If not, why not? Present each group in a format that gives appropriate identity and inverse elements. For those that are not groups, indicate what is necessary to make them so, if anything.
   a. The set of even integers under the operation of addition.
   b. The set of rational numbers under the operation of addition.
   c. The set of real numbers under the operation of multiplication.
   d. The set of positive rational numbers under the operation of multiplication.
   e. The set formed by the reciprocals of all positive integers n, under multiplication.
   f. The set of all integers under subtraction.

7. What do you conclude about the second and fourth transformation groups described in the examples following Table 2.1?

8. Demonstrate that for a finite transformation group each element of the group appears exactly once in each row and column of its Cayley table.

9. Determine which of the following sets are groups:
   a. Rotations of a circle through 0, $2\pi/3$, $4\pi/3$.
   b. Rotations of a circle through $\pi/2$, π, $3\pi/2$, 2π.
   c. Reflection of a square about a diagonal, rotation about its center by π, and the identity element.
   d. The reflections of a circle in two perpendicular diameters and the identity element.
   e. The reflection of a circle in its center and the identity element.

10. Show that every group of rotations of a circle is commutative.

2.2 Linear Transformations

Rigid-body, general affine and projective transformations we express algebraically as systems of linear equations. In this section, we review the forms and properties of these equations, their direct relationship to matrix algebra, and their role in describing geometric transformations. Our modus operandi is to show that linear transformations do indeed constitute a group and, therefore, meet the qualifications we discussed in the previous section.

Systems of Linear Equations

A set of equations of the form

$$a_{11}x^1 + a_{12}x^2 + \ldots + a_{1n}x^n = c_1$$
$$a_{21}x^1 + a_{22}x^2 + \ldots + a_{2n}x^n = c_2$$
$$\vdots$$
$$a_{r1}x^1 + a_{r2}x^2 + \ldots + a_{rn}x^n = c_r$$

is a system of linear equations in the n unknowns $x^1, x^2, \ldots, x^n$. The number of equations in this system is r, where r may be less than, equal to, or greater than n. For the present, we assume that the coefficients a_{ij} and the constants c_i are real numbers. If all the $c_i = 0$, then we have a system of homogeneous linear equations.

Here is an alternative notation scheme for a system of linear equations that is simple:

$$\sum_{j=1}^{n} a_{ij}x^j = c_i \quad \text{for } i = 1, 2, \ldots, r$$

The so-called *Einstein convention* recognizes the presence of a repeated index in a term to indicate summation and assumes that $r = n$, unless otherwise indicated. We ordinarily work in two- or three-dimensional space, so $n = 2$ or 3. Using the Einstein convention for a system of homogeneous linear equations, we write

$$a_{ij}x^j = 0$$

What could be simpler?

The solution space for a system of linear equations depends on both n and r. The geometric interpretation of a solution space is direct and obvious. For example, if $n = 3$ and $r = 1, 2$, or 3, then the solution space is a plane, a line, or a point in space, respectively.

The transformations of most importance to us here we express as sets of linear equations, or more precisely as sets of linear substitutions describing point transformations. Although projective transformations are the most general and inclusive

THEORY OF TRANSFORMATIONS

of a whole class of linear transformations, we will use the affine transformations as our starting point because they have more familiar geometric interpretations, and relatively simple restrictions of general affine geometry lead directly to the Euclidean geometries of similarity and congruence.

If x', y', z' are linear functions of x, y, z of the form

$$\begin{aligned} x' &= a_{11}x + a_{12}y + a_{13}z + a_{14} \\ y' &= a_{21}x + a_{22}y + a_{23}z + a_{24} \\ z' &= a_{31}x + a_{32}y + a_{33}z + a_{34} \end{aligned} \quad (2.1)$$

then this system describes an affine transformation. The constants a_{14}, a_{24}, a_{34} represent translations parallel to the coordinate axes, and we can account for them independently of the effects of the other coefficients, as we shall soon see. This means that to explore the more interesting aspects of affine transformations we must study the homogeneous linear transformation given by

$$\begin{aligned} x' &= a_{11}x + a_{12}y + a_{13}z \\ y' &= a_{21}x + a_{22}y + a_{23}z \\ z' &= a_{31}x + a_{32}y + a_{33}z \end{aligned} \quad (2.2)$$

We easily solve this system of equations for x, y, z as follows: Let Δ represent the determinant of the coefficients of Equations 2.2, where

$$\Delta = \begin{vmatrix} a_{11} & a_{12} & a_{13} \\ a_{21} & a_{22} & a_{23} \\ a_{31} & a_{32} & a_{33} \end{vmatrix}$$

If $\Delta \neq 0$, then the system of Equations 2.2 has a unique solution:

$$\begin{aligned} x &= a'_{11}x' + a'_{12}y' + a'_{13}z' \\ y &= a'_{21}x' + a'_{22}y' + a'_{23}z' \\ z &= a'_{31}x' + a'_{32}y' + a'_{33}z' \end{aligned} \quad (2.3)$$

In the plane these equations become

$$\begin{aligned} x &= a'_{11}x' + a'_{12}y' \\ y &= a'_{21}x' + a'_{22}y' \end{aligned} \quad (2.4)$$

Let's see how we affect some familiar geometric figures of the plane by subjecting them to an affine transformation. We start with a line in the plane

$$Ax + By + C = 0 \quad (2.5)$$

Substituting from Equations 2.4 we obtain

$$A(a'_{11}x' + a'_{12}y') + B(a'_{21}x' + a'_{22}y') + C = 0$$

or

$$(Aa'_{11} + Ba'_{21})x' + (Aa'_{12} + Ba'_{22})y' + C = 0$$

If we let $A' = (Aa'_{11} + Ba'_{21})$, $B' = (Aa'_{12} + Ba'_{22})$, and $C' = C$, then

$$A'x' + B'y' + C' = 0 \tag{2.6}$$

As you can see, Equation 2.6 has exactly the same form as Equation 2.5, and we correctly conclude that lines transform into lines under affine transformations.

Now let us try applying an affine transformation to a second-degree curve,

$$Ax^2 + 2Bxy + Cy^2 + 2Dx + 2Ey + F = 0 \tag{2.7}$$

Once again, we perform the linear substitutions from Equations 2.4 to obtain

$$A(a'_{11}x' + a'_{11}y')^2 + 2B(a'_{11}x' + a'_{12}y')(a'_{21}x' + a'_{22}y')$$
$$+ C(a'_{21}x' + a'_{22}y')^2 + 2D(a'_{11}x' + a'_{12}y')$$
$$+ 2E(a'_{21}x' + a'_{22}y') + F = 0$$

If we expand this rather messy equation and rearrange terms appropriately, we obtain

$$A'(x')^2 + 2B'x'y' + C'(y')^2 + 2D'x' + 2E'y' + F' = 0 \tag{2.8}$$

where the $A', B', \ldots, F'$ are combinations of $A, B, \ldots, F$ and the coefficients a'_{ij} of the linear transformation. The form of Equation 2.8 is identical to Equation 2.7, so that an affine transformation maps a second-degree curve into another second-degree curve. Do you see that by the very nature of these linear substitutions an affine transformation maps any curve of degree n into another curve of the same degree?

As promised above, we now return to see how we account for the translation components a_{14}, a_{24}, a_{34} of Equations 2.1. We limit our discussion to the space of the plane, so that from Equation 2.1 if $a_{13}, a_{23}, a_{31}, a_{32}, a_{33}, a_{34} = 0$, then

$$\begin{aligned} x' &= a_{11}x + a_{12}y + a_{14} \\ y' &= a_{21}x + a_{22}y + a_{24} \end{aligned} \tag{2.9}$$

We interpret this transformation as the succession of

$$\begin{aligned} x' &= a_{11}x + a_{12}y \\ y' &= a_{21}x + a_{22}y \end{aligned} \tag{2.10}$$

and the translation

$$\begin{aligned} x'' &= x' + a_{14} \\ y'' &= y' + a_{24} \end{aligned} \tag{2.11}$$

Equations 2.10 represent the homogeneous or centered affine group, and all the unique characteristics of the group we determine by the a_{ij} coefficients. The term "centered" indicates that the origin or center is invariant under all transformations of the group. This is easily verified by letting $x = 0$, $y = 0$ in Equation 2.10.

THEORY OF TRANSFORMATIONS

We must now demonstrate that the set of homogeneous affine transformations does indeed form a group. We begin by showing that a succession of linear substitutions on the homogeneous affine set is a closed binary operation. Let H_1 and H_2 denote two such transformations, so that

$$H_1 \begin{cases} x' = a_{11}x + a_{12}y \\ y' = a_{21}x + a_{22}y \end{cases} \text{ and } H_2 \begin{cases} x'' = b_{11}x' + b_{12}y' \\ y'' = b_{21}x' + b_{22}y' \end{cases}$$

If the succession, or binary operation, $H_1 \circ H_2$ is also a member of the set, then the binary operation is closed. Substitute the equations for H_1 into those for H_2, producing

$$H_1 \circ H_2 \begin{cases} x'' = b_{11}(a_{11}x + a_{12}y) + b_{12}(a_{21}x + a_{22}y) \\ y'' = b_{21}(a_{11}x + a_{12}y) + b_{22}(a_{21}x + a_{22}y) \end{cases}$$

and by rearranging terms obtain

$$H_1 \circ H_2 \begin{cases} x'' = (a_{11}b_{11} + a_{21}b_{12})x + (a_{12}b_{11} = a_{22}b_{12})y \\ y'' = (a_{11}b_{21} + a_{21}b_{22})x + (a_{21}b_{21} + a_{22}b_{22})y \end{cases} \quad (2.12)$$

The form of this set of equations is identical to that of Equations 2.10, and we conclude that the succession $H_1 \circ H_2$ is closed.

It is now appropriate for a brief but important digression to introduce and review matrix algebra.

Matrices and Linear Transformations

In order to write the coefficients of H_1 and H_2 in matrix form, let

$$\mathbf{M}_1 = \begin{bmatrix} a_{11} & a_{12} \\ a_{21} & a_{22} \end{bmatrix} \text{ and } \mathbf{M}_2 = \begin{bmatrix} b_{11} & b_{12} \\ b_{21} & b_{22} \end{bmatrix}$$

If $\mathbf{p} = \begin{bmatrix} x & y \end{bmatrix}^T$, $\mathbf{p}' = \begin{bmatrix} x' & y' \end{bmatrix}^T$, and $\mathbf{p}'' = \begin{bmatrix} x'' & y'' \end{bmatrix}^T$, using matrix notation

$$\mathbf{p}' = \mathbf{M}_1 \mathbf{p} \text{ and } \mathbf{p}'' = \mathbf{M}_2 \mathbf{p}'$$

Substituting appropriately, we find

$$\mathbf{p}'' = \mathbf{M}_2 \mathbf{M}_1 \mathbf{p} \quad (2.13)$$

It is a simple exercise to show that Equation 2.13 produces the same result as Equations 2.12. $\mathbf{M}_2 \mathbf{M}_1$ represents the succession, or product, of two homogeneous affine transformations $\mathbf{M}_1$ and $\mathbf{M}_2$.

We will now use matrices to simplify the work that remains. Let us return to our demonstration. We may safely assume that the results of additional affine transformations on $\mathbf{p}$ are determined by the matrix product $\mathbf{M}_n \mathbf{M}_{n-1} \ldots \mathbf{M}_2 \mathbf{M}_1 \mathbf{p}$. From the properties of matrix multiplication we infer that the product of a succession of homogeneous affine transformations is associative; for example $(\mathbf{M}_3\mathbf{M}_2)\mathbf{M}_1\mathbf{p}$ and $\mathbf{M}_3(\mathbf{M}_2\mathbf{M}_1)\mathbf{p}$ are equivalent to $\mathbf{M}_3\mathbf{M}_2\mathbf{M}_1\mathbf{p}$.

The identity transformation associated with the affine group is simply $x' = x$, $y' = y$. This means that $a_{11} = a_{22} = 1$ and $a_{21} = a_{12} = 0$, so that the identity matrix is

$$\mathbf{I} = \begin{bmatrix} 1 & 0 \\ 0 & 1 \end{bmatrix}$$

Finally, we must demonstrate that each homogeneous affine transformation has an inverse. Again, we shall resort to using properties of matrices. First, we find the inverses of the functions given in Equations 2.10 by solving for x and y:

$$\begin{aligned} x &= \frac{a_{22}}{a_{11}a_{22} - a_{21}a_{12}} x' - \frac{a_{12}}{a_{11}a_{22} - a_{21}a_{12}} y' \\ y &= \frac{-a_{21}}{a_{11}a_{22} - a_{21}a_{12}} x' + \frac{a_{11}}{a_{11}a_{22} - a_{21}a_{12}} y' \end{aligned} \quad (2.14)$$

These equations hold if $a_{11}a_{22} - a_{21}a_{12} \neq 0$. Obviously, the form of this set of equations is identical to Equations 2.10, so it is a valid affine transformation, and these equations do represent the inverse of that transformation. Therefore, we find that to each point x, y not only does there correspond a point x', y', but also there is only one, and the transformation from x', y' to x, y is an affine transformation. Furthermore, if

$$\mathbf{M} = \begin{bmatrix} a_{11} & a_{12} \\ a_{21} & a_{22} \end{bmatrix}$$

and

$$\mathbf{M}^{-1} = \begin{bmatrix} \dfrac{a_{21}}{a_{11}a_{22} - a_{21}a_{12}} & \dfrac{-a_{12}}{a_{11}a_{22} - a_{21}a_{12}} \\ \dfrac{-a_{21}}{a_{11}a_{22} - a_{21}a_{12}} & \dfrac{a_{11}}{a_{11}a_{22} - a_{21}a_{12}} \end{bmatrix}$$

provided that the $\det \mathbf{M} \neq 0$. We observe that the elements of $\mathbf{M}^{-1}$ are precisely the coefficients of Equations 2.14. We now assert that Equations 2.2 or 2.10 describe a homogeneous affine transformation group if and only if the determinant of the coefficients does not vanish. The results above apply equally to spaces of two, three, or more dimensions. This completes the demonstration that the homogeneous affine transformations form a group. Using a similar approach, we can show how other linear transformations conform to our notion of a group.

But wait; we seem to have some unfinished business here. What does happen if the determinant vanishes (i.e., $\det \mathbf{M} = 0$)? Which brings us to …

Theory of Transformations

Singular and Nonsingular Linear Transformations

The restriction on the matrix of transformation coefficients such that $\det \mathbf{M} \neq 0$ was no idle threat. If $\det \mathbf{M} = 0$, then we no longer have a one-to-one transformation but a many-to-one mapping of a space onto a subset of itself, and it has no inverse. We call this mapping a singular linear transformation. Here is an example: Let

$$x' = x + 2y$$
$$y' = 3x + 6y$$

The determinant of this set of equations is equal to zero, and we find that all points lying on the line $y = -x/2$ map onto $x' = 0$, $y' = 0$, the origin, and that there is no way to sort them out and get them back with an inverse transformation! Obviously it is preferable to work with the more tractable nonsingular linear transformations.

Here is a brief discussion of an interesting consequence of some of the properties of homogeneous linear transformations. It is an interpretation of nonsingular linear homogeneous transformations of the plane.

Every nonsingular linear homogeneous transformation of the plane is a product of a reflection of the plane with respect to the line $y = x$, expansion (or contraction) of the plane, and shears parallel to a coordinate axis. This transformation is given by

$$x' = ax + by$$
$$y' = cx + dy$$

and whose transformation matrix is

$$\mathbf{M} = \begin{bmatrix} a & b \\ c & d \end{bmatrix}$$

where $\det \mathbf{M} \neq 0$.

It is easy to see that we could define $\mathbf{M}$ as the product of other transformation matrices of the form

$$\mathbf{A} = \begin{bmatrix} 0 & 1 \\ 1 & 0 \end{bmatrix}, \mathbf{B} = \begin{bmatrix} k & 0 \\ 0 & 1 \end{bmatrix}, \mathbf{C} = \begin{bmatrix} 1 & 0 \\ 0 & k \end{bmatrix},$$

$$\mathbf{D} = \begin{bmatrix} 1 & k \\ 0 & 1 \end{bmatrix}, \text{ and } \mathbf{E} = \begin{bmatrix} 1 & 0 \\ k & 1 \end{bmatrix}$$

Matrix $\mathbf{A}$ represents a reflection of the plane in the line $y = x$, $\mathbf{B}$ and $\mathbf{C}$ represent expansions or contractions of the plane, and $\mathbf{D}$ and $\mathbf{E}$ represent shears parallel to a coordinate axis.

Section 2.2 Exercises

1. Give a geometric interpretation of each of the following linear transformations of the plane. Assume all constants are non-zero.
   a. $x' = x, \quad y' = ay$
   b. $x' = ax, \quad y' = by$
   c. $x' = ax, \quad y' = ay$
   d. $x' = x, \quad y' = ax + y$
   e. $x' = (x - y)/\sqrt{2}, \quad y' = (x + y)/\sqrt{2}$

2. Show that the following is a singular linear transformation, and give a geometric interpretation of the mapping.
   $$x' = x + y, \quad y' = y + z, \quad z' = x + 2y + z$$

Note: Regarding exercises 3 through 5, this textbook does not discuss methods for solving systems of linear equations. However, you should be familiar with a variety of methods for doing this from other studies.

3. Solve the following system of linear equations:
   $$2x - 4y + 5z = 10$$
   $$2x - 11y + 10z = 36$$
   $$4x - y + 5z = -6$$

4. Solve the following system of linear equations:
   $$2x - y + z = 1$$
   $$x + y - z = 2$$
   $$3x - y + z = 0$$

5. Solve the following system of linear equations:
   $$x + 3y + z = 6$$
   $$3x - 2y - 8z = 7$$
   $$4x + 5y - 3z = 17$$

6. Prove that under nonsingular linear homogeneous transformations of the plane:
   a. The image of the origin is the origin.
   b. The images of parallel lines are parallel lines.

7. Describe the effect on the plane of the transformation
   $$\mathbf{M} = \begin{bmatrix} 6 & 3 \\ 2 & 1 \end{bmatrix}$$

2.3 Geometric Invariants

Consider the group of all homogeneous linear substitutions for the coordinates of a point in space. As we did earlier, we write these as

$$x' = a_{11}x + a_{12}y + a_{13}z$$
$$y' = a_{21}x + a_{22}y + a_{23}z$$
$$z' = a_{31}x + a_{32}y + a_{33}z$$

The nine coefficients a_{ij} are not at all independent of one another, because their relationships will depend on the conditions of invariance that we establish. For example, we may choose relationships that arise in the isometric orthogonal substitutions, as is the case for rotation about the origin, where $x'^2 + y'^2 + z'^2 = x^2 + y^2 + z^2$. (This says that the distance from any point to the origin is invariant under a rotation.) Substitution produces the following six relationships for the nine coefficients:

$$a_{11}^2 + a_{21}^2 + a_{31}^2 = 1$$
$$a_{12}^2 + a_{22}^2 + a_{32}^2 = 1$$
$$a_{13}^2 + a_{23}^2 + a_{33}^2 = 1$$
$$a_{12}a_{13} + a_{22}a_{23} + a_{32}a_{33} = 0$$
$$a_{11}a_{13} + a_{21}a_{23} + a_{31}a_{33} = 0$$
$$a_{11}a_{12} + a_{21}a_{22} + a_{31}a_{32} = 0$$

This leads to a solution for the so-called transposed linear substitutions:

$$x = a_{11}x' + a_{21}y' + a_{31}z'$$
$$y = a_{12}x' + a_{22}y' + a_{32}z'$$
$$z = a_{13}x' + a_{23}y' + a_{33}z'$$

Now we consider functions of these coordinate variables (perhaps defining lines, planes, curves or surfaces). We limit the discussion to homogeneous functions; for example, the linear forms $f = Ax + By + Cz$, and the quadratic forms

$$g = Ax^2 + By^2 + Cz^2 + 2Dxy + 2Exz + 2Fyz + Gx + Hy + Iz$$

or forms of higher dimension, simultaneous systems of linear or higher forms, and so on. Given a set of points and some linear, quadratic, or higher form that is a function of these point coordinates, then any other function of the coordinates in the specified form that remains unchanged under a certain set of linear substitutions (transformations) is a geometric invariant of the set. The principal objective of the theory of invariants is to discover those functions and forms that various linear transformations preserve. Notice that possible geometric invariants under a particular transformation may include metric properties (e.g., length, angle, area, and volume), relational properties (between, inside, outside, etc.), and form (circle-to-circle, line-to-line, etc.).

Each transformation group preserves some geometric properties (leaves them invariant) and fails to preserve others. Transformations producing metric invariants, that is distance-preserving transformations, form perhaps the largest, most important group we will study. However, there are also many important nonmetric invariants to consider. These two distinct characteristics of transformations pose many interesting problems. For example, we might ask if a particular transformation or transformation group produces the metric invariants that preserve distance, or angles, or areas. Other transformations we try to characterize by their nonmetric invariants, such as those preserving betweenness, the degree of a curve or surface, and whether or not a closed curve remains closed. In this way we explore and learn to use various sets, usually groups, of transformations to produce invariants of a well-defined type. Affine invariants are characteristic of affine transformations, as are projective and topological invariants characteristic of projective and topological transformations. These processes we will explore thoroughly in the succeeding sections and chapters. This section strives for an intuitive appreciation of five important kinds of geometric invariants: those that fix points, those that preserve lines, those that preserve distance and angle (the metric transformations), and those that preserve orientation of a figure.

In the remaining following sections, we consider the invariant metric properties of Euclidean geometry. Then, by imposing a sequence of generalizations, we gradually reduce the number of geometric invariants until we are left with only the invariant properties of projective transformations.

Fixed Points

A point is a fixed point under a transformation of the plane or space if the transformation maps it onto itself. For all rotations about the origin, the origin itself is the fixed point, similarly for expansion or contraction about the origin. There are no fixed points under translation transformations. However, because we can rotate a geometric figure about any point in the plane, that point is a fixed point under that rotation. Again, the same generalization is true for expansion and contraction.

Line-Preserving Transformations

A transformation α, having the property such that if l is a line and $\alpha(l)$ is also a line, is a collineation, meaning that α preserves the line l. We shall take the view that a line consists of a set of points, and $\alpha(l)$ is the set of all points $\alpha(P)$, where any point P is on the line l. Given a collineation α, then $\alpha(l)$ is a line, and point $\alpha(P)$ is on line $\alpha(l)$ if and only if point P is on line l (Figure 2.21). Translations and rotations are examples of collineation transformations. A collineation β that fixes every point on a line m is an axial collineation, where m is the axis of β. A rotation in space is an axial collineation, and so is a reflection in a line.

Theory of Transformations

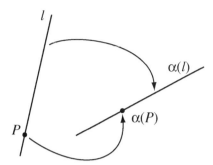

Figure 2.21 Line-preserving transformations.

Distance-Preserving Transformations

All distance-preserving transformations are also angle preserving and area- and volume-preserving. The most restricted Euclidean metric geometry, the geometry of congruent figures and their measurement, assumes that all motions of a figure must leave its size and shape invariant. Sometimes mathematicians call this assumption or postulate the principle of relativity for Euclidean geometry. This property depends on two underlying, fundamental invariants: distance and angle.

Translation and rotation are distance-preserving metric transformations. A square figure, for example, retains all its geometric properties relating to size and shape under these transformations. There are, however, certain metric transformations that preserve neither distance nor angle, yet preserve area or volume. These are equiareal or equivolume affine transformations.

Angle-Preserving Transformations

A conformal transformation preserves the angle measures of a geometric figure. It is indeed in the metric family of transformations, although it may not necessarily preserve the distance measures. The simplest example is the expansion or contraction of a figure, giving rise to the geometry of similarity (Figure 2.22).

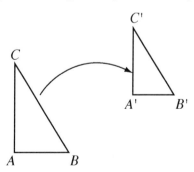

Figure 2.22 Angle-preserving transformations.

Conformal transformations extend to stereographic projections, to intersecting curves (Figure 2.23), and to nets of curves or surfaces.

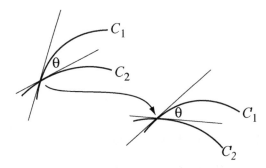

Figure 2.23 The angle between intersecting curves is invariant under translation and rotation.

Orientation-Preserving Transformations

Given a square, $ABCD$, we can denote and traverse its perimeter in two ways, in either a clockwise or counterclockwise sense (Figure 2.24). Thus, a square has two possible orientations. A transformation that retains or preserves the original orientation is a *direct* transformation; one that reverses orientation is an *opposite* transformation.

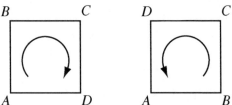

Figure 2.24 Two orientations of a square.

In Figure 2.25 notice that a reflection in the line m in the plane, or in the plane Π in space, is an operation that reverses orientation. Translations and rotations, on the other hand, preserve orientations.

2.4 Isometries

An *isometry* (from the Greek *isos* and *metron*...equal measure) is a rigid-body motion of the points of a geometric figure. The defining property of an isometry is that it preserves distances between points on the figure. If an isometry α maps points P and Q to P' and Q', then $PQ = P'Q'$ (Figure 2.26). In other words, a transformation α is an isometry if $P' = \alpha(P)$, $Q' = \alpha(Q)$ and $P'Q' = PQ$.

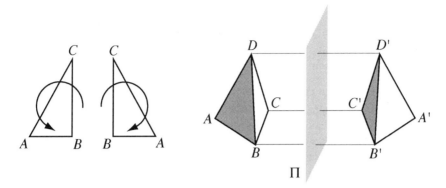

Figure 2.25 Reflection is an orientation-reversing transformation.

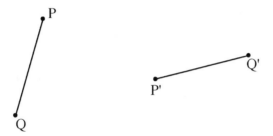

Figure 2.26 A simple isometry.

We often describe these transformations as motions or rigid-body motions because they resemble physical movements. The analogy is useful; however, remember no real motion is actually taking place, and there is no preferred path described for the transformation from P to P'. In fact, there is no path at all. Because these transformations preserve distance, it is easy to demonstrate that motions send each geometric figure into a congruent one, so sometimes we call them congruent transformations. Given an isometry α and two sets of points $\{P_i\}, \{Q_i\}$; if $\alpha\{P_i\} = \{Q_i\}$, then $\{P_i\}$ and $\{Q_i\}$ are congruent.

The general equations for an isometry in the plane are

$$x' = a_{11}x + a_{12}y + a_{14}$$
$$y' = a_{21}x + a_{22}y + a_{24}$$
(2.15)

where $\begin{vmatrix} a_{11} & a_{12} \\ a_{21} & a_{22} \end{vmatrix} = \pm 1$, and terms a_{14} and a_{24} represent translations parallel to the coordinate axes. We interpret the coefficients a_{ij} as we progress through this section.

The general equations for an even isometry in the plane, neglecting translations, are $x' = a_{11}x + a_{12}y$ and $y' = a_{21}x + a_{22}y$, where $a_{11} = a_{22}$, $a_{12} = -a_{21}$, and $a_{11}^2 + a_{12}^2 = 1$.

We interpret an even isometry as one that is orientation preserving. However, we allow for the possibility of orientation-reversing or odd isometries, and this leads to a further generalization or refinement of the conditions on the above equations. We constrain the coefficients so that their determinant

$$|a_{ij}| = \pm 1 \begin{cases} + \text{ even isometry} \\ - \text{ odd isometry} \end{cases}$$

Isometries are the first of four major transformation groups to explore here. The other three are similarities, affinities and projectivities. We recognize the isometries as a special subgroup of affine transformations. There are four types of isometries in the plane (with corresponding isometries in space), each either orientation-preserving or orientation-reversing and thus either a direct or an indirect isometry, respectively:

1. Reflection, indirect
2. Rotation, direct
3. Translation, direct
4. Glide reflection, indirect

We will show that a rotation is a product of two reflections in intersecting lines, a translation is a product of two reflections in parallel lines, and a glide reflection is a product of three reflections. We can express all isometries as products of reflections, and, conversely, products of reflections produce one of these isometries. So, it turns out that reflection is the basic building block of all isometries.

Reflections in the Plane

We begin with a synthetic definition of a reflection in the plane and later extend this to a three-dimensional space. In due course, we will develop appropriate analytic expressions. We denote a reflection by σ_m, where m identifies the line or axis of reflection. In Figure 2.27a, the point $\sigma_m(A)$ is the reflected image of point A in line m, and conversely we can think of A as the reflected image of $\sigma_m(A)$. We find that, given a point A and a line of reflection m, a reflection operator σ_m produces an image point B such that the line segment constructed between A and B is perpendicular to m and bisected by it (Figure 2.27b). Notice that $\sigma_m(A) = B$, $\sigma_m(B) = A$, and that $\sigma_m^2 = 1$; therefore σ_m is an involution. Furthermore, σ_m is one-to-one because if $\sigma_m(A) = \sigma_m(B)$, then $\sigma_m(\sigma_m(A)) = \sigma_m(\sigma_m(B))$ and $A = B$. Finally, if A is on m, then $\sigma_m(A) = A$.

THEORY OF TRANSFORMATIONS 57

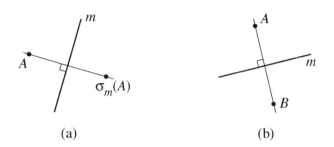

Figure 2.27 Reflection in a line.

A reflection is orientation reversing. The reflection of triangle *ABC* in line *m* produces triangle $A'B'C'$ (Figure 2.28). A line *m* is a line of symmetry for any set of points $\{P_i\}$ if $\sigma_m\{P_i\} = \{P_i\}$. This is certainly true for the set of six points defining the vertices of the two triangles *ABC* and $A'B'C'$.

In Figure 2.29, we see that if a line is parallel to the axis of reflection, then so is its image. Also, if a line is perpendicular to the axis of reflection, then so is its image; and the line and its image are collinear.

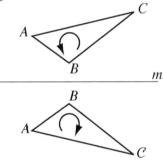

Figure 2.28 A reflection is orientation-reversing.

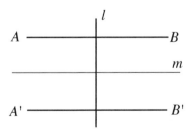

Figure 2.29 Reflection of lines parallel and perpendicular to the axis of reflection.

Again, although we have yet to demonstrate it, reflection is the basis of all isometries and therefore has all the properties we would expect of the isometries. Other invariants of reflection to consider are: parallelness, perpendicularity, betweenness and midpoints. To reinforce and expand your conception of isometries in general and reflections in particular, try to prove the following four assertions:

1. If three noncollinear points A, B, and C and the isometries α and β are such that $\alpha(A) = \beta(A)$, $\alpha(B) = \beta(B)$ and $\alpha(C) = \beta(C)$, then $\alpha = \beta$.
2. If an isometry fixes two points, then it is either a reflection or the identity.
3. If an isometry fixes two distinct points on a line, then it fixes all the points on the line.
4. If an isometry fixes three noncollinear points, then the isometry must be the identity.

Inversion in a Point and Halfturns in the Plane

Consider the construction in Figure 2.30. Given three arbitrary points A, B, and C in the plane and a fourth point P, map point A' so that A, P, A' are collinear and $AP = PA'$, similarly for B' and C'. This geometric transformation is an inversion, and point P is the center of inversion. As you can see by comparing triangles ABC and $A'B'C'$ this transformation is orientation pre-serving and an isometry.

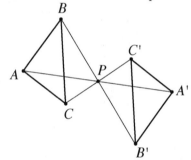

Figure 2.30 Inversion in a point.

An inversion produces a halfturn. It is as though we create the image of a figure by a rigid body rotation of 180° about the center of inversion. A halfturn is nothing more than a special rotation (special because it is always 180°). Although we are anticipating later developments of this sequel, we will proceed a bit farther because these notions seem to offer a natural transition to products of reflections. An inversion has many features in common with a reflection. However, the correspondence breaks down when we recall that inversion is orientation preserving and reflection is not. In fact, we will soon see that an inversion is the product of two reflections subject to certain conditions.

The product of two halfturns about the same center is the identity transformation $\sigma_P \sigma_P = i$, so we characterize it as an involutory rotation. However, the product of two halfturns about different centers is a translation (Figure 2.31), which is obviously orientation preserving and an isometry.

Theory of Transformations

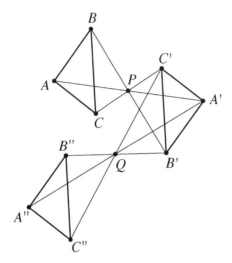

Figure 2.31 The product of two halfturns.

Furthermore, referring to Figure 2.32, we discover the rather surprising fact that given any point A and two centers of inversion P and Q, the directed distance AA'' is exactly twice PQ.

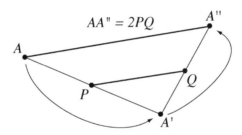

Figure 2.32 The product of two halfturns moves A twice the directed distance from P to Q.

We find that the product of three halfturns is a halfturn and yields yet another surprising result. Given three noncollinear points P, Q, and R as centers of inversion and any arbitrary point A, then $\sigma_R \sigma_Q \sigma_P(A) = \sigma_S(A)$, where point S completes the parallelogram $PQRS$! (Figure 2.33)

Here are the equations for various kinds of reflections and inversions in the plane. The derivations for some of these are requested in the exercises that conclude this section.

Inversion in the origin (Figure 2.34):

$x' = -x$
$y' = -y$

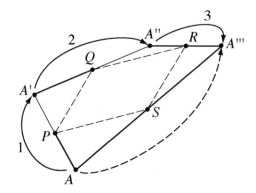

Figure 2.33 The product of three halfturns.

Inversion about *a*, *b* (Figure 2.34):

$x' = -x + 2a$
$y' = -y + 2b$

Reflection in the *y* axis (Figure 2.35):

$x' = -x$
$y' = y$

Reflection in the *x* axis (Figure 2.35):

$x' = x$
$y' = -y$

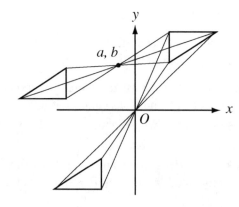

Figure 2.34 Inversion in the origin and inversion in an arbitrary point *a*, *b*.

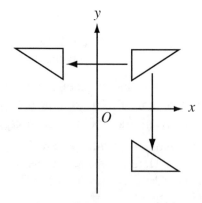

Figure 2.35 Reflection in the *x* and *y* axis.

THEORY OF TRANSFORMATIONS

Reflection in a line through the origin (Figure 2.36):

$x' = x\cos 2\theta + y\sin 2\theta$
$y' = x\sin 2\theta - y\cos 2\theta$

where θ is the inclination of m.

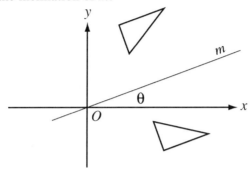

Figure 2.36 Reflection in a line through the origin.

Reflection in an arbitrary line (Figure 2.37):

A. $x' = x\cos 2\theta + y\sin 2\theta + 2d\cos\psi$ and $y' = x\sin 2\theta - y\cos 2\theta + 2d\sin\psi$
where θ is the inclination of m and ψ is the angle of a directed vector from the origin and perpendicular to m.

B. If m is given by $ax + by + c = 0$, then

$$x' = x - 2a(ax + by + c)/(a^2 + b^2)$$
$$y' = y - 2b(ax + by + c)/(a^2 + b^2)$$

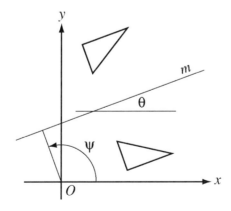

Figure 2.37 Reflection in an arbitrary line.

The Product of Two Reflections in the Plane

Given two parallel lines l and m, then $\sigma_m\sigma_l$ is a translation through twice the directed distance from l to m (Figure 2.38). The first reflection reverses the orientation of ABC, and the second reflection restores the original orientation. The distance between l and m is d, and $d = DC' + C'D'$. Because $CD = DC'$ and $D'C'' = C'D'$, then $CD + DC' + C'D' + D'C'' = 2d$. If line b is perpendicular to l and m and intersects them at L and M, then $\sigma_m\sigma_l = \sigma_M\sigma_L$. We may now assert that every translation is a product of two reflections in parallel lines; conversely, a product of two reflections in parallel lines is a translation.

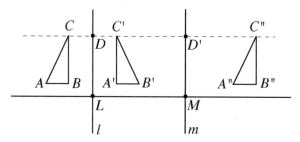

Figure 2.38 The product of two reflections in parallel lines.

Here is a demonstration of an interesting equivalence between certain products of parallel reflections. Given parallel lines l, m, and n, an arbitrary point A, and a line b through A and perpendicular to l, m, and n,, there are unique lines p and q such that $\sigma_m\sigma_l = \sigma_n\sigma_p = \sigma_q\sigma_n$ (Figure 2.39). From $\sigma_m\sigma_l = \sigma_n\sigma_p$ we obtain $\sigma_n\sigma_m\sigma_l = \sigma_p$, and we solve for the distance x_p:

$$a + 2e = 2(a + b + e - \frac{b}{2} - x_p) \text{ and } x_p = (a+b)/2 = d$$

We discover that p is the unique line such that the directed distance from p to n equals the directed distance from l to m.

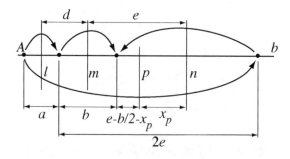

Figure 2.39 Equivalent products of parallel reflections: Part 1.

Theory of Transformations 63

Finally, from $\sigma_m\sigma_l = \sigma_q\sigma_n$ we obtain $\sigma_q\sigma_m\sigma_l = \sigma_n$, and we solve for the distance x_q (Figure 2.40).

$$a+b+2\left(e-\frac{b}{2}+x_q\right) = 2\left(\frac{a}{2}+d+e\right) \text{ and } x_q = d$$

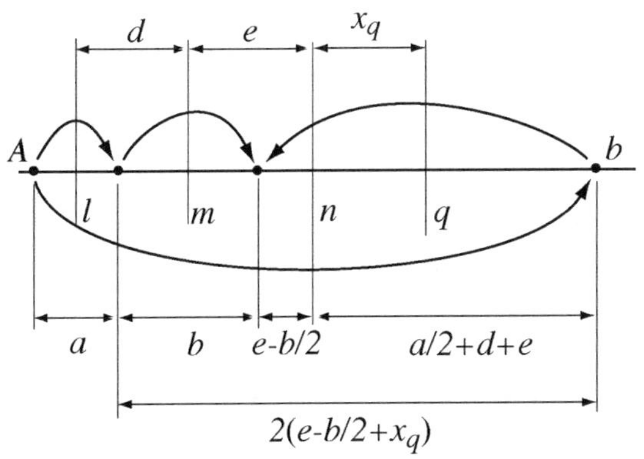

Figure 2.40 Equivalent products of parallel reflections: Part 2.

Similar to our discovery above, we find that q is also a unique line, and its directed distance from n to q also equals the directed distance from l to m.

If two lines l and m intersect at a point O (Figure 2.41), then the product of the two reflections $\sigma_m\sigma_l$ is a rotation $\rho_{O\theta}$, where the Greek letter rho denotes rotation, the first subscript O denotes the center of rotation and θ is the angle through which the points or figures are rotated. Notice that $\theta/2$ is the angle between l and m. This product is orientation preserving and an isometry.

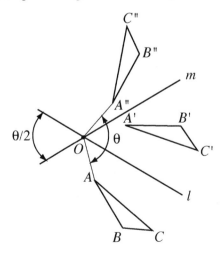

Figure 2.41 The product of two reflections in two intersecting lines.

The converse is also true: every rotation is a product of two reflections in intersecting lines. In general, for l and m intersecting, $\sigma_m \sigma_l \neq \sigma_l \sigma_m$ (Figure 2.42).

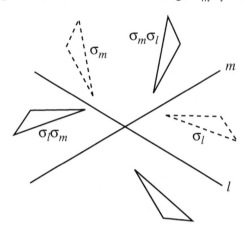

Figure 2.42 Noncommutative products of reflection: $\sigma_m \sigma_l \neq \sigma_l \sigma_m$.

However, we find that a halfturn σ_P is the product of two reflections in any two perpendicular lines r and s: thus $\sigma_P = \sigma_s \sigma_r$. Furthermore, we can do these reflections in either order, so that $\sigma_s \sigma_r = \sigma_r \sigma_s$ (Figure 2.43).

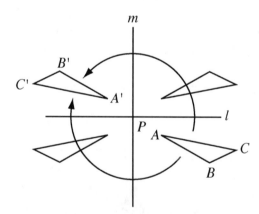

Figure 2.43 Halfturns as the product of reflections in perpendicular lines.

To find the product of rotations about different centers expressed as a product of reflections, we proceed as follows: Given two rotations about different centers in the plane, $\sigma_b \sigma_a = \rho_{A\alpha}$ and $\sigma_d \sigma_c = \rho_{B\beta}$ (Figure 2.44), find a simple formula for their product, $\rho_{B\beta} \rho_{A\alpha}$. To begin, nothing prevents us from rotating the sets $\{a,b\}$ and $\{c,d\}$ about A and B, respectively, so that lines b and c are collinear with a line AB through A and B. Then we have

Theory of Transformations

$$\rho_{B\beta}\rho_{A\alpha} = \sigma_d\sigma_c\sigma_b\sigma_a$$

where $\sigma_c = \sigma_b$ and $\sigma_c\sigma_b = \iota$. Thus

$$\rho_{B\beta}\rho_{A\alpha} = \sigma_d\sigma_a$$

If a and d are not parallel, then we assert that

$$\rho_{c,(\alpha+\beta)} = \rho_{B\beta}\rho_{A\alpha}$$

where $C = \rho_{B,\beta/2}AB \cap \rho_{A,-\alpha/2}AB$.

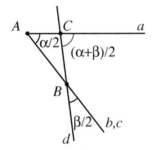

Figure 2.44 The product of rotations about different centers.

The Product of Three or More Reflections in the Plane

The product of three reflections in parallel lines is equal to one reflection. So if l, m, and m are parallel lines and $P' = \sigma_n\sigma_m\sigma_n P$, then $\sigma_n\sigma_m\sigma_l = \sigma_b$ and $P' = \sigma_b P$, where b is the perpendicular bisector of PP' (Figure 2.45).

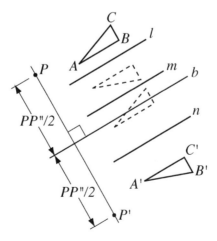

Figure 2.45 The product of three reflections in parallel lines.

In other words, we can replace any product of three reflections in parallel lines by a single equivalent reflection. As you can see, both transformations are isometries and are orientation reversing. A similar argument holds for the product of three reflections in concurrent lines (Figure 2.46).

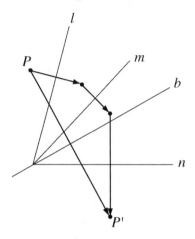

Figure 2.46 The product of three reflections in concurrent lines.

We are now ready to consider the product of three reflections where l, m, and m are neither concurrent nor mutually parallel. We will explore only the special case where l and m are perpendicular to n. If l and m are distinct, non collinear lines perpendicular to line n, then the product $\sigma_n \sigma_m \sigma_l$ is a *glide reflection* with axis n (Figure 2.47). A glide reflection is an orientation-reversing isometry.

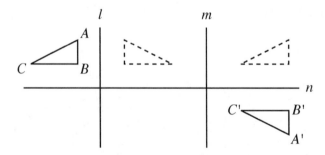

Figure 2.47 Glide reflection.

If an isometry is a product of an even number of reflections, then it is an *even isometry*; if it is a product of an odd number of reflections, then it is an *odd isometry*. Although it is outside the scope of this text, we can prove that any isometry is a product of reflections; therefore, any isometry is either even or odd. (The term parity refers to the mathematical property of an integer as being either even or odd.)

Theory of Transformations

Now consider the following assertions and try to develop their proofs:

1. We can reduce an even isometry of four or more reflections to a product of two reflections.
2. We can reduce an odd isometry of five or more reflections to a product of three reflections.
3. No isometry is both even and odd.

Isometries in Space

Let's extend what we know about isometries and reflections in the plane to a space of three dimensions. Most of the concepts of plane isometries have direct analogs in space. Here are the equations of the principal reflections in space:

Inversion in the origin (Figure 2.48):

$x' = -x$
$y' = -y$
$z' = -z$

Inversion in the point (a, b, c):

$x' = -x + 2a$
$y' = -y + 2b$
$z' = -z + 2c$

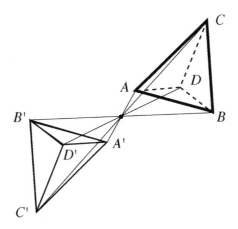

Figure 2.48 Inversion in the origin.

Reflection in the x axis:

$x' = x$
$y' = -y$ and similarly for the y and z axis.
$z' = -z$

Reflection in the $z = 0$ plane:

$x' = x$

$y' = y$ and similarly for the $x = 0$ and $y = 0$ planes.

$z' = -z$

Again, as for plane isometries, reflection is the basic building block. Given a plane Π and a point P in space, P' is the image of P under σ_Π; that is, $P' = \sigma_\Pi P$. The plane Π is the perpendicular bisector of line PP' (Figure 2.49).

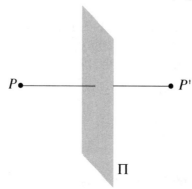

Figure 2.49 Reflection in the plane.

Reflection in a plane is an orientation-reversing isometry (Figure 2.50).

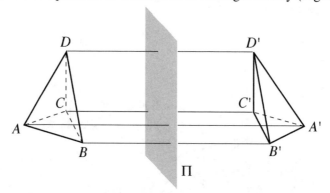

Figure 2.50 Reflection in a plane is orientation-reversing.

Other properties we easily infer, for example:

1. σ_Π is an involution, because $\sigma_\Pi^2 = \iota$.
2. σ_Π is a collineation; it maps all straight lines onto straight lines.
3. σ_Π fixes all points in the plane Π.
4. σ_Π fixes all lines perpendicular to Π.
5. If a line l is parallel to Π, then $\sigma_\Pi l$ is parallel to Π.

THEORY OF TRANSFORMATIONS 69

Given two parallel planes Π and Δ separated by distance d, and any point P, the transformation $\sigma_\Delta \sigma_\Pi$ is a translation through twice the distance separating the two planes. Algebraically, we write $P' = \sigma_\Delta \sigma_\Pi P$, where $PP' = 2d$ (Figure 2.51).

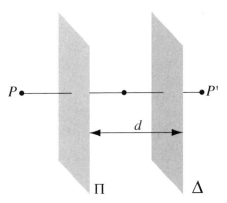

Figure 2.51 The product of two reflections in parallel planes.

We can express every translation in space as a product of two reflections in parallel planes. Clearly, a translation transformation is an orientation-preserving isometry (Figure 2.52).

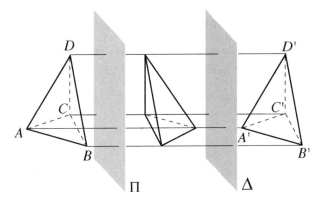

Figure 2.52 The product of two reflections in parallel planes is orientation-preserving.

Other properties include:

1. $\sigma_\Delta \sigma_\Pi$ fixes all lines perpendicular to Π and Δ.
2. $\sigma_\Delta \sigma_\Pi$ fixes all planes perpendicular to Π and Δ.
3. $\sigma_\Delta \sigma_\Pi$ fixes no points.

If planes Π and Δ are mutually perpendicular, then $\sigma_\Delta \sigma_\Pi$ is a halfturn about the line of intersection of the two planes. This transformation is a line reflection (Figure 2.53).

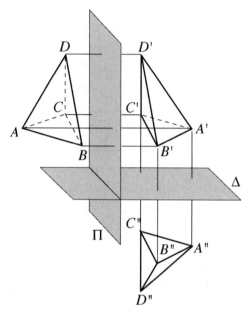

Figure 2.53 Reflection in two perpendicular planes.

If planes Π and Δ intersect in line l, then $\sigma_\Delta \sigma_\Pi$ is a rotation about line l through a directed angle equal to twice the directed angle between Π and Δ (Figure 2.54). Thus, $\sigma_\Delta \sigma_\Pi = \rho_{l\theta}$.

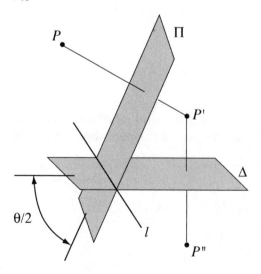

Figure 2.54 Reflection in two intersecting planes.

Here are some additional isometries in space:

1. Given two parallel planes Γ and Δ, each perpendicular to plane Π, then $\sigma_\Pi \sigma_\Delta \sigma_\Gamma$ is a glide reflection with axis Π.

2. Given two intersecting planes Γ and Δ, perpendicular to plane Π, then $\sigma_\Pi \sigma_\Delta \sigma_\Gamma$ is a rotary reflection about the point common to the three planes.

3. The product of a translation along a given line l and a rotation about the same line is a twist or glide rotation.

4. Given a point M in space, then σ_M is the transformation such that $\sigma_M P = P'$, where M is the midpoint of PP'. We call this transformation a point reflection or a central inversion.

5. Given three mutually perpendicular planes Π, Δ, and Γ, then $\sigma_\Gamma \sigma_\Delta \sigma_\Pi$ is an inversion in M, where M is the common point of intersection of the three planes.

Finally, we assert without proof that we can reduce any isometry in space to one of the following types: a single reflection, a translation, a rotation, a glide reflection, a glide rotation, or a rotary reflection. If we ignore the translations in space, then the homogenous equations for isometries in space are:

$$x' = a_{11}x + a_{12}y + a_{13}z$$
$$y' = a_{21}x + a_{22}y + a_{23}z$$
$$z' = a_{31}x + a_{32}y + a_{33}z$$

where, again, $|a_{ij}| = \pm 1$. For example, consider the simple case of reflection in the origin (an inversion). Here we have $a_{11} = a_{22} = a_{33} = -1$ and $a_{ij} = 0$ for $i \neq j$, and

$$\begin{vmatrix} -1 & 0 & 0 \\ 0 & -1 & 0 \\ 0 & 0 & -1 \end{vmatrix} = -1$$

where the evaluation of the determinant tells us that the transformation is an odd, or orientation-reversing, isometry.

Next, we consider reflection in the $y = 0$ plane. This means that $a_{11} = a_{33} = 1$, $a_{22} = -1$, and $a_{ij} = 0$ for $i \neq j$, so that

$$\begin{vmatrix} 1 & 0 & 0 \\ 0 & -1 & 0 \\ 0 & 0 & 1 \end{vmatrix} = -1$$

The determinant tells us that this transformation is also orientation reversing. This interesting mathematical property is further explored in Chapters 3, 4, and 5.

The Group of Isometries

The inverse of any isometry or motion is another isometry or motion, and the result of any two isometries is again another isometry (closure). Identity and associativity conditions are assumed. We conclude that the set of all isometries is a group, satisfying inverse, identity, closure and associativity conditions. Now let's consider subgroups of isometries. Recall from the preceding discussion that there are six types of isometries, all based upon products of reflection. These are: a single reflection, translation, rotation, glide reflection, glide rotation, and rotary reflection.

Does the set of rotations and translations form a group? Closure is the only condition we will need to investigate; we will assume the other three conditions to apply. We demonstrated earlier that the product of two rotations about different points is, indeed, another rotation, unless the sum of the two angles is 2π, in which case the result is a translation. This leaves just three other possibilities to consider: the product of two translations, the product of a rotation followed by a translation, and the product of a translation followed by a rotation. The first possibility poses no problem, because we know that the product of two translations is another translation. For the second possibility, we proceed as follows (Figure 2.55). Given a center O and angle of rotation α, where l is a line through O in the direction of translation, construct m and n perpendicular to l with m through O and m and n separated by a distance $d/2$. Lay out line p through O such that $\sigma_m \sigma_p = \rho_{o\alpha}$. This produces the translation $\sigma_n \sigma_m = \tau_{nm}$. It follows that

$$\tau_{nm}\rho_{0\alpha} = \sigma_n \sigma_m \sigma_m \sigma_p$$
$$= \sigma_n \sigma_p$$

but $\sigma_n \sigma_p$ is a rotation through α about a different, displaced center. We develop the third possibility, similarly. In both cases, the result is a rotation, a member of the set of all rotations and translations. Thus, we demonstrate that closure and the set of all rotations and translations do form a group: a subgroup of the group of isometries.

This leads us to a slightly different point of view. We may choose instead to emphasize the fact that every direct isometry is the product of two reflections and is, therefore, either a translation or a rotation, forming the group of direct isometries.

We can also show that the set of all translations forms a group (a subgroup of the group of direct isometries). This set has the closure property because the product of two translations is another translation. Obviously, the inverse of a translation is simply its negation.

There are many other subgroups of the group of direct isometries. We now have a greater appreciation for the work of Felix Klein who regarded geometry as the study of those properties of figures left invariant under all mappings of a group. Indeed, for every group of transformations, we can construct a particular geometry. The following discussion introduces us to a possible geometry of a square by developing the group of transformations that map a square onto itself.

THEORY OF TRANSFORMATIONS

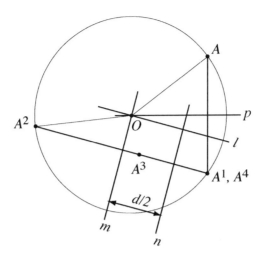

Figure 2.55 Rotation followed by translation.

We can map a square onto itself by any of the eight isometries shown here (Figure 2.56):

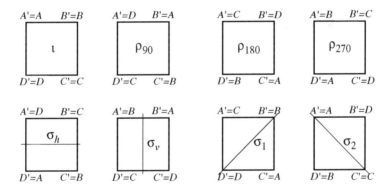

Figure 2.56 A group of transformations mapping a square onto itself.

Here we see the identity transformation, three rotations, and four reflections. These are the only isometries that map a square onto itself. No translations or glide reflection can do this. No other isometry leaves the square invariant.

If these eight mappings form a group, then the product of any two of them must itself be one of the original eight (closure). And, the inverse of each mapping must also belong to this set of eight. We use a Cayley table to demonstrate the closure property.

Notice some of the interesting properties of Table 2.6. First, in every row and every column, every transformation occurs exactly once. This is a property of all Cayley tables. Second, the transformations generating the upper left quadrant of the table form a subgroup. Third, the group is noncommutative. Try to verify this.

Table 2.6 A Cayley demonstrating the closure property of the group of transformations mapping a square onto itself.

	ι	σ_{90}	σ_{180}	σ_{270}	σ_h	σ_v	σ_{d_1}	σ_{d_2}
ι	ι	σ_{90}	σ_{180}	σ_{270}	σ_h	σ_v	σ_{d_1}	σ_{d_2}
σ_{90}	σ_{90}	σ_{180}	σ_{270}	ι	σ_{d_1}	σ_{d_2}	σ_v	σ_h
σ_{180}	σ_{180}	σ_{270}	ι	σ_{90}	σ_v	σ_h	σ_{d_2}	σ_{d_1}
σ_{270}	σ_{270}	ι	σ_{90}	σ_{180}	σ_{d_2}	σ_{d_1}	σ_h	σ_v
σ_h	σ_h	σ_{d_2}	σ_v	σ_{d_1}	ι	σ_{180}	σ_{270}	σ_{90}
σ_v	σ_v	σ_{d_1}	σ_h	σ_{d_2}	σ_{180}	ι	σ_{90}	σ_{270}
σ_{d_1}	σ_{d_1}	σ_h	σ_{d_2}	σ_v	σ_{90}	σ_{270}	ι	σ_{180}
σ_{d_2}	σ_{d_2}	σ_v	σ_{d_1}	σ_h	σ_{270}	σ_{90}	σ_{180}	ι

Section 2.4 Exercises

1. Prove that a reflection preserves distance.
2. Prove that a reflection preserves parallelism.
3. If the equation of line m is $x = y$, find the image of the following points under a reflection in m.
   a. $(3,3)$ c. $(5,1)$ e. (x,y)
   b. $(3,-3)$ d. (a,b)
4. Derive the equations for reflection in a line through the origin (see discussion above on equivalent products of parallel reflections).
5. Derive the equations for reflection in an arbitrary line.
6. Prove that inversion in a point is an isometry.
7. Given that σ_P, σ_Q, σ_R are halfturns, show that $\sigma_R \sigma_Q \sigma_P = \sigma_P \sigma_Q \sigma_R$.
8. Show that the product of rotations about the vertices of a triangle ABC by twice the angles at the vertices is the identity, ι.
9. Show that the midpoint of any point and its image under a glide reflection lies on the axis.
10. Show that a glide reflection is the composite of a reflection in some line k followed by a halfturn about some point off k.

Theory of Transformations

11. Show that in space a reflection in a plane preserves distance.
12. Show that, in general, the product of two reflections in parallel lines or planes is not commutative; that is, $\sigma_n \sigma_m \neq \sigma_m \sigma_n$.
13. Identify the orientation-reversing mappings of the square onto itself.
14. Find the group of transformations that maps a cube onto itself.
15. Identify the orientation-reversing mappings of a cube onto itself.

2.5 Similarities

After the isometries, the most elementary transformation of a geometric figure is a similarity, which alters the size but preserves the shape of a figure. This is the first step we take in generalizing the isometries, and in doing this we produce a geometry with fewer invariants; obviously distance is no longer invariant. The geometry of similarities is, nonetheless, of central importance to Euclidean geometry. From the properties of similarities, we can construct all the trigonometries, for example. Two figures are similar if corresponding lengths have the same ratio, so that one is either an enlargement or a reduction of the other (Figure 2.57), where

$$\frac{AB}{DE} = \frac{BC}{EF} = \frac{CA}{FD} \text{ and } \angle A = \angle D, \quad \angle B = \angle E, \quad \angle C = \angle F.$$

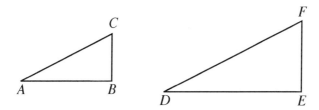

Figure 2.57 Similar triangles.

Or, we can say that a similarity transformation is a one-to-one mapping of the plane onto itself such that it multiplies each distance by the same ratio, k. Furthermore, similarities preserve angles. Geometric figures may be similar yet have opposite orientations (Figure 2.58).

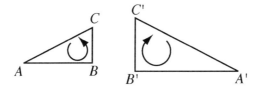

Figure 2.58 Similar but oppositely oriented triangles.

A similarity transformation sends triangles into similar triangles, polygons into similar polygons, polyhedra into similar polyhedra, and, in general, any figure into a similar figure. The image of any angle is an angle of equal size, and we may note, in particular, that this transformation preserves perpendicularity.

Notice that a similarity is a composite transformation, consisting of a dilation component and an isometry. Sometimes we use the term homothety instead of dilation; you will find both terms in the literature of transformations (see Chapter 6). This means that a similarity may change the size of a figure as well as introduce a "motion" of some sort (a single reflection, a translation, and/or a rotation). Similar figures, thus, may occupy widely separated spatial positions and may differ in their rotational orientation (Figure 2.59).

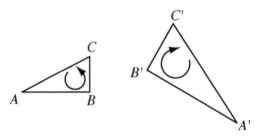

Figure 2.59 Similarities: A product of a dilation and isometry.

Anticipating later work, here is a cursory preview of the Cartesian equations for a similarity in the plane:

$$x' = ax - by + c$$
$$y' = \pm(bx + ay) + d \tag{2.16}$$

where $\sqrt{a^2 + b^2} = k$ and $k \neq 0$ $\begin{cases} k > 0 \text{ for a direct similarity} \\ k < 0 \text{ for an opposite similarity} \end{cases}$, and where k is the ratio of expansion or contraction, and c and d are the components of a translation. This is a different, somewhat simpler (in that it is specialized) notation scheme for the coefficients. It is a way to emphasize the special relationships among them.

If we restrict our investigation to the dilation component of a similarity, then we must rewrite Equations 2.16 as follows:

$$x' = kx + c$$
$$y' = ky + d \tag{2.17}$$

This set of equations cannot produce a rotation, whereas Equations 2.16 can. So the transformation of Equations 2.17 yields similar figures whose corresponding lines are parallel if $k > 0$. This rather specialized similarity is a homothety, mentioned earlier. Of course, we may further simplify these equations by neglecting the translation coefficients and writing the homogeneous equations:

THEORY OF TRANSFORMATIONS

$$x' = kx$$
$$y' = ky \quad (2.18)$$

These equations tell us that this dilation transformation multiplies the distance of any point $P(x,y)$ from the origin by the factor $|k|$. What about dilation about some arbitrary point? Here is a straightforward solution:

Given a dilation factor k about an arbitrary point $P(x_P, y_P)$ in the plane, find the image of any figure, say $\triangle ABC$ (Figure 2.60), translate this figure so that P is coincident with the origin, and we have $\tau_{P,O}\triangle ABC = \triangle DEF$.

Apply the dilation factor k so that $\delta_{O,k}\triangle DEF = \triangle D'E'F'$

Then reverse the initial translation: $\tau_{O,P}\triangle D'E'F' = \triangle A'B'C'$

This, of course, is the same as $\delta_{P,k}\triangle ABC = \triangle A'B'C'$

This means that $\delta_{P,k} = \tau_{O,P}\delta_{O,k}\tau_{P,O}$

We describe this transformation algebraically as follows:

1. Translate P to the origin: $\begin{cases} x' = x - x_P \\ y' = y - y_P \end{cases}$

2. Apply k: $\begin{cases} x' = k(x - x_P) \\ y' = k(y - y_P) \end{cases}$

3. Reverse the initial translation: $\begin{cases} x' = k(x - x_P) + x_P \\ y' = k(y - y_P) + y_P \end{cases}$

4. Simplify: $\begin{cases} x' = kx + (1-k)x_P \\ y' = ky + (1-k)y_P \end{cases}$

A further development of this transformation appears in Chapter 6 in the guise of vectors, matrices, and homogeneous coordinates.

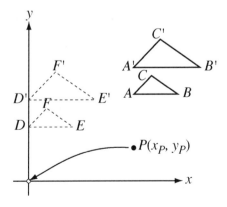

Figure 2.60 Dilation about an arbitrary point in the plane.

Dilation in the Plane

Let's consider a strictly graphical, geometric approach to understanding dilations, remembering that a similarity is nothing more than the product of a dilation and an isometry. To begin, we work in the plane. Given a figure, say triangle ABC, a center of dilation P, and a scale factor k, then $PA'/PA = PB'/PB = PC'/PC = k$. Figure 2.61 shows us that for $k > 1$ the image triangle $A'B'C'$ lies beyond the original, relative to the center of dilation. If the image lies between P and the original, then it must be that $0 < k < 1$. In both cases, the image retains the orientation of the original. However, if $k < 0$, producing $\Delta A''B''C''$ in Figure 2.61, then the center of dilation lies between the image and the original and the image has a reverse orientation.

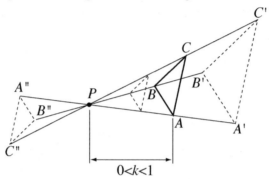

Figure 2.61 Dilation of a triangle about a point.

Here is one way to denote a dilation: δ_{Pk}, where P is the center of the dilation and k is the ratio or scale factor. Consider the following properties of dilations:

1. Dilations are line preserving.
2. Dilations are angle preserving.
3. The lengths of corresponding line segments are in the ratio $1:k$.
4. The ratio of the lengths of two image line segments is equal to the ratio of the lengths of the corresponding original line segments. Referring to Figure 2.61, again, we find, for example, that $AB/BC = A'B'/B'C'$.
5. The area of a figure and its image under a dilation are in the ratio $1:k^2$.
6. The inverse of a dilation δ_{Pk} is $\delta_{P,1/k}$.
7. $\delta_{Pk}^{-1} = \delta_{P,1/k}$
8. $\delta_{o,-1} = \sigma_o$ a dilation with a scale factor of -1 is the same as a halfturn or an inversion in the center of dilation.
9. $\delta_{o,1} = \iota$ a dilation with a scale factor of 1 is the identity transformation.
10. If $k > 0$, then δ_{Pk} maps parallel lines onto parallel lines.
11. Lines through the center of a dilation map onto themselves.

THEORY OF TRANSFORMATIONS

Two circles in the plane present an interesting condition; see the following discussion on two circles and two centers of dilation. From this and the preceding discussion, it should be easy to demonstrate that all circles in the plane are similar.

Given two circles in the plane, if the circles are not concentric, then they have two distinct centers of dilation (Figure 2.62). These two centers, O and O', are on the line through the centers of the circles. To find O and O', construct any two parallel diameters, say AB and $A'B'$. Lines AA' and BB' intersect at one center of dilation, and lines AB' and $A'B$ intersect at the other. The algebraic derivation of these centers is straightforward. Try it. (We could, of course, dispense with circles altogether in this discussion. Given any two parallel line segments, the identical conditions and results prevail.)

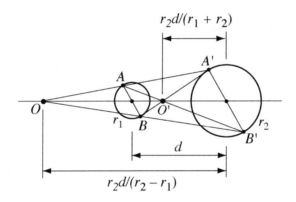

Figure 2.62 Two circles and two centers of dilation.

The Product of Two Dilations with the Same Center

Given a figure, say triangle ABC and two dilations δ_{Ok_1} and δ_{Ok_2} about the same center O, we find that $\delta_{Ok_2}\delta_{Ok_1} = \delta_{Ok_1 k_2}$ (Figure 2.63).

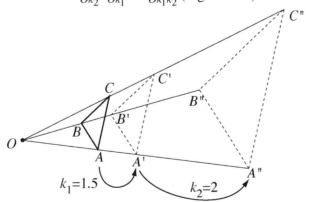

Figure 2.63 The product of two dilations with the same center.

Furthermore, dilations about the same center are commutative, so that $\delta_{O,k_2}\delta_{O,k_1} = \delta_{O,k_1}\delta_{O,k_2}$. Algebraically we demonstrate this property as follows:

$$\text{if } \begin{Bmatrix} x' = k_1 x \\ y' = k_1 y \end{Bmatrix} \text{ and } \begin{Bmatrix} x'' = k_2 x' \\ y'' = k_2 y' \end{Bmatrix}, \text{ then } \begin{Bmatrix} x'' = k_1 k_2 x \\ y'' = k_1 k_2 y \end{Bmatrix} \quad (2.18)$$

The Product of Two Dilations about Different Centers

The product of two dilations k_1 and k_2 with centers O_1 and O_2 is an equivalent dilation of ratio $k_1 k_2$, whose center is collinear with the centers of δ_1 and δ_2 and divides the line segment connecting their centers in the ratio $(k_2 - 1)/k_2(k_1 - 1)$. So that $\delta_{O_2,k_2}\delta_{O_1,k_1} = \delta_{O',k_1 k_2}$, where O_1, O', O_2 are collinear. In Figure 2.64, we first produce $\Delta A'B'C' = \delta_{O_1,k_1} \Delta ABC$. Next, we construct $\Delta A''B''C'' = \delta_{O_2,k_2} \Delta A'B'C'$. Finally, we see that $\Delta A''B''C'' = \delta_{O',k_1 k_2} \Delta ABC$. In this example $k_1 = 1.5$, $k_2 = 2.0$ and $k_1 k_2 = 3.0$.

Dilations in Space

Dilations in space are directly analogous to the dilations in the plane. The simple geometric object in Figure 2.65 is a good illustration. For $k > 0$ the dilation is orientation preserving, and for $k < 0$ it is orientation reversing. For dilation about the origin, neglecting isometries, the following simple homogeneous transformation equations apply: $x' = kx$, $y' = ky$, and $z' = kz$

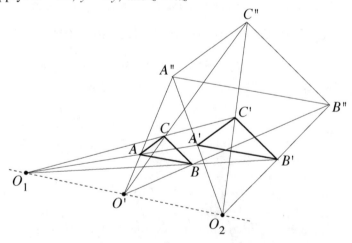

Figure 2.64 The product of two dilations with different centers.

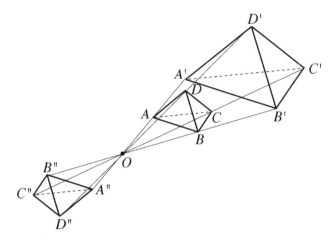

Figure 2.65 Dilation in space.

If we want to include dilation about an arbitrary point and the isometries, then, by analogy we extend the equations developed in the discussion above on the dilation about an arbitrary point in the plane to include the third dimension. There are some minor complications along the way to achieving this extension, and Chapter 4 *et seq* address this problem with the thoroughness it requires.

The Group of Similarities

We have focused on the dilation component of similarity transformations, but now we must again consider the similarity transformation as a composite of a dilation and an isometry. The identity transformation results if $k = 1$ and the isometry produces no "motion." We know every isometry has an inverse, but so does every dilation $\delta_{O,k}^{-1} = \delta_{O,1/k}$. Closure is easy to demonstrate, because we have already seen that the product of two dilations is another dilation: $\delta_{O,k_2}\delta_{O,k_1} = \delta_{O,k_1 k_2}$ and $\delta_{O_2,k_2}\delta_{O_1,k_1} = \delta_{O',k_1 k_2}$. Although by no means does the preceding constitute a formal proof, with some intuitive support we may now assert that the set of all similarity transformations is a group, the similarity group. To this we might add that the set of all isometries is a subgroup of the similarity group.

Our experience with standard school geometry gives us a somewhat limited meaning to the concept of similarity. We are familiar with similar polygons, mostly triangles, but beyond this lies *terra incognita*. However, as we progress in this sequel we will broaden the application of similarity to find that any two figures of the same shape are similar. (This must remain intuitive, however, because mathematicians have as yet no precise definition for the term "shape.")

Section 2.5 Exercises

1. What does $\delta_{O,-1} = \sigma_O$ mean?
2. Prove that if $k > 0$, then δ_{Pk} maps parallel lines onto parallel lines.
3. Show that all line segments are similar.
4. Show that all ellipses with the same eccentricity are equal.
5. Show that the product of two dilations about the same center is commutative.
6. Is the product of two dilations about different centers commutative? Demonstrate your answer with a geometric construction.
7. Derive an algebraic equation for the product of two dilations with different centers (x_1, y_1) and (x_2, y_2), and ratios k_1, k_2. Find the coordinates of the center of the equivalent dilation, (x_P, y_P).

2.6 Affinities

Our studies up to this point in the sequel are of classic elementary geometry, albeit from a new point of view. With affine geometry, we begin a subject first introduced in the eighteenth century by Leonhard Euler, a subject not usually part of school geometry. An affine property is one that is preserved by all affine transformations, but not by the more general projective transformations. Collinearity and parallelism are the two most important invariant properties of affinities.

Recall that isometries preserve distance between corresponding pairs of points in the plane or in space. Therefore, angles and other metric properties are invariant, and we obtain the familiar Euclidean geometry of congruent figures. When we give up the requirement that distances must be invariant but retain invariance of angles, the resulting linear substitutions are similarities. If we go one step further and give up both distance and angle invariance, then, algebraically, this means that we relax almost all of the earlier restrictions we placed upon the coefficients of the transformation equations. Consider the homogeneous linear transformation

$$\begin{aligned} x' &= a_{11}x + a_{12}y + a_{13}z \\ y' &= a_{21}x + a_{22}y + a_{23}z \\ z' &= a_{31}x + a_{32}y + a_{33}z \end{aligned} \tag{2.19}$$

This time the only restriction we place on the coefficients is $\det a_{ij} \neq 0$, or

$$\begin{vmatrix} a_{11} & a_{12} & a_{13} \\ a_{21} & a_{22} & a_{23} \\ a_{31} & a_{32} & a_{33} \end{vmatrix} \neq 0 \tag{2.20}$$

THEORY OF TRANSFORMATIONS

Equations 2.19 and the conditions placed on the coefficients by Equation 2.20 define an affine transformation or affinity. Note, we ignore the displacements parallel to the three coordinate axes, because these produce an isometry that can be superimposed at any time without affecting the underlying affine properties that we are interested in here.

Here is one way to form an intuitive, geometric interpretation of a homogeneous affine transformation in all its generality: Imagine a unit cube aligned with the principal coordinate axes with one vertex at the origin. Label the three edges coincident at O as OA, OB, OC (Figure 2.66). Next, deform the "framework" (i.e., edges) of the cube by arbitrarily and independently expanding and/or contracting each of these three edges while at the same time altering their directions (i.e., they should no longer be aligned with the principal axes). Finally, construct a parallelepiped based upon the three altered edges OA', OB', OC'. The resulting parallelepiped is an affine image of the original cube. Lines go into lines, parallel lines go into parallel lines, and so parallelepipeds go into parallelepipeds. We also notice that a scale change in three independent directions is possible. Clearly, this transformation preserves neither distances nor angles. (The effect of affine transformations on distance is a most curious and mathematically interesting phenomenon that we will explore in more depth later in this section.)

After we study vector spaces in Chapter 3 we will have a greater selection of mathematical tools to apply to the problems of affine geometry and transformations. This includes basis vectors, oblique coordinate systems, and the notion of a metric tensor to give us a more powerful way of defining distance. Here affine geometry is particularly important because more general nonlinear transformations become linear in the limit of very small displacements. This means that any geometry expressed with infinitesimal distances (for example, differential geometry) is necessarily affine. We will see a strong hint of this later, in Chapter 3.

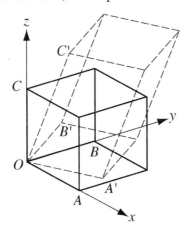

Figure 2.66 Affine deformation of a cube.

Affine Geometry in the Plane

It is best to simplify conditions as much as we can at the outset of our study of affinities so we can focus on the central characteristics and problems of these transformations. So first, let us restrict ourselves for the moment to affine geometry in the plane. We rewrite Equations 2.19 and 2.20 to reflect this and change the notation on the coefficients to make it somewhat easier to track their effects. Thus

$$x' = ax + by$$
$$y' = cx + dy \tag{2.21}$$

where $ad - bc \neq 0$.

What can we learn about a geometry given by these equations and the conditions placed upon their coefficients? We start by asking questions that are more specific. First, how are lines in the plane transformed? We must do some preliminary work first by solving for x and y in Equations 2.21 to obtain the inverse relationships:

$$x = \frac{d}{\Delta}x' - \frac{b}{\Delta}y'$$
$$y = -\frac{c}{\Delta}x' + \frac{a}{\Delta}y' \tag{2.22}$$

where $\Delta = ad - bc$.

The slope-intercept form of equation for a line in the plane is $y = Ax + B$, where A is the slope and B the y intercept. If we substitute from Equations 2.22 into this equation for a line, then we obtain

$$y' = \frac{dA + c}{a + bA}x' + \frac{ad - bc}{a + bA}B \tag{2.23}$$

We immediately see from Equation 2.23 that an affine transformation sends lines into lines, since its algebraic form is the same as $y = Ax + B$. If we let A' and B' denote the slope and intercept, respectively, of the transformed line, then

$$y' = A'x' + B'$$

where $A' = \dfrac{dA + c}{a + bA}$ and $B' = \dfrac{ad - bc}{a + bA}$.

Now we can ask if any lines are fixed. We begin by equating slopes A and A' and intercept B and B'. For the slopes we find

$$A = \frac{dA + c}{a + bA}$$

or
$$A = \frac{-(a - d) \pm \sqrt{(a - d)^2 + 4bc}}{2b} \tag{2.24}$$

This tells us that there are two families of parallel lines, those with slopes, say A_1 and A_2, corresponding to the two possible solutions for Equation 2.24. If the original line has slope A_1, then its image will also have slope A_1. (Of course, this also applies to slope A_2.) This means that, in this situation, the original and image lines are parallel.

Equating the intercepts B and B', we find that

$$B' = \frac{ad-bc}{a+bA} B \qquad (2.25)$$

and this tells us that if the y intercept of the original line is zero, then the y intercept of the image line will also equal zero. There is another possibility. If $(ad-bc)/(a+bA) = 1$, then, again, the image and original have identical intercepts, and they are collinear or fixed; but, special conditions on the coefficients a, b, c, and d are obviously required. Furthermore, only one, if any, of the two families of parallel lines can possibly satisfy Equation 2.25, either the family with slope A_1 or the family with slope A_2, but not both.

Figure 2.67 illustrates one way to interpret the foregoing algebraic statement of affine characteristics and from them derive an affine geometric construction technique. Here are the ground rules for this transformation: Select a line l in the plane, a direction not parallel to it, represented by line m, and a scale factor k, where $k \neq 0$. Then lines through pairs of corresponding points P and P' are parallel to the direction given by line m and $(P'P*)/(PP*) = k$, where $P*$ is the intersection of the line PP' with l. This construction is a perspective affinity or an axial stretching. The line l is the axis of affinity, the line m the direction of affinity, and k the scale factor of the affinity.

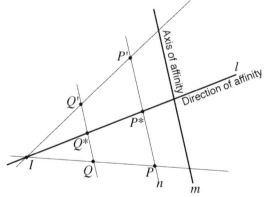

Figure 2.67 Affine construction.

Given any point P, we find its image P' by first constructing a line n through P parallel to the direction of affinity, m. Call the intersection of this line with the axis of affinity l point $P*$. From $P*$ lay off P' along n so that $P'P* = kPP*$.

The intersection of a line and its image lies on the axis of affinity. For example, in Figure 2.67 the line through points P,Q and the line through its image defined by P',Q' intersect at point l on the axis.

This transformation fixes all points on l, so it fixes l. It also fixes lines parallel to the direction of affinity, but not pointwise. This means that a perspective affinity is line-preserving, that corresponding lines intersect on the axis of affinity, and that parallel lines map onto parallel lines. The inverse of a perspective affinity has the same axis and direction, but its scale factor is $1/k$. The following discussion presents a specific example of a perspective affinity in both algebraic and geometric terms.

Let line $y = Ax$, the y axis is the direction of affinity, and k is the scale factor. The line passes through the origin for simplicity, since this produces homogeneous transformation equations (Figure 2.68). The shows figure, it is easy to see that

$$x' = x$$
$$y' = A(k+1)x - ky$$

where the general affine coefficients are $a = 1$, $b = 0$, $c = a(k+1)$ and $d = -k$. So, for the determinant of their coefficients we have

$$\begin{vmatrix} 1 & 0 \\ A(k+1) & -k \end{vmatrix} = -k \neq 0$$

To find fixed points, set $x' = x$ and $y' = y = A(k+1)x - ky$. By doing this we find that $y + ky = A(k+1)x$ or $y = ax$, confirming that all points on the axis of affinity are, indeed, fixed.

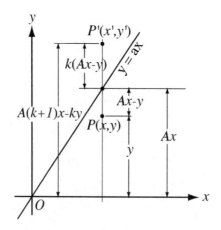

Figure 2.68 An example of perspective affinity.

Theory of Transformations

Shear

Another special affine transformation is a shear fixing either the x or y axis. If it fixes the x axis, then the defining equations are:

$$x' = x + k_y y$$
$$y' = y \quad (2.26)$$

where k_y is the shear coefficient. Denote this transformation by v_x. Figure 2.69 illustrates a shear transformation of a point in the plane. The transformation fixes points on the x axis because $y = 0$, and moves all other points horizontally a directed distance proportional to their distance from the x axis.

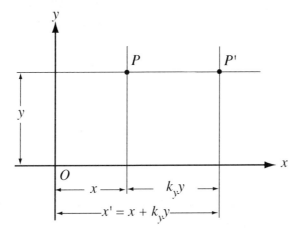

Figure 2.69 Shear transformation relative to the x axis.

Similarly, a shear transformation fixing the y axis (denoted by v_y) is:

$$x' = x$$
$$y' = k_x x + y \quad (2.27)$$

Under shear, lines transform into lines, parallel lines transform into parallel lines, and lines parallel to the axis of shear remain parallel to it. Figure 2.70 shows three collinear points and their images under a shear transformation

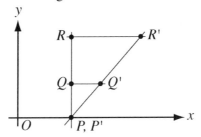

Figure 2.70 Points under a shear transformation relative to the x axis

The product of two shears $v_y v_x$ is given by

$$x' = x + k_y y$$
$$y' = k_x x + (k_x k_y + 1) y$$

Taking the product in the reverse order, $v_x v_y$, we obtain

$$x' = (k_x k_y + 1) x + k_y$$
$$y' = k_x x + y$$

In general, the product of two shears is not commutative; thus $v_x v_y \neq v_y v_x$.

Curiously, a shear preserves area, so it is in a class of affine transformations that we call equiaffine or equiareal (Chapter 6). It is very easy to see why the area of a figure doesn't change under a shear transformation. For example, the triangle in Figure 2.71, after undergoing the transformation v_x, still has the same base and altitude and, therefore, the same area.

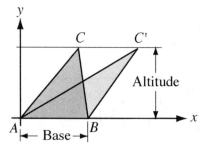

Figure 2.71 Area-preserving shear transformation.

Strain

Strain transformations behave like unidirectional (or nonisotropic) dilations. Here is an example:

$$x' = k_x x$$
$$y' = y \tag{2.28}$$

where k_x is now the strain coefficient. This is a strain, ε_x, in the x direction (Figure 2.72). It fixes the origin; it fixes the y axis pointwise; and it fixes the x axis and all lines parallel to it, though not pointwise.

Strain in the y direction, ε_y, is given by

$$x' = x$$
$$y' = k_y y \tag{2.29}$$

THEORY OF TRANSFORMATIONS

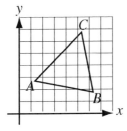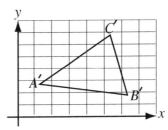

$A' = \varepsilon_x A$
$B' = \varepsilon_x B$
$C' = \varepsilon_x C$

Figure 2.72 Strain.

There is something special about these strains—they are orthogonal. Their principal lines of action are mutually perpendicular. This type of transformation is closely related to the mechanics of materials and the extension or compression of structural elements under load. Later, in Chapter 6, we will discuss nonorthogonal strains and strains not aligned with the coordinate axes.

The product of the two strains $\varepsilon_y \varepsilon_x$ is

$$x' = k_x x$$
$$y' = k_y y \tag{2.30}$$

If the strains are equal, then the product is a simple dilation about the origin with the ratio $k = k_x = k_y$. The product of these two strains is commutative. What about the product of two strains, both of which are in the same direction, say $\varepsilon_{x_2} \varepsilon_{x_1}$?

Equiareal Transformations

An affine transformation whose determinant has the value ± 1 is an equiareal transformation. It is the only affine transformation that preserves the area of a triangle. In fact, any two figures that correspond under an equiareal transformation will have the same area. We can go even farther than this and assert that an affine transformation multiplies the area of all figures by a factor equal to the absolute value of the determinant of the transformation. We prove this for triangles in the discussion below, where we consider the effect of an affine transformation on the area of a triangle. Finally, note that we use the term affinely equivalent in the sense implied by the following example: Two figures are affinely equivalent if there is an affine transformation that maps one onto the other.

Analytic geometry gives the following formula for the area of a triangle (Figure 2.73):

$$A = \frac{1}{2}\begin{vmatrix} x_1 & y_1 & 1 \\ x_2 & y_2 & 1 \\ x_3 & y_3 & 1 \end{vmatrix} = \frac{1}{2}\left[x_1(y_2 - y_3) - x_2(y_1 - y_3) + x_3(y_1 - y_2) \right]$$

where $(x_1, y_1), (x_2, y_2), (x_3, y_3)$ are the coordinates of the vertices of the triangle. If we transform these coordinates so that $(x_1', y_1') = \alpha(x_1, y_1)$, $(x_2', y_2') = \alpha(x_2, y_2)$ and $(x_3', y_3') = \alpha(x_3, y_3)$, then the area of the transformed triangle formed by these transformed points is given by

$$A' = \frac{1}{2} \begin{vmatrix} x_1' & y_1' & 1 \\ x_2' & y_2' & 1 \\ x_3' & y_3' & 1 \end{vmatrix}$$

Make the linear substitutions given by Equations 2.21 to obtain

$$A' = \frac{1}{2} \begin{vmatrix} ax_1 + by_1 & cx_1 + dy_1 & 1 \\ ax_2 + by_2 & cx_2 + dy_2 & 1 \\ ax_3 + by_3 & cx_3 + dy_3 & 1 \end{vmatrix}$$

Expanding this determinant, we find

$$A' = \frac{1}{2}(ad - bc)\left[x_1(y_2 - y_3) - x_2(y_1 - y_3) + x_3(y_1 - y_2)\right]$$

This tells us that $A' = (ad - bc)A$ or $A'/A = (ad - bc)$: The ratio of the area of the transformed triangle to the area of the original is equal to the determinant of the homogeneous linear transformation—the homogeneous affine transformation.

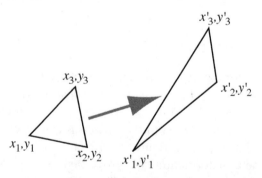

Figure 2.73 The effect of an affine transformation on the area of a triangle.

Distance in Affine Geometry

Under affine transformations, we compare distances only when they are on the same or parallel lines. In other words, we can compare the lengths of two segments only if they are collinear or parallel. It is not possible to compare two segments or distances that are not on the same or parallel lines using the congruence

THEORY OF TRANSFORMATIONS

theorems of synthetic Euclidean geometry. A consequence of this is that there is no two- or three-dimensional distance system in affine geometry.

We can construct a distance system on each family of parallel lines and compare segments on them, but we cannot compare two segments if they belong to different families. However, an affine transformation preserves the ratios of the lengths of two parallel line segments. In Figure 2.74, (a) shows the original pair of parallel line segments, (b) shows the same segments after a shear transformation, and (c) shows the same line segments after a more general affine transformation. Thus, $AB/CD = A'B'/C'D'$. In the figure, a grid is superimposed to show the effect of the transformation on the underlying space.

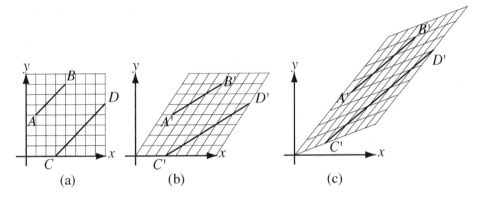

Figure 2.74 Distance in affine geometry.

The square of the distance s between two points P_1, P_2 in the plane is given by the familiar function

$$s^2 = (x_1 - x_2)^2 + (y_1 - y_2)^2$$

The distance between the images P_1', P_2' of the original points is

$$s' = (x_1' - x_2')^2 + (y_1' - y_2')^2$$

which, after appropriate substitution, becomes

$$s'^2 = (a^2 + c^2)(x_1 - x_2)^2$$
$$+ (ab + cd)(x_1 - x_2)(y_1 - y_2)$$
$$+ (b^2 + d^2)(y_1 - y_2)^2$$

If the transformation is an isometry, then $a^2 + c^2 = b^2 + d^2 = 1$ and $ab + cd = 0$, and we immediately see that $s' = s$. For a similarity we must have $a^2 + c^2 = b^2 + d^2$ and $ab + cd = 0$, so that $s' = s\sqrt{a^2 + c^2}$, as we should expect. And for a dilation $a = d = k$, $b = c = 0$, so that $s' = ks$.

The relationship between the distance functions becomes increasingly more complex as we begin to relax the restrictions of the coefficients. For example, consider a shear transformation in the x direction. Recall the transformation equations

$$x' = x + k_y y \text{ and } y' = y$$

The square of the distance between corresponding pairs of points is

$$s'^2 = (x_1 - x_2)^2 + 2k_y(x_1 - x_2)(y_1 - y_2) + (k_y^2 + 1)(y_1 - y_2)^2$$

If the original pair of points lies on a line parallel to the x axis, then so does its image. This means that $y_1 - y_2 = 0$ and $s' = s$. If the original pair of points lies on a line parallel to the y axis, then its image lies on a line whose slope is $1/k_y$ and where $s' = (y_1 - y_2)\sqrt{k_y^2 + 1}$; the distance is proportional to the difference in the y coordinates of the original points. This problem reappears later, in Chapter 6, in a slightly different form and resolution.

The Affine Group

We already know that an affine transformation is a one-to-one transformation because it is given as a linear substitution. From the development of Equations 2.22, we see that the inverse of an affine transformation is another affine transformation. Now, what about the product of two affinities? Given two successive homogeneous affine transformations

$$\begin{aligned} x' &= ax + by \\ y' &= cx + dy \end{aligned} \text{ and } \begin{aligned} x'' &= a'x' + b'y' \\ y'' &= c'x' + d'y' \end{aligned} \quad (2.31)$$

in terms of the original point, their product is

$$\begin{aligned} x'' &= a'(ax + by) + b'(cx + dy) = (aa' + cb')x + (ba' + db')y \\ y'' &= c'(ax + by) + d'(cx + dy) = (ac' + cd')x + (bc' + dd')y \end{aligned} \quad (2.32)$$

Equations 2.32 clearly have the form of an affine transformation, and the determinant of this set $\begin{vmatrix} aa' + cb' & ba' + db' \\ ac' + cd' & bc' + dd' \end{vmatrix}$ is the product of the determinants of Equations 2.31: $\begin{vmatrix} a & b \\ c & d \end{vmatrix}$ and $\begin{vmatrix} a' & b' \\ c' & d' \end{vmatrix}$. If each of these is nonzero, then their product is nonzero. So, we conclude that the product of two successive affine transformations is an affine transformation whose determinant is the product of their determinants. Finally, we may now assert that the set of all affine transformations is a group, the affine group. There are many important subgroups, among them three metric groups: the rigid body transformations, the similarities, and the equiareal group. Among the nonmetric subgroups we must include the strains and shears (see Exercises 5 and 6).

Theory of Transformations

Review

A brief review of affine transformations is now appropriate. We will confine our review to affine transformations in the plane given by the pair of homogeneous linear equations

$$x' = ax + by$$
$$y' = cx + dy$$

where $ad - bc \neq 0$. By imposing various restrictions on the coefficients, we obtain the special instances of affine transformations. Table 2.7 lists the transformation and the corresponding restrictions on the coefficients.

Table 2.7 Affine transformations in the plane given by $\begin{Bmatrix} x' = ax + by \\ y' = cx + dy \end{Bmatrix}$

Transformation	Restriction of Coefficients
General affine (nonsingular)	$ad - bc \neq 0$ $\begin{Bmatrix} +, \text{ orientation-preserving} \\ -, \text{ orientation-reversing} \end{Bmatrix}$
General affine (singular)	$ad - bc = 0$
Identity	$a = 1, b = c = 0, d = 1$
Isometry	$\begin{cases} a^2 + c^2 = 1 \\ b^2 + d^2 = 1 \\ ab + cd = 1 \end{cases}$
Similarity	$\begin{cases} a^2 + c^2 = b^2 + d^2 \\ ab + cd = 0 \end{cases}$
Equiareal	$|ad - bc| = 1$
Shear, x-direction	$a = 1, b = k_y, c = 0, d = 1$
Shear, y-direction	$a = 1, b = 0, c = k_x, d = 1$
Strain, x-direction	$a = k_x, b = c = 0, d = 1$
Strain, y-direction	$a = 1, b = c = 0, d = k_y$
Dilation	$a = k, b = c = 0, d = k$
Point reflection	$a = -1, b = c = 0, d = -1$

Many important geometric properties of linear transformations in the plane depend on the value of the determinant $ad - bc$. If the value of the determinant is zero, then we say that the transformation is singular; if all of $a, b, c, d = 0$, then all points map onto a single point; if $ad - bc = 0$ and not all of a, b, c, d are zero, then all points map onto a single line.

Parallel Projection

The notion of parallel projection allows us to make an easy transition from the more familiar world of affine geometry to that of projective geometry. Consider first parallel projection in the plane. Given two lines in the plane, l and m, and a direction of projection represented by line n, where n is not collinear with l or m, then the parallel projection of any point A on l onto m is point A' such that the line AA' is parallel to n (Figure 2.75). If A', B', C' are images of the parallel projections of A, B, and C, respectively, onto line m, then $AA' \parallel BB' \parallel CC' \parallel n$.

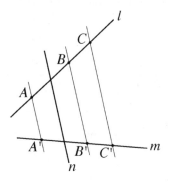

Figure 2.75 Parallel projection in the plane.

A similar interpretation applies to a parallel projection in space (Figure 2.76). Given two planes Γ and Π, we project any point A on Γ onto Π by first selecting a direction of projection and then finding the image of A at the intersection with Π of a line through and parallel to the direction of projection. Again, if A', B', C' on Π are the parallel projections of A, B, C on Γ, then $AA' \parallel BB' \parallel CC'$.

Parallel projection is a special kind of affine transformation. We can interpret any parallel projection in the plane as an example of the affine construction shown in Figure 2.67. The demonstration of this assertion is requested in Exercise 7. Although we will not demonstrate it here, it should be clear that all the geometric properties preserved by affine transformations are also preserved by parallel projections.

Theory of Transformations

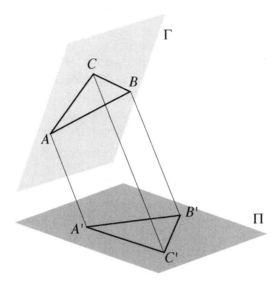

Figure 2.76 Parallel projection in space.

Section 2.6 Exercises

1. Show that a perspective affinity does not preserve angles.
2. Find the inverse of the shear v_x.
3. Find the inverse of the strain ε_x.
4. Show that the image of a conic curve under an affine transformation is a conic curve of the same type.
5. Show that strains form a subgroup.
6. Show how the shears form a subgroup.
7. Show that any parallel projection in the plane can be represented by a construction such as that shown in Figure 2.67.
8. Show that an ellipse is the image of a unit circle under an affine transformation.
9. Give an example of an equiareal collineation that is neither an isometry nor a shear.
10. Find the inverse of $x' = 3x + y$, $y' = -x + 2y$, and show that it is affine.

2.7 Projectivities

After studying affine transformations, it must seem that not many invariant geometric properties are left to be given up to further generalizations. But of course there are, and the most important is that of parallelism. Affine transformations preserve parallelism, projective transformations do not.

We find the roots of projective geometry in the work of the artists and architects of the fifteenth century. Brunelleschi, an Italian architect of that era, was probably the first to discuss a theory of perspective drawing and its geometric interpretation and implications. Mathematics done over the next four hundred years added to and refined these few early theorems. Then, in the late nineteenth century, Felix Klein created a firm algebraic foundation for what we now call projective geometry.

Although the principles of projective geometry apply to a space of any dimension, we will limit our initial explorations to plane projective geometry. Here we begin to study the geometric properties of figures that are invariant under what we call central projection.

The Equations of Projective Transformations

Projective geometry is the most universal of all the geometries that are characterized by linear transformations. It is the study of properties invariant under linear fractional transformations. The following rather formidable equations produce a projective transformation, or projectivity, in three-dimensional space:

$$x' = \frac{a_{11}x + a_{12}y + a_{13}z + a_{14}}{a_{41}x + a_{42}y + a_{43}z + a_{44}}$$

$$y' = \frac{a_{21}x + a_{22}y + a_{23}z + a_{24}}{a_{41}x + a_{42}y + a_{43}z + a_{44}} \quad (2.33)$$

$$z' = \frac{a_{31}x + a_{32}y + a_{33}z + a_{34}}{a_{41}x + a_{42}y + a_{43}z + a_{44}}$$

where for every point x, y, z there is a corresponding point x', y', z'. There are two important restrictions on these equations. First, the denominator cannot equal zero, and second, each function must have the same denominator. As the point x, y, z approaches the plane $a_{41}x + a_{42}y + a_{43}z + a_{44} = 0$, the corresponding point x', y', z' moves to infinity. What happens if the denominators are different? Finally, if we set $a_{41}, a_{42}, a_{43} = 0$ and $a_{44} = 1$, then, of course, we produce the system of linear transformations we call the affine transformations.

As we have done for other transformations, we can rewrite Equations 2.33 for the projective geometry of two-dimensional space and, in the process, simplify them as follows

Theory of Transformations

$$x' = \frac{ax+by+h}{ex+fy+g}$$
$$y' = \frac{cx+dy+k}{ex+fy+g}$$

(2.34)

where $adg - afk + bek - bcg + cfh - deh \neq 0$. This restriction on the coefficients is analogous to $ad - bc \neq 0$ for the affine transformation. If we let $g = 1$ and $e, f, h, k = 0$, then Equations 2.34 reduce to the homogeneous affine transformation, and the restriction on the coefficients becomes $ad - bc \neq 0$, as expected. The restriction on the coefficients ensures that the projective transformation is indeed a one-to-one mapping with an inverse that is also a projective transformation.

Later, in Chapter 6, when we have the power of vectors and matrix methods at our disposal, as well as an understanding of homogeneous coordinates, we will return to the study of the analytic description and nature of projectivities. For now, however, we will explore the geometric construction and interpretation of a central projection.

Central Projection of a Plane Figure

To perform a central projection transformation, we will need (Figure 2.77): a point E (the center of projection), a projection plane Π, and a plane of origin, Γ containing the figure to be transformed. If we cast a ray from E through each point of the original figure, then each ray intersects the projection plane in a point, the projected image of the original point. In the original figure, we find $AB \parallel CD$; this parallel relationship, in general, breaks down under a central projection and is usually not preserved. (Only special arrangements of the figure, the planes and center of projection preserve the parallel property.) Thus, $A'B'$ is not parallel to $C'D'$. If segment AB and CD are each extended indefinitely far to F and G, respectively, then the angle between rays EF and EG approaches zero, and EF and EG approach being perpendicular to Π. The projected points F', G' approach concurrence. This is an example of the effect of perspective on parallel lines as they recede from us along our line of sight; railroad tracks are an oft-quoted instance of this. Clearly, this transformation preserves collinearity, but not parallels.

Central projection preserves other geometric properties, for example, cross-ratios of intersecting line segments and the degree of a curve. (We will not examine these now.) Points and straight lines are invariant under projection, but of course, angles and the distance between two points are not.

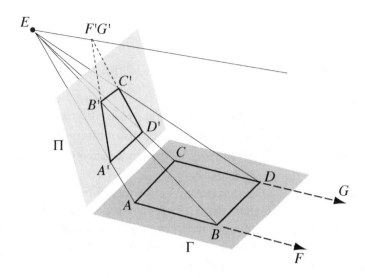

Figure 2.77 Central projection in plane projective geometry.

Central Projection of a Figure in Space

Here we go a step beyond the preceding example. Instead of a plane figure, let the original be an arbitrary figure in space (Figure 2.78). We construct its projected image exactly the same way as for a plane figure. A ray is cast from the center of projection E through each point of the original figure and through the projection plane Π. This produces image points at the intersections of the rays with Π. We may choose to place the projection plane either beyond the object or between the object and the center of projection.

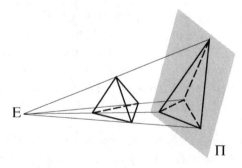

Figure 2.78 Central projection of a figure in space.

The center of projection in certain applications might represent the eye of the viewer of a computer graphic scene. In this case it is important to position E, Π, and the object so that the angles between the rays and the line-of-sight perpendicular to Π are not too large, otherwise a distortion occurs relative to what is ordinarily seen. (In part, this happens because the plane of projection of the eye is not a plane but a curved surface.)

If we are careful in how we position a coordinate system and choose the projection plane and center of projection, then we can simplify Equations 2.33. Here is an example: We let the plane $z = 0$ represent the projection plane Π and the point $(0,0,z_E)$ represent the center of projection (Figure 2.79).

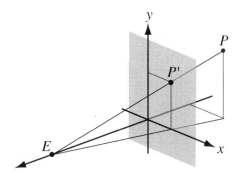

Figure 2.79 Central projection of points in space relative to a coordinate system.

From similar triangles, we obtain the following transformation equations:

$$x' = \frac{z_E x}{z_E - z}$$

$$y' = \frac{z_E y}{z_E - z}$$

$$z' = 0$$

where $a_{11}, a_{22}, a_{44} = z_E$, $a_{43} = -1$ and all the other coefficients are zero.

2.8 Topological Transformations

We now make a brief but significant departure from the world of linear transformations to that of topological transformations, which in all their generality are distinctly nonlinear. The properties of figures that are invariant under transformations that are continuous both ways form the study of topology. What does this mean? First, the ground rules: A topological transformation may stretch, bend, twist and compress a figure, but may not tear it, puncture it, nor cause it to be self-intersecting. So, "cubical" and "spherical" are not topological properties, because we can continuously deform either of these surfaces into the other. However, the "knottedness" of a closed loop of string incorporating a square knot is a topological property, because we cannot undo the knot using only the allowable topological deformations.

What other kinds of properties are possibly left invariant under such a deformational onslaught? Surprisingly, there are several, and they are of great interest and importance to geometers. For example, topological transformations preserve linear order and cyclic order. Lines, parabolas and the branches of a hyperbola belong to a class of topologically equivalent figures we call simple arcs. The property of being a closed or open curve or surface is a topological invariant. One-sidedness (a Möbius strip or Klein bottle, for example) or two-sidedness are topologically invariant properties of surfaces. However, topology is a geometry in which size and shape have no meaning.

If a topological transformation is a one-to-one mapping of the plane onto itself with equations

$$x' = f(x, y)$$
$$y' = g(x, y)$$

and its inverse has equations

$$x = \psi(x', y')$$
$$y = \phi(x', y')$$

where functions f, g, ψ, ϕ are continuous and unique, then we call it a homeomorphism. Homeomorphisms happen to form a group, and we can show that affine transformations are a subgroup. However, we would quickly go far beyond the scope of this text were we to attempt to elaborate these distinctions. We will explore some special topological transformations in Chapter 7, and we will meet them under the guise of nonlinear transformations.

3 Vector Spaces

Any set of vectors over the field of real numbers that is closed under addition and scalar multiplication defines a real vector space. What does the term "vector" mean in this context? The answer develops early in the first section, which introduces linear spaces, vectors, linear dependence, and vector products, properties and geometry. (Many readers will now see vectors from a somewhat different point of view from what an earlier experience with them might have given, particularly as they relate to geometry, transformations and invariance.) The second section is a discussion of basis vectors, beginning with their definition and proceeding with such related topics as change of basis, oblique coordinates, orthogonal bases and matrices, orthogonal transformations, and transformations relative to different bases. The third section introduces eigenvalues, eigenvectors, and the so-called characteristic equation that defines them, their geometric interpretation, symmetric transformations, the diagonalization of matrices, quadratic forms and conics. The fourth and final section of the chapter introduces tensors and their relationship to transformations. Here a distinction is made between contravariant and covariant tensors. This section concludes with a brief look at what is probably one of the most important concepts in differential geometry and general relativity—the metric tensor.

3.1 Introduction to Linear Vector Spaces

Recall that one of the central problems we must resolve is this: How do we formulate propositions and develop computational techniques to represent and analyze geometric objects and physical phenomena in a way that is independent of any underlying arbitrarily-chosen coordinate system? Vectors and tensors allow us to do just that, to study invariant geometric and algebraic properties not dependent upon a particular coordinate system.

The points of space taken together comprise the stage upon which we describe and analyze geometric objects. To each point we may assign a scalar, characterized by a single number whose value at different points is a function of the point coordinates; a vector, characterized by a set of numbers with the same dimensionality, n, as that of the space; or a tensor, characterized by an m-dimensional array of numbers. Thus, in a space of n dimensions, a single number describes a scalar, n

numbers describe a vector, and an array of n^m numbers describes a tensor. These numbers are the components of the vector or tensor, and their values depend on the particular coordinate system chosen. If we move from one coordinate system to another via a transformation, then the components of the vector or tensor in the new system relate to those of the original in a definite way, as governed by the transformation and depending only on the partial derivatives of the new coordinates with respect to the old. These characteristics allow us to associate invariant magnitudes with vectors and tensors. Now, let's begin our study of some basic concepts so that later we can use vectors and tensors to better understand geometric transformations.

Linear Vector Spaces

The notion of a free vector is one we commonly interpret visually as a directed line segment or arrow that we can transport in space so that it always remains parallel to its original direction. We infer and compute many properties of free vectors by using geometric constructions without reference to a coordinate system in much the same way that we approach problems in synthetic Euclidean geometry. In many free vector operations, we treat them almost surreptitiously as if they originated at a common starting point. The nature of a particular problem in physics or engineering makes this computationally convenient, if not a logical necessity. When we deal with vectors we must be prepared for a rather interesting duality in their nature: free and fixed, imparting a certain vagueness surrounding any clear visual or precise concrete interpretation we may try to give them. This is more apparent once we discover that vectors (and scalars, for that matter) are special kinds of tensors. However, we are digressing. When we create a vector algebra that treats vectors as if they originated at a common point, then we work with a linear vector space. In such a space, not only do all vectors have a common starting point (in the arrow metaphor, the tail end), but also every vector combines under addition with any other vector according to the parallelogram law. Now we will see just what that means.

We will use boldface lowercase Roman letters to denote vectors, such as **a**, **b**, **c**, ..., **p**, **q**, **r**, **s**, **t**.. To simplify matters we assume that all vectors have the origin of a coordinate system as their initial points.

We must now define the two operations on vectors for them to qualify as constituents of a linear vector space:

1. Addition of two vectors: Add **p** and **q** to obtain their sum, **p** + **q**.

2. Product of a vector and scalar: Multiply any vector **p** by any real number k to obtain their product, k**p**.

The set of all vectors is closed with respect to these two operations, because both the sum of two vectors and the product of a vector and a scalar are themselves vectors. These operations have the following properties:

Vector Spaces

1. Commutativity: $\mathbf{a}+\mathbf{b}=\mathbf{b}+\mathbf{a}$.
2. Associativity: $(\mathbf{a}+\mathbf{b})+\mathbf{c}=\mathbf{a}+(\mathbf{b}+\mathbf{c})$.
3. Identity Element: $\mathbf{a}+\mathbf{0}=\mathbf{a}$.
4. Inverse: $\mathbf{a}-\mathbf{a}=\mathbf{0}$.
5. Identity under scalar multiplication: $k\mathbf{a}=\mathbf{a}$, when $k=1$.
6. $c(d\mathbf{p})=(cd)\mathbf{p}$
7. $(c+d)\mathbf{p}=c\mathbf{p}+d\mathbf{p}$
8. $k(\mathbf{p}+\mathbf{q})+k\mathbf{p}+k\mathbf{q}$

Any set of elements satisfying the two operations with the eight properties mentioned above forms a linear space or linear vector space. Of course, we are primarily concerned with the set of vectors that satisfies these requirements. This brings us back to the definition presented in the introductory paragraph to this chapter, namely: Any set of vectors over the field of real numbers that is closed under addition and scalar multiplication is a real vector space. There are of course many other operations, such as the inner product and the vector product, but they are not pertinent to this definition of a linear vector space.

The set of all vectors of the form $\mathbf{r}=(r_1, r_2, \ldots, r_n)$, where $r_1, r_2, \ldots, r_n$, are real numbers, obviously constitutes a vector space since we can easily show that it is closed under addition and scalar multiplication. (Notice that we have backed into the definition of a vector as an ordered n-tuple of real numbers.)

We can form linear combinations of vectors and show that the set of all linear combinations of a given set forms a vector space. Let $\mathbf{x}_1, \mathbf{x}_2, \ldots, \mathbf{x}_n$ be any n vectors; then $a_1\mathbf{x}_1 + a_2\mathbf{x}_2 + \ldots + a_n\mathbf{x}_n$ (where $a_1, a_2, \ldots, a_n$ are scalars) is a linear combination of the vectors $\mathbf{x}_1, \mathbf{x}_2, \ldots, \mathbf{x}_n$. However, we can go much farther. Let

$$\mathbf{s} = a_1\mathbf{x}_1 + a_2\mathbf{x}_2 + \ldots + a_n\mathbf{x}_n$$
$$\mathbf{t} = b_1\mathbf{x}_1 + b_2\mathbf{x}_2 + \ldots + b_n\mathbf{x}_n$$

Then the vectors $\mathbf{s}$ and $\mathbf{t}$ are linear combinations of the vectors $\mathbf{x}_1, \mathbf{x}_2, \ldots, \mathbf{x}_n$, so that $\mathbf{s}+\mathbf{t} = (a_1+b_1)\mathbf{x}_1 + (a_2+b_2)\mathbf{x}_2 + \ldots + (a_n+b_n)\mathbf{x}_n$

Furthermore, we have $k\mathbf{s} = (ka_1)\mathbf{x}_1 + (ka_2)\mathbf{x}_2 + \ldots + (ka_n)\mathbf{x}_n$, which is also a linear combination of $\mathbf{x}_1, \mathbf{x}_2, \ldots, \mathbf{x}_n$. We say that the space of all linear combinations of a given set of vectors is the space generated by that set.

It is time for some examples. Given just a single vector $\mathbf{x}$, then the space generated by all scalar multiples of $\mathbf{x}$ is a straight line collinear with $\mathbf{x}$ (we assume $\mathbf{x} \neq \mathbf{0}$). Given two vectors $\mathbf{p}$ and $\mathbf{q}$, where $\mathbf{q}$ is not a scalar multiple of $\mathbf{p}$, then the space generated by their linear combinations is the plane containing $\mathbf{p}$ and $\mathbf{q}$. In Figure 3.1 we see that $\mathbf{r} = a_1\mathbf{p} + a_2\mathbf{q}$, and from the parallelogram law of vector addition that the vectors $\mathbf{p}$, $\mathbf{q}$, and $\mathbf{r}$ are coplanar. We could easily continue this process, generating spaces of three or more dimensions merely by increasing the number of vectors in the generating set. We do this, but not without certain conditions (as in the last example above, where we did not permit $\mathbf{p}$ and $\mathbf{q}$ to be scalar multiples of one another). This leads us to the idea of linear dependence of vectors.

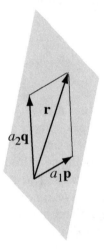

Figure 3.1 The set of linear combinations of p and q generates a plane.

Linear Dependence

Linear dependence is an idea common to many areas of mathematics, applying both to functions and to vectors, for example, and to linear differential equations and linear algebraic equations, as two more specific examples. Mathematicians are able to demonstrate that the linear dependence of functions is a special case of the linear dependence of vectors. Here is how we apply this notion to vectors: Vectors $\mathbf{x}_1, \mathbf{x}_2, \ldots, \mathbf{x}_n$ are linearly dependent if and only if there are real numbers $a_1, a_2, \ldots, a_n$ not all equal to zero such that

$$a_1\mathbf{x}_1 + a_2\mathbf{x}_2 + \ldots + a_n\mathbf{x}_n = 0 \tag{3.1}$$

If Equation 3.1 is true only if $a_1, a_2, \ldots, a_n$ are all zero, then $\mathbf{x}_1, \mathbf{x}_2, \ldots, \mathbf{x}_n$ are linearly independent. If the vectors $\mathbf{x}_1, \mathbf{x}_2, \ldots, \mathbf{x}_n$ are linearly dependent, then we can represent any one of them as a linear combination of the others. Conversely, if one of the vectors $\mathbf{x}_1, \mathbf{x}_2, \ldots, \mathbf{x}_n$ is a linear combination of the others, then the vectors are linearly dependent. This is easy to prove. In Equation 3.1 let $a_1 \neq 0$, then

$$\mathbf{x}_1 = -\frac{a_2}{a_1}\mathbf{x}_2 - \frac{a_3}{a_1}\mathbf{x}_3 - \ldots - \frac{a_n}{a_1}\mathbf{x}_n \tag{3.2}$$

Vector Spaces

Another way of saying this is that the vectors $x_1, x_2, \ldots, x_n$ are linearly dependent if and only if one of them belongs to the space generated by the remaining $n-1$ vectors. From this observation, we define the dimension of a linear space as equal to the maximum number of linearly independent vectors that it can contain. We will make good use of this fact in the next section, where we study basis vectors. Figure 3.2 presents a clear picture of how this works in the plane. Here we have three arbitrary coplanar vectors **p**, **q**, and **r**. We easily find coefficients a_1 and a_2 so that $a_1\mathbf{p} + a_2\mathbf{q} = \mathbf{r}$ or $a_1\mathbf{p} + a_2\mathbf{q} - \mathbf{r} = 0$.

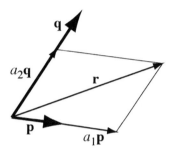

Figure 3.2 Linear dependence of three vectors in the plane.

Two vectors are dependent if and only if they are parallel (or collinear); any three vectors are dependent if and only if they are coplanar; any four vectors are dependent in a space of three dimensions; any n vectors are dependent in a space of $n-1$ dimensions. Finally, in a space of three dimensions, we make the following brash assertion (try to demonstrate its truth.): A set of three vectors **r**, **s**, and **t** is linearly dependent if and only if the determinant

$$\begin{vmatrix} r_1 & r_2 & r_3 \\ s_1 & s_2 & s_3 \\ t_1 & t_2 & t_3 \end{vmatrix} = 0 \qquad (3.3)$$

where the elements of the determinant are the separate components of the vectors. If the vectors $x_1, x_2, \ldots, x_n$ are linearly independent, then it is impossible to represent any one of them as a linear combination of the other $n-1$ vectors.

Vectors

Certain properties of vectors and vector algebra are worth reviewing before we move on to other subjects. So, first let us consider the commutativity of vector addition, which tells us that $\mathbf{a} + \mathbf{b}$ is the same as $\mathbf{b} + \mathbf{a}$. It does not matter in what order we add the two vectors. This is the algebraic embodiment of the parallelogram law of vector addition (Figure 3.3). If $\mathbf{a} + \mathbf{b} = \mathbf{r}$, then $\mathbf{b} + \mathbf{a} = \mathbf{r}$ as well. It does not matter which of the two paths we take; the resultant **r** is the same.

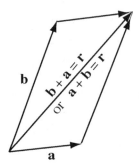

Figure 3.3 Parallelogram law of vector addition.

A vector has components, but it has no "natural" components (although it may seem that way when we express components as projections upon the axes of some conveniently installed coordinate system). Figure 3.4 is an example of this for a space of two dimensions. Here $\mathbf{p} = \mathbf{x} + \mathbf{y}$, where $\mathbf{x}$ and $\mathbf{y}$ are vectors of appropriate magnitude and collinear with the x and y axes. Alternatively, $\mathbf{p} = \mathbf{a} + \mathbf{b}$, where $\mathbf{a}$ and $\mathbf{b}$ are linearly independent vectors. We are free to express any vector in a space of n dimensions as a linear combination of n independent vectors. Thus,

$$\mathbf{p} = a_1 \mathbf{x}_1 + a_2 \mathbf{x}_2 + \ldots + a_n \mathbf{x}_n \tag{3.4}$$

where $a_1, a_2, \ldots, a_n$ are real numbers (i. e., scalars). We call the vectors $a_1 \mathbf{x}_1$, $a_2 \mathbf{x}_2$, ..., $a_n \mathbf{x}_n$ the components of $\mathbf{p}$. We do not require special conditions, such as that the $\mathbf{x}_1, \mathbf{x}_2, \ldots, \mathbf{x}_n$ vectors are mutually orthogonal or that they must be unit vectors.

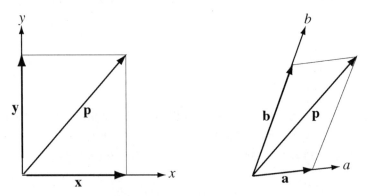

Figure 3.4 Vector components.

As you see, every vector has a potentially infinite number of sets of components. However, the magnitude of a vector is invariant under all isometries, even though these transformations dramatically affect the components. The invariance of a vector's direction under an isometry is another matter, for direction is often described relative to some coordinate system, which is, of course, subject to change.

Vector Spaces

In metric spaces, a vector has a magnitude, a single number that is invariant under appropriate metric transformations (isometries). When we use the arrow or directed line segment model of a vector, then the length of the segment is what we take to be the magnitude. In a rectangular Cartesian coordinate system, with the components of a vector given as a set of linearly independent vectors collinear with the coordinate axes, the magnitude is found simply by the application of the Pythagorean theorem (Figure 3.5). Thus

$$|\mathbf{r}| = \sqrt{r_x^2 + r_y^2 + r_z^2} \qquad (3.5)$$

where $|\mathbf{r}|$ is a positive real number.

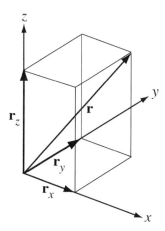

Figure 3.5 Vector magnitude.

Do vectors have to have a magnitude? No, there are nonmetric spaces where the use of vectors is not only appropriate but also necessary but where the notion of length or magnitude in the ordinary sense does not apply.

Given any vector $\mathbf{a}$, we compute its corresponding unit vector as follows:

$$\hat{\mathbf{a}} = \frac{\mathbf{a}}{|\mathbf{a}|} \qquad (3.6)$$

where $|\hat{\mathbf{a}}| = 1$. If $\mathbf{a}$ is a vector in a three-dimensional Cartesian space, then the components of $\hat{\mathbf{a}}$ are $a_x/|\mathbf{a}|, a_y/|\mathbf{a}|, a_z/|\mathbf{a}|$.

The direction of a vector $\mathbf{a}$ in a three-dimensional Cartesian system is given by the components of its unit vector $\mathbf{a}/|\mathbf{a}|$. We often refer to these particular components as direction numbers $l, m,$ and n, or direction cosines (Figure 3.6). The angles corresponding to these we denote as α, β, γ, so that

$$l = \cos\alpha = a_x/|\mathbf{a}|, \quad m = \cos\beta = a_y/|\mathbf{a}|, \quad n = \cos\gamma = a_z/|\mathbf{a}|$$

Since these are components of a unit vector they are obviously related by means of the Pythagorean theorem; thus $|\hat{\mathbf{a}}| = l^2 + m^2 + n^2 = 1$.

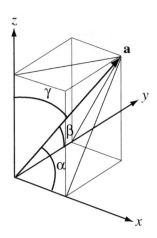

Figure 3.6 Direction cosines.

Scalar Product

The scalar product of two vectors **p** and **q** is the sum of the products of their corresponding components. It is a single real number. In three-dimensional Cartesian space this product is

$$\mathbf{p} \bullet \mathbf{q} = p_x q_x + p_y q_y + p_z q_z \tag{3.7}$$

Its value is independent of the coordinate system and invariant under isometric transformations. Furthermore, if **p** = **q**, then

$$\mathbf{p} \bullet \mathbf{p} = |\mathbf{p}|^2 \tag{3.8}$$

and the magnitude of **p** is $\sqrt{\mathbf{p} \bullet \mathbf{p}}$. To anticipate later developments, this is how to write the inner product using tensor notation:

$$\mathbf{p} \bullet \mathbf{q} = p^j q^k \delta_{jk} \tag{3.9}$$

where $\delta_{jk} = \begin{cases} 1 \text{ if } j = k \\ 0 \text{ if } j \neq k \end{cases}$ and δ_{jk} is the Kronecker delta.

The scalar product produces useful geometric information. Since proofs are readily available in almost any text on elementary vector algebra, we will assert that:

1. If $\mathbf{p} \bullet \mathbf{q} = 0$, then **p** and **q** are mutually perpendicular.

2. The angle θ between two vectors **p** and **q** is given by

$$\theta = \cos^{-1}(\mathbf{p} \bullet \mathbf{q} / |\mathbf{p}||\mathbf{q}|) \tag{3.10}$$

Vector Spaces

Vector Product

The vector product of two vectors **p** and **q** is another vector **r**. We write this product as

$$\mathbf{p} \times \mathbf{q} = \mathbf{r} \tag{3.11}$$

where **r** is perpendicular to the plane containing **p** and **q**.

The component form for the product of two vectors in Cartesian space is

$$\mathbf{p} \times \mathbf{q} = (p_y q_z - p_z q_y)\mathbf{u}_x + (p_z q_x - p_x q_z)\mathbf{u}_y + (p_x q_y - p_y q_x)\mathbf{u}_z \tag{3.12}$$

which is the expansion of a determinant:

$$\mathbf{p} \times \mathbf{q} = \begin{vmatrix} \mathbf{u}_x & \mathbf{u}_y & \mathbf{u}_z \\ p_x & p_y & p_z \\ q_x & q_y & q_z \end{vmatrix} \tag{3.13}$$

where $\mathbf{u}_x, \mathbf{u}_y, \mathbf{u}_z$ are unit vectors in the x, y, and z directions, respectively.

Recall the right-hand rule. It describes the direction of the vector **r** generated by the product $\mathbf{p} \times \mathbf{q}$ (Figure 3.7). Imagine rotating **p** into **q** and then curling the fingers of your right hand in this same sense. When you do this the extended thumb of your right hand points in the direction of the resultant **r**. This gives a visual and intuitive appreciation for the vector product that corresponds to the analytical representation.

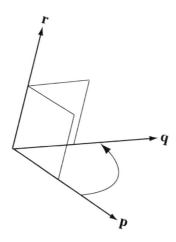

Figure 3.7 Vector product $\mathbf{p} \times \mathbf{q} = \mathbf{r}$ and the right-hand rule.

As with the scalar product, the vector product also conveys geometric information. For example, the area of a triangle OAB is $|\mathbf{a} \times \mathbf{b}|/2$, where **a** is the directed line segment OA and **b** is the directed segment OB (Figure 3.8). (Again, rigorous proof is available in any text on vector algebra.).

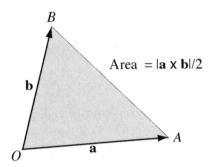

Figure 3.8 The area of a triangle as a vector product.

Finally, the angle θ between vectors **p** and **q** is

$$\theta = \sin^{-1}(|\mathbf{p} \times \mathbf{q}|/|\mathbf{p}||\mathbf{q}|) \tag{3.14}$$

Compare this to Equation 3.10. From this equation we see that if $\mathbf{p} \times \mathbf{q} = \mathbf{0}$, then $\theta = 0$; this means that **p** and **q** are parallel. Here is a summary some of the more important vector properties:

1. $\mathbf{a} + \mathbf{b} = \mathbf{b} + \mathbf{a}$
2. $(\mathbf{a} + \mathbf{b}) + \mathbf{c} = \mathbf{a} + (\mathbf{b} + \mathbf{c})$
3. $\mathbf{a} + \mathbf{0} = \mathbf{a}$, $(\mathbf{0}$ is the null vector$)$
4. $\mathbf{a} - \mathbf{a} = \mathbf{0}$
5. $k(l\mathbf{a}) = (kl)\mathbf{a}$, (k,l are scalars)
6. $k\mathbf{a} = \mathbf{a}$, $(k = 1)$
7. $(k+l)\mathbf{a} = k\mathbf{a} + l\mathbf{a}$
8. $k(\mathbf{a} + \mathbf{b}) = k\mathbf{a} + k\mathbf{b}$ $k(\mathbf{a} + \mathbf{b}) = k\mathbf{a} + k\mathbf{b}$
9. $\mathbf{a} \bullet \mathbf{b} = |\mathbf{a}||\mathbf{b}\cos\theta|$, where θ = the angle between **a** and **b**.
10. $\mathbf{a} \bullet \mathbf{a} = |\mathbf{a}|^2$
11. $\mathbf{a} \bullet \mathbf{b} = \mathbf{b} \bullet \mathbf{a}$
12. $\mathbf{a} \bullet (\mathbf{b} + \mathbf{c}) = \mathbf{a} \bullet \mathbf{b} + \mathbf{a} \bullet \mathbf{c}$
13. $(k\mathbf{a}) \bullet \mathbf{b} = \mathbf{a} \bullet (k\mathbf{b}) = k(\mathbf{a} \bullet \mathbf{b})$
14. If $\mathbf{a} \bullet \mathbf{b} = 0$, then **a** and **b** are perpendicular.
15. If $\mathbf{a} \times \mathbf{b} = \mathbf{c}$, then **c** is perpendicular to both **a** and **b**.

Vector Spaces

16. $\mathbf{a} \times \mathbf{b} = \begin{vmatrix} \mathbf{u}_x & \mathbf{u}_y & \mathbf{u}_z \\ a_x & a_y & a_z \\ b_x & b_y & b_z \end{vmatrix}$

17. $\mathbf{a} \times \mathbf{b} = -\mathbf{b} \times \mathbf{a}$

18. $\mathbf{a} \times (\mathbf{b} + \mathbf{c}) = \mathbf{a} \times \mathbf{b} + \mathbf{a} \times \mathbf{c}$

19. $(k\mathbf{a}) \times \mathbf{b} = \mathbf{a} \times (k\mathbf{b}) = k(\mathbf{a} \times \mathbf{b})$

20. If $\mathbf{a} \times \mathbf{b} = 0$, then $\mathbf{a}$ and $\mathbf{b}$ are parallel.

21. $\mathbf{a} \times \mathbf{a} = 0$

Vector Geometry

Now let's put vectors to work doing what they are very good at—geometry. We start with some easy applications with which you are probably familiar. In later chapters, you will see increasingly more sophisticated and powerful applications and will come to appreciate the special relationships of vectors, matrices, and geometric transformations.

In a Cartesian coordinate system, we use fixed or position vectors to define points, where the vector components represent corresponding point coordinates. This means we can use the terms point and vector interchangeably. For example, in Figure 3.9 we speak of the points (vectors) $\mathbf{p}$ and $\mathbf{q}$.

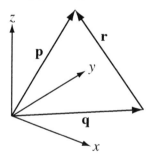

Figure 3.9 Distance between two points.

The distance $|\mathbf{r}|$ between points $\mathbf{p}$ and $\mathbf{q}$ is

$$|\mathbf{r}| = |\mathbf{p} - \mathbf{q}| \qquad (3.15)$$

and from Equations 3.5 and 3.8 $|\mathbf{r}| = \sqrt{(\mathbf{p} - \mathbf{q}) \bullet (\mathbf{p} - \mathbf{q})}$, or

$$|\mathbf{r}| = \sqrt{(p_x - q_x)^2 + (p_y - q_y)^2 + (p_z - q_z)^2} \qquad (3.16)$$

This is the Pythagorean theorem at work via vector algebra.

The vector equation of a line through $\mathbf{r}_0$ and parallel to $\mathbf{t}$ (Figure 3.10) is

$$\mathbf{r} = \mathbf{r}_0 + u\mathbf{t} \qquad (3.17)$$

where $\mathbf{r}$ is a point on the line and u is a real number.

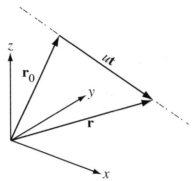

Figure 3.10 Vector representation of a line.

In terms of components, we have

$$\begin{aligned}r_x &= r_{0x} + ut_x \\ r_y &= r_{0y} + ut_y \\ r_z &= r_{0z} + ut_z\end{aligned} \qquad (3.18)$$

The vector equation of a plane through a point $\mathbf{r}_0$ and parallel to the two vectors $\mathbf{s}$ and $\mathbf{t}$ (Figure 3.11) is

$$\mathbf{r} = \mathbf{r}_0 + u\mathbf{s} + v\mathbf{t} \qquad (3.19)$$

where $\mathbf{r}$ is a point on the plane and u and v are real numbers.

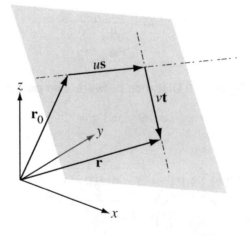

Figure 3.11 Vector representation of a plane.

Vector Spaces

In terms of components we have

$$r_x = r_{0x} + us_x + vt_x$$
$$r_y = r_{0y} + us_y + vt_y \qquad (3.20)$$
$$r_z = r_{0z} + us_z + vt_z$$

Section 3.1 Exercises

1. Show that the set of all vectors lying in the first quadrant of the x, y plane is not a linear space.
2. Find nontrivial linear relations for the following sets of vectors.
   a. $\mathbf{p} = (1, 0, -2)$, $\mathbf{q} = (3, -1, 3)$, $\mathbf{r} = (5, -2, 8)$
   b. $\mathbf{p} = (2, 0, 1)$, $\mathbf{q} = (0, 5, 1)$, $\mathbf{r} = (6, -5, 4)$
   c. $\mathbf{p} = (3, 0)$, $\mathbf{q} = (1, 4)$, $\mathbf{r} = (2, -1)$
3. Given that $\mathbf{p}$ and $\mathbf{q}$ are linearly independent vectors in the plane, find the value of k that makes each of the following pairs of vectors collinear.
   a. $k\mathbf{p} + 2\mathbf{q}$, $\mathbf{p} - \mathbf{q}$
   b. $(k+1)\mathbf{p} + \mathbf{q}$, $2\mathbf{q}$
   c. $k\mathbf{p} + \mathbf{q}$, $\mathbf{p} + k\mathbf{q}$
4. Show that three vectors in space are linearly dependent if and only if they lie in one plane.
5. Are vectors $\mathbf{r} = (1, -1, 0)$, $\mathbf{s} = (0, 2, -1)$, $\mathbf{t} = (2, 0, -1)$ linearly dependent? Why?
6. Show that the magnitude of a unit vector is equal to one.
7. Find the magnitude and direction numbers for each of the following vectors.
   a. $\mathbf{a} = (3, 4)$ d. $\mathbf{d} = (1, 4, -3)$
   b. $\mathbf{b} = (0, -2)$ e. $\mathbf{e} = (x, y, z)$
   c. $\mathbf{c} = (-3, -5, 0)$
8. Find the inner product of the following pairs of vectors.
   a. (0, −2), (1, 3)
   b. (4, −1), (2, 1)
   c. (1, 0), (0, 4)
   d. (3, 0, −2), (0, −1, −3)
   e. (5, 1, 7), (−2, 4, 1)

9. Find the angle between each pair of vectors in Exercise 8.
10. Find the vector product for each of the following pairs of vectors.
    a. (3, –1, 2), (2, 0, 2)
    b. (4, 1, –5), (3, 6, 2)
    c. (2, –1, 3), (–4, 2, –6)
    d. (0, 1, 0), (1, 0, 0)
    e. (0, 0, 1), (1, 0, 0)
11. Prove that the vectors $\left(-\frac{1}{3}, \frac{2}{3}, \frac{2}{3}\right)$, $\left(\frac{2}{3}, -\frac{1}{3}, \frac{2}{3}\right)$, $\left(-\frac{2}{3}, -\frac{2}{3}, \frac{1}{3}\right)$ are mutually perpendicular.
12. Prove that the area of triangle OAB is given by $|\mathbf{a} \times \mathbf{b}|/2$, where $\mathbf{a}$ is directed line segment OA and $\mathbf{b}$ is OB.
13. Show that in three-dimensional Cartesian space the inequality $|\mathbf{a}+\mathbf{b}| \leq |\mathbf{a}|+|\mathbf{b}|$ demonstrates that the sum of the lengths of any two sides of a triangle is greater than the length of the third side.
14. Show that if $|\mathbf{a}+\mathbf{b}| = |\mathbf{a}|+|\mathbf{b}|$, then $\mathbf{a}$ and $\mathbf{b}$ are linearly dependent.
15. Write the vector equation of a line through two points $\mathbf{p}_0$ and $\mathbf{p}_1$.
16. Write the vector equation of a plane containing the three noncollinear points $\mathbf{p}_0, \mathbf{p}_1, \mathbf{p}_2$.
17. Compare and verify Equations 3.10 and 3.14.

3.2 Basis Vectors

We have identified two kinds of vectors: fixed and free. Free vectors, remember, are completely independent of any coordinate system, and for this reason we call them objective. Fixed vectors, on the other hand, always relate either to some implied frame of reference or to an explicitly defined one. The notion of using vectors to represent points is a powerful one, as we will demonstrate many times in this text. We have already discovered that all vectors have components, but only the components of fixed vectors are also coordinates.

Unless indicated otherwise, we will use Cartesian systems. Rectangular Cartesian coordinates are a special case of the more general Cartesian systems. For each dimension of a general Cartesian system there is a family of parallel straight lines, each with a uniform scale or metric. All families may have the same scale, or each may be different (Figure 3.12). The families need not be mutually orthogonal, and, of course, we must assign one point to the role of the origin.

Vector Spaces

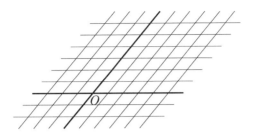

Figure 3.12 Two-dimensional Cartesian coordinate system.

A set of linearly independent vectors $e_1, e_2, ..., e_n$ forms the basis of a Cartesian space S_n of n dimensions, if we can express any position vector r in S_n as a linear combination of these basis vectors. Let's restrict this discussion to spaces of two or three dimensions, keeping in mind that analogous arguments and results apply to higher dimensions. In three dimensions, a set of basis vectors e_1, e_2, e_3 (Figure 3.13) emanating from a common point O, the origin, defines three families of parallel lines and forms a Cartesian system. The three lines X_1, X_2, X_3 concurrent at O and collinear with e_1, e_2, e_3, respectively, define the coordinate axes.

For any point (position vector) r in this system, we find coordinates (components), as follows: Construct a parallelepiped with O at one vertex, r as a body diagonal, and concurrent edges lying collinear with the basis vectors. The three directed line segments corresponding to edges OA, OB, and OC define the vector components of r in this basis system. Denote these as a, b, and c, respectively. The very nature of this construction technique ensures that the parallelogram law of vector addition applies. Thus,

$$r = a + b + c \tag{3.21}$$

What are the coordinates of r relative to this basis? To answer this we begin with these definitions: Let

$$r_1 = \frac{|a|}{|e_1|}, \; r_2 = \frac{|b|}{|e_2|}, \; r_3 = \frac{|c|}{|e_3|} \tag{3.22}$$

Remember, we do not insist that basis vectors must be unit vectors. Now, using these definitions we rewrite Equation 3.21 to obtain

$$r = r_1 e_1 + r_2 e_2 + r_3 e_3 \tag{3.23}$$

where we identify $r_1, r_2,$ and r_3 as the components or coordinates of r with respect to the frame of reference defined by the basis vector e_1, e_2, and e_3. Sometimes we speak of these as the parallel coordinates of the point r. These coordinates coincide with the coordinates of the affine three-dimensional space if the basis vectors are unit vectors. Notice that this system of basis vectors as shown in Figure 3.13 is a right-handed one, but it could just as well be left-handed.

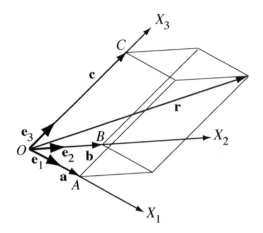

Figure 3.13 A three-dimensional basis.

We have just constructed an oblique Cartesian coordinate system of the most general sort. Such a system may offer computational or other advantages if the principle vectors defining a particular problem are not orthogonal. Such a system becomes the stage upon which we may perform various feats of affine geometry. Now lets look at how we can indeed do geometry in an oblique system.

First, let vectors $\mathbf{e}_1$ and $\mathbf{e}_2$ define an oblique coordinate system in the xy plane. Then every point in the plane has a position vector given by $\mathbf{p} = \alpha \mathbf{e}_1 + \beta \mathbf{e}_2$ in this system. The oblique coordinates of a point are thus (α, β) (Figure 3.14). The magnitudes of $\mathbf{e}_1$ and $\mathbf{e}_2$ establish the scaling along the two axes.

Using the properties of similar triangles, we write the equation of a line through the points $P_1(\alpha_1, \beta_1)$ and $P_2(\alpha_2, \beta_2)$ as $(\alpha - \alpha_1)(\beta_2 - \beta_1) = (\beta - \beta_1)(\alpha_2 - \alpha_1)$. For example, the equation of the line through $P_1(1,0)$ and $P_2(0,1)$ is $\alpha + \beta = 1$.

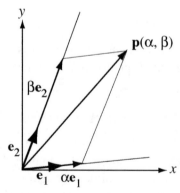

Figure 3.14 Oblique coordinate system.

Vector Spaces

The x, y coordinates of any point with oblique coordinates α, β are, in matrix terms,

$$\begin{bmatrix} x \\ y \end{bmatrix} = \begin{bmatrix} e_{1,x} & e_{2,x} \\ e_{1,y} & e_{2,y} \end{bmatrix} \begin{bmatrix} \alpha \\ \beta \end{bmatrix}$$

where $(e_{1,x}, e_{1,y}), (e_{2,x}, e_{2,y})$ are rectangular Cartesian components of $\mathbf{e}_1$ and $\mathbf{e}_2$.

After performing the indicated matrix multiplication we obtain

$$x = \alpha e_{1,x} + \beta e_{2,x}$$
$$y = \alpha e_{1,y} + \beta e_{2,y}$$

This is just what we should expect: a homogeneous affine transformation.

If we write the general equation for a conic (see Equation 2.7) in matrix form and perform the necessary substitutions, then we obtain

$$\begin{bmatrix} x & y & 1 \end{bmatrix} \begin{bmatrix} A & B & D \\ B & C & E \\ D & E & F \end{bmatrix} \begin{bmatrix} x \\ y \\ 1 \end{bmatrix} = 0$$

or

$$\begin{bmatrix} \alpha \\ \beta \\ 1 \end{bmatrix}^T \begin{bmatrix} e_{1,x} & e_{1,y} & 0 \\ e_{2,x} & e_{2,y} & 0 \\ 0 & 0 & 1 \end{bmatrix} \begin{bmatrix} A & B & D \\ B & C & E \\ D & E & F \end{bmatrix} \begin{bmatrix} e_{1,x} & e_{2,x} & 0 \\ e_{1,y} & e_{2,y} & 0 \\ 0 & 0 & 1 \end{bmatrix} \begin{bmatrix} \alpha \\ \beta \\ 1 \end{bmatrix} = 0$$

This matrix product yields $A'\alpha^2 + 2B'\alpha\beta + C'\beta^2 + 2D'\alpha + 2E'\beta + F' = 0$, where the $A', B', \ldots, F'$ terms are various functions of $e_{1,x}$, $e_{1,y}$, $e_{2,x}$, $e_{2,y}$, $A, B, \ldots, F$, all algebraically too messy to present here. We see that the form of a conic equation in the oblique system is identical to the form in the rectangular one. Furthermore, the discriminant in the oblique system, $A'C' - B'^2$ allows us to distinguish between the ellipse, parabola and hyperbola, as we would in the rectangular system.

Finally, we find that the square of the distance between the two points $P_1(\alpha_1, \beta_1)$ and $P_2(\alpha_2, \beta_2)$ is

$$\left(\overline{P_1 P_2}\right)^2 = |\mathbf{p}_2 - \mathbf{p}_1|^2$$
$$= |(\alpha_2 - \alpha_1)\mathbf{e}_1 + (\beta_2 - \beta_1)\mathbf{e}_2|^2$$
$$= (\alpha_2 - \alpha_1)^2 \mathbf{e}_1 \bullet \mathbf{e}_1 + 2(\alpha_2 - \alpha_1)(\beta_2 - \beta_1) \mathbf{e}_1 \bullet \mathbf{e}_2$$
$$+ (\beta_2 - \beta_1)^2 \mathbf{e}_2 \bullet \mathbf{e}_2$$

If $\mathbf{e}_1, \mathbf{e}_2$ are mutually perpendicular unit vectors, then $\mathbf{e}_1 \bullet \mathbf{e}_1 = \mathbf{e}_2 \bullet \mathbf{e}_2 = 1$ and $\mathbf{e}_1 \bullet \mathbf{e}_2 = 0$, and the expression for distance simplifies to the familiar Pythagorean theorem of rectangular Cartesian coordinates.

Change of Basis

In a space of n dimensions, any n linearly independent vectors may define its basis. There is no limit to the number of basis systems we can construct in a space. Every basis of a space determines a unique coordinate system and a corresponding set of coordinates for any point in the space. So it is natural to ask: How do we change from one basis to another? We might also ask: Given the components of a vector relative to one basis, what are its components relative to another? We simply apply what we have learned about linear transformations to answer these questions.

Given a basis $\mathbf{e}_1, \mathbf{e}_2$, we change to another basis $\bar{\mathbf{e}}_1, \bar{\mathbf{e}}_2$ using a set of linear transformation equations (Figure 3.15):

$$\bar{\mathbf{e}}_1 = a\mathbf{e}_1 + b\mathbf{e}_2$$
$$\bar{\mathbf{e}}_2 = c\mathbf{e}_1 + d\mathbf{e}_2 \qquad (3.24)$$

where $ad - bc \neq 0$.

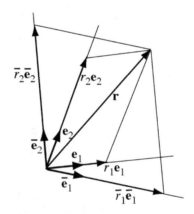

Figure 3.15 Change of basis in two dimensions.

The equation expressing the components of any vector $\mathbf{r}$ in basis $\bar{\mathbf{e}}_1, \bar{\mathbf{e}}_2$ is

$$\mathbf{r} = r_1 \mathbf{e}_1 + r_2 \mathbf{e}_2 \qquad (3.25)$$

as we might expect by analogy with Equation 3.23.

If we solve Equation 3.24 for $\mathbf{e}_1$ and $\mathbf{e}_2$, we obtain

$$\mathbf{e}_1 = \frac{d}{\Delta}\bar{\mathbf{e}}_1 - \frac{b}{\Delta}\bar{\mathbf{e}}_2 \text{ and } \mathbf{e}_2 = -\frac{c}{\Delta}\bar{\mathbf{e}}_1 + \frac{a}{\Delta}\bar{\mathbf{e}}_2 \qquad (3.26)$$

where $\Delta = ad - bc$.

Vector Spaces

Substitute from Equations 3.26 into 3.25 for $\mathbf{e}_1$ and $\mathbf{e}_2$ and rearrange terms as necessary so that

$$\mathbf{r} = \left(\frac{d}{\Delta}r_1 - \frac{c}{\Delta}r_2\right)\overline{\mathbf{e}}_1 + \left(-\frac{b}{\Delta}r_1 + \frac{a}{\Delta}r_2\right)\overline{\mathbf{e}}_2 \qquad (3.27)$$

Because of the objective (invariant) nature we attribute to $\mathbf{r}$, we expect to describe the components of $\mathbf{r}$ in terms of the new basis $\overline{\mathbf{e}}_1, \overline{\mathbf{e}}_2$, and so we write

$$\mathbf{r} = \overline{r}_1 \overline{\mathbf{e}}_1 + \overline{r}_2 \overline{\mathbf{e}}_2 \qquad (3.28)$$

Now, compare Equations 3.27 and 3.28. It must be true that

$$\begin{aligned}\overline{r}_1 &= \frac{d}{\Delta}r_1 - \frac{c}{\Delta}r_2 \\ \overline{r}_2 &= -\frac{b}{\Delta}r_1 + \frac{a}{\Delta}r_2\end{aligned} \qquad (3.29)$$

Compare the matrix $\mathbf{B}$ of coefficients that transforms the $\overline{\mathbf{e}}$ basis to the $\mathbf{e}$ basis, to the matrix $\mathbf{C}$ of coefficients that transforms the components of $\mathbf{r}$ from the $\mathbf{e}$ basis to the $\overline{\mathbf{e}}$ basis:

$$\mathbf{B} = \begin{bmatrix} \dfrac{d}{\Delta} & -\dfrac{b}{\Delta} \\ -\dfrac{c}{\Delta} & \dfrac{a}{\Delta} \end{bmatrix} \qquad \mathbf{C} = \begin{bmatrix} \dfrac{d}{\Delta} & -\dfrac{c}{\Delta} \\ -\dfrac{b}{\Delta} & \dfrac{a}{\Delta} \end{bmatrix}$$

We immediately see that $\mathbf{C}$ is the transpose of $\mathbf{B}$, or $\mathbf{C} = \mathbf{B}^T$, so that their elements correspond according as

$$c_{ij} = b_{ji} \qquad (3.30)$$

If we denote the matrix that transforms the $\mathbf{e}$ basis into the $\overline{\mathbf{e}}$ basis as $\mathbf{A}$, then clearly $\mathbf{B} = \mathbf{A}^{-1}$. This gives us

$$\mathbf{C} = \left[\mathbf{A}^{-1}\right]^T \qquad (3.31)$$

The relationships expressed in Equations 3.24, 3.26 and 3.29 are deceptively simple but, nonetheless, very powerful. They show how the equations transforming basis vectors relate to those transforming the vector components. The notation scheme we have been using obscures somewhat the generality of these relationships. In the next section we will see how tensors let us describe the transformations between basis systems and between vector components in them in a generalized and systematic way that is applicable to a Cartesian space of n dimensions.

Basis Vectors and Coordinate Systems

Given the familiar linear transformation

$$x' = ax + by$$
$$y' = cx + dy \quad (3.32)$$

where x and y are rectangular Cartesian coordinates and $ad - bc \neq 0$. we interpret x' and y' as coordinates of a Cartesian system whose x' axis corresponds to the line $cx + dy = 0$, and whose y' axis corresponds to the line $ax + by = 0$. The slope of the x' axis is $-c/d$, and of the y' axis $-a/b$ (Figure 3.16).

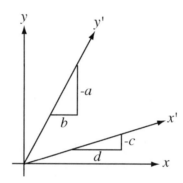

Figure 3.16 Cartesian coordinate system transformation.

The direction cosines of the x' axis are $d/\sqrt{c^2 + d^2}$ and $-c/\sqrt{c^2 + d^2}$, and the y' axis $b/\sqrt{a^2 + b^2}$ and $-a/\sqrt{a^2 + b^2}$. Write Equations 3.32 in matrix form as

$$\begin{bmatrix} x' \\ y' \end{bmatrix} = \begin{bmatrix} a & b \\ c & d \end{bmatrix} \begin{bmatrix} x \\ y \end{bmatrix} \quad (3.33)$$

The inverse of the matrix of transformation coefficients $\mathbf{A}$ is

$$\mathbf{A}^{-1} = \begin{bmatrix} \dfrac{d}{\Delta} & -\dfrac{b}{\Delta} \\ -\dfrac{c}{\Delta} & \dfrac{a}{\Delta} \end{bmatrix} \quad (3.34)$$

so

$$x = \frac{d}{\Delta} x' - \frac{b}{\Delta} y'$$
$$y = -\frac{c}{\Delta} x' + \frac{a}{\Delta} y' \quad (3.35)$$

Our next task is to construct a basis corresponding to the x', y' system (Figure 3.17), and we now have enough information to do this.

Vector Spaces

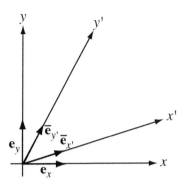

Figure 3.17 A basis corresponding to a transformed coordinate system.

We write the following matrix equation

$$\begin{bmatrix} \bar{\mathbf{e}}_x \\ \bar{\mathbf{e}}_y \end{bmatrix} = \begin{bmatrix} \dfrac{d}{\Delta} & -\dfrac{c}{\Delta} \\ -\dfrac{b}{\Delta} & \dfrac{a}{\Delta} \end{bmatrix} \begin{bmatrix} \mathbf{e}_x \\ \mathbf{e}_y \end{bmatrix} \qquad (3.36)$$

where the matrix of coefficients is nothing more than the transpose of $\mathbf{A}^{-1}$. Do the indicated matrix multiplication to obtain

$$\begin{aligned} \bar{\mathbf{e}}_x &= \frac{d}{\Delta}\mathbf{e}_x - \frac{c}{\Delta}\mathbf{e}_y \\ \bar{\mathbf{e}}_y &= -\frac{b}{\Delta}\mathbf{e}_x + \frac{a}{\Delta}\mathbf{e}_y \end{aligned} \qquad (3.37)$$

It is a simple matter to show that the direction cosines of $\bar{\mathbf{e}}_x$ and $\bar{\mathbf{e}}_y$ are identical to those of the x' and y' axes, so we have indeed found a set of basis vectors collinear to these axes. Again, we simplify all of this rather awkward algebra when we employ tensor notation, coming soon in Section 3.4. It is now time for an example. Figure 3.18 offers graphic support for what follows.

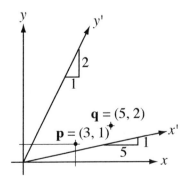

Figure 3.18 Affine transformation of a rectangular Cartesian coordinate system.

Given the linear transformation $x' = 2x - y$ and $y' = -x + 5y$ and an arbitrary point in the initial rectangular Cartesian coordinate system, say $\mathbf{p} = (3,1)$, we first check. $ad - bc = 9 \neq 0$, then compute $x' = 5$, $y' = 2$. How do we plot this point? If we retain the initial x, y coordinate system, then we simply map the point to a new location, $\mathbf{q} = (5,2)$. However, we may instead take the point of view that we have transformed the coordinate system itself and now wish to map the point in the resulting new system. It is, of course, an affine transformation, where parallel lines remain parallel, although we must now account for any change of scale along each axis x', y' that such a transformation might induce.

First, we plot the new axes as shown in the figure. The slopes of the x' and y' axes are 0.2 and 2.0, respectively. Next, we establish a scale factor for the x' axis, say k'_x, and one for the y' axis, say k'_y. For the initial rectangular Cartesian coordinate system the x and y scales are uniform and equal. We compute the untransformed length x'_s corresponding to x' along the x' axis, so that $x'_s = k'_x x'$. We do the same for the y' axis; thus, $y'_s = k'_y y'$.

Now, if $x' = 1$ and $y' = 0$ (corresponding to a unit length along the x' axis, from the origin), then $k'_x = x'_s$. using the Pythagorean theorem (Figure 3.19) we find

$$k'_x = \frac{\sqrt{c^2 + d^2}}{ad - bc} \tag{3.38}$$

Similarly, for the y' axis we assume $x' = 0$ and $y' = 1$ to find

$$k'_y = \frac{\sqrt{a^2 + b^2}}{ad - bc} \tag{3.39}$$

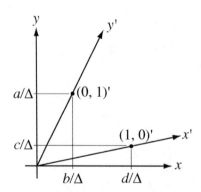

Figure 3.19 Scaling parameters for the affine transformation of a rectangular Cartesian coordinate system.

Vector Spaces

For our example we find $k'_x = \sqrt{26}/9$ and $k'_y = \sqrt{5}/9$, and if $x' = 5$ and $y' = 2$, then $x'_s = 5\sqrt{26}/9$ and $y'_s = 2\sqrt{5}/9$. This tells us that 5 units of length along the x' axis correspond to approximately 2.833 units of length along the x axis. Similarly, 2 units along the y' axis correspond to approximately 0.497 units along the y axis. Using these values, we plot the points corresponding to $x' = 5$ and $y' = 2$ on their respective axes (Figure 3.20).

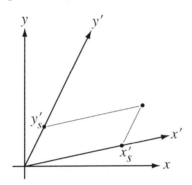

Figure 3.20 Parallel projection of p onto the x' and y' axes.

The line through points x', y' and $x', 0$ is parallel to the y' axis, and the line through points x', y' and $0, y'$ is parallel to the x' axis. We can prove that such pairs of lines are indeed parallel, by proceeding as follows:

The slopes of the x', y' axes of the transformed coordinate system are $-c/d$ and $-a/b$, respectively. Given any point x, y, we can demonstrate that the slopes of the two lines determined by it and its transformed images on the x', y' axes are parallel to these axes (Figure 3.21). The line AB through (x, y) and $(x', 0)$ is parallel to the y' axis, and the line AC through (x, y) and $(0, y')$ is parallel to the x' axis.

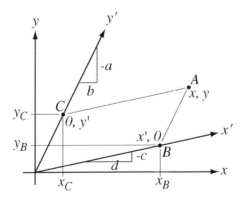

Figure 3.21 Parallel projection of point coordinates onto axes under affine transformations.

Let's find the slope of line AB in the rectangular system, recognizing that a similar process applies to line AC. The general equation for the slope is $\Delta y / \Delta x$. Let (x_B, y_B) be the coordinates in the rectangular system corresponding to $(x',0)$. Thus

$$\frac{\Delta y}{\Delta x} = \frac{y - y_B}{x - x_B} \tag{3.40}$$

Since the direction cosines of the x' axis are $d/\sqrt{c^2 + d^2}$ and $-c/\sqrt{c^2 + d^2}$, and the length of the segment from the origin to the point x_B, y_B is $k'_x(ax + by)$, where $k'_x = \sqrt{c^2 + d^2}/(ad - bc)$, then

$$\begin{aligned} x_B &= dk'_x(ax+by)/\sqrt{c^2+d^2} = d(ax+by)/(ad-bc) \\ y_B &= ck'_x(ax+by)/\sqrt{c^2+d^2} = -c(ax+by)/(ad-bc) \end{aligned} \tag{3.41}$$

From here it is only a matter of substitution into $\Delta y/\Delta x$, above, and some careful algebra to show that $\Delta y/\Delta x = -a/b$, which is the slope of the y' axis.

We can also demonstrate the objectivity of a vector under an affine transformation. However, $\bar{\mathbf{e}}_x$ and $\bar{\mathbf{e}}_y$ are not necessarily unit vectors. Substituting from Equations 3.37 we find $|\bar{\mathbf{e}}_x|^2 = \bar{\mathbf{e}}_x \bullet \bar{\mathbf{e}}_x = \sqrt{c^2+d^2}/(ad-bc)$ so that $|\bar{\mathbf{e}}_x|$ is a unit vector if and only if $\sqrt{c^2+d^2}/(ad-bc) = 1$.

In the transformed system we express $\mathbf{r}$ as $\mathbf{r} = \bar{r}_x|\bar{\mathbf{e}}_x|\hat{\mathbf{e}}_x + \bar{r}_y|\bar{\mathbf{e}}_y|\hat{\mathbf{e}}_y$ where $\hat{\mathbf{e}}_x$ and $\hat{\mathbf{e}}_y$ are unit vectors. The terms $\bar{r}_x|\bar{\mathbf{e}}_x|$ and $\bar{r}_y|\bar{\mathbf{e}}_y|$ are measures of the parallel projections of $\mathbf{r}$ onto the x' and y' axes (Figure 3.22). This follows from the parallelogram law of vector addition. If $|\bar{\mathbf{e}}_x|$ and $|\bar{\mathbf{e}}_y|$ specify the units of measurement along each axis, then $\bar{r}_x$ and $\bar{r}_y$ represent the parallel projections.

Reciprocal Basis Vectors

Two sets of basis vectors, $\mathbf{e}_1, \mathbf{e}_2, \mathbf{e}_3$ and $\mathbf{e}^1, \mathbf{e}^2, \mathbf{e}^3$, are reciprocal if and only if the following relations are true:

$$\begin{array}{lll} \mathbf{e}_1 \bullet \mathbf{e}^1 = 1 & \mathbf{e}_1 \bullet \mathbf{e}^2 = 0 & \mathbf{e}_1 \bullet \mathbf{e}^3 = 0 \\ \mathbf{e}_2 \bullet \mathbf{e}^1 = 0 & \mathbf{e}_2 \bullet \mathbf{e}^2 = 1 & \mathbf{e}_2 \bullet \mathbf{e}^3 = 0 \\ \mathbf{e}_3 \bullet \mathbf{e}^1 = 0 & \mathbf{e}_3 \bullet \mathbf{e}^2 = 0 & \mathbf{e}_3 \bullet \mathbf{e}^3 = 1 \end{array} \tag{3.42}$$

Vector Spaces

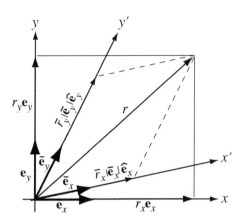

Figure 3.22 Objectivity of a vector under an affine transformation.

The superscript index has significance beyond its most apparent role here, namely that of helping to denote a distinction between the two sets of basis vectors, and we will discuss this point later.

The Kronecker delta, δ_i^j, allows us to compress these nine equations into the much more compact form of

$$\mathbf{e}_i \bullet \mathbf{e}^j = \delta_i^j \tag{3.43}$$

Remember, $\delta_i^j = 1$ if $i = j$; otherwise $\delta_i^j = 0$. For example, the equations $\mathbf{e}_1 \bullet \mathbf{e}^2 = 0$ and $\mathbf{e}_1 \bullet \mathbf{e}^3 = 0$ tell us that $\mathbf{e}_1$ is perpendicular to both $\mathbf{e}^2$ and $\mathbf{e}^3$. Furthermore, we notice that a set of basis vectors and their reciprocals are identical if

$$\mathbf{e}_i \bullet \mathbf{e}_j = \delta_{ij} \tag{3.44}$$

There is no reason to use superscripts or to distinguish between reciprocal sets of basis vectors that satisfy this condition, since they are orthonormal unit vectors. The rectangular Cartesian basis system has this property.

Let us look at what happens in a two-dimensional vector space (Figure 3.23). We begin with two unit vectors $\hat{\mathbf{e}}_1, \hat{\mathbf{e}}_2$ forming a basis on the x^1, x^2 axes. We then construct a new set of axes x_1, x_2 such that x_1 is perpendicular to x^2 and x_2 is perpendicular to x^1. This means that

$$\hat{\mathbf{e}}_1 \bullet \hat{\mathbf{e}}^2 = 0$$
$$\hat{\mathbf{e}}_2 \bullet \hat{\mathbf{e}}^1 = 0$$
$$\hat{\mathbf{e}}_1 \bullet \hat{\mathbf{e}}^1 = |\hat{\mathbf{e}}_1||\hat{\mathbf{e}}^1|\sin\theta$$

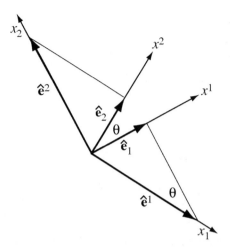

Figure 3.23 Reciprocal basis vectors.

From the figure we see that $\sin\theta = |\hat{\mathbf{e}}_1|/|\hat{\mathbf{e}}^1|$, so that

$$\hat{\mathbf{e}}_1 \bullet \hat{\mathbf{e}}^1 = |\hat{\mathbf{e}}_1||\hat{\mathbf{e}}^1|\frac{|\hat{\mathbf{e}}_1|}{|\hat{\mathbf{e}}^1|}$$

$$= |\hat{\mathbf{e}}_1|^2 \quad \text{(similarly } \hat{\mathbf{e}}_2 \bullet \hat{\mathbf{e}}^2 = 1 \text{)}$$

$$= 1$$

Thus $\hat{\mathbf{e}}_1, \hat{\mathbf{e}}_2$ and $\hat{\mathbf{e}}^1, \hat{\mathbf{e}}^2$ are reciprocal sets of basis vectors.

Finally, notice that the scale of the $|\hat{\mathbf{e}}_i|$ is different from that of the $|\hat{\mathbf{e}}^j|$. Since we began with the assumption that $\hat{\mathbf{e}}_1, \hat{\mathbf{e}}_2$ were unit vectors, it should be clear that $|\hat{\mathbf{e}}^j| > 1$; to put it another way, we see that $|\hat{\mathbf{e}}^1| = 1/\sin\theta > 1$.

Orthogonal Basis Vectors and Matrices

If the angle between two vectors is a right angle, then the vectors are mutually orthogonal. We recall the condition that characterizes orthogonal vectors in three-dimensional space: $\mathbf{p} \bullet \mathbf{q} = 0$, or $p_x q_x + p_y q_y + p_z q_z = 0$.

The most common system of basis vectors is one in which each basis vector in it is mutually perpendicular to the others, so that $\mathbf{e}_i \bullet \mathbf{e}_j = \delta_{ij}$, an orthogonal system corresponding to a rectangular Cartesian coordinate system (Figure 3.24). If it is also true that the $|\mathbf{e}_i| = 1$, then it is a system of orthonormal basis vectors.

Vector Spaces

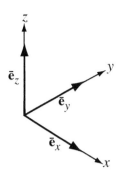

Figure 3.24 Orthogonal basis vectors.

If e_i is an orthonormal basis, and if a_{ij} represents elements of the transformation matrix from this basis to another, $\bar{e}_i$, then a necessary and sufficient condition for the $\bar{e}_i$ basis to be orthonormal is that a_{ij} must be an orthogonal matrix.

Given a real, square matrix $\mathbf{A}$ of order n, if $\mathbf{AA}^T = \mathbf{I}$, then $\mathbf{A}^T = \mathbf{A}^{-1}$ and $\mathbf{A}$ is an orthogonal matrix. Let a_{ij} denote the elements of $\mathbf{A}$; then if $\mathbf{A}$ is orthogonal

$$\sum_{k=1}^{n} a_{ik} a_{jk} = \delta_{ij} \tag{3.45}$$

Here are some characteristics of orthogonal transformation matrices:

1. The row vectors of $\mathbf{A}$ are mutually orthogonal unit vectors.
2. The column vectors of $\mathbf{A}$ are mutually orthogonal unit vectors.
3. The inverse of an orthogonal matrix is an orthogonal matrix.
4. If matrices $\mathbf{A}$ and $\mathbf{B}$ are orthogonal matrices, their product is orthogonal.
5. If $\mathbf{A}$ is an orthogonal matrix, then $|\mathbf{A}| = \pm 1$.
6. If $|\mathbf{A}| = +1$, then $\mathbf{A}$ is an orientation-preserving proper orthogonal matrix.
7. If $|\mathbf{A}| = -1$, then $\mathbf{A}$ is an improper orthogonal matrix. It is orientation-reversing (e.g., reflection across a line through the origin).
8. A linear transformation $\mathbf{A}$ is orthogonal if it preserves scalar products: $\mathbf{Ap} \bullet \mathbf{Aq} = \mathbf{p} \bullet \mathbf{q}$.
9. If $\mathbf{A}$ is an orthogonal matrix, then the sum of the products of the elements of any row (or column) with the corresponding elements of any other row (or column) equals zero.
10. If $\mathbf{A}$ is an orthogonal matrix, then the sum of the squares of the elements of any row (or column) equals one.

Section 3.2 Exercises

1. Describe each of the following matrices: orthogonal, proper, improper?

   a. $\begin{bmatrix} \sqrt{2}/2 & \sqrt{2}/2 \\ -\sqrt{2}/2 & \sqrt{2}/2 \end{bmatrix}$
   b. $\begin{bmatrix} 1 & 1/2 \\ 2 & 0 \end{bmatrix}$
   c. $\begin{bmatrix} 3 & 1 \\ 5 & 2 \end{bmatrix}$
   d. $\begin{bmatrix} -\sqrt{3}/2 & 1/2 \\ 1/2 & \sqrt{3}/2 \end{bmatrix}$

2. Describe each of the following matrices: orthogonal, proper, improper?

   a. $\begin{bmatrix} 0 & 0 & 1 \\ 1 & 0 & 0 \\ 0 & -1 & 0 \end{bmatrix}$,
   b. $\begin{bmatrix} 1/2 & -\sqrt{3}/2 & 0 \\ \sqrt{3}/2 & 1/2 & 0 \\ 0 & 0 & 1 \end{bmatrix}$,
   c. $\begin{bmatrix} \sqrt{3}/2 & \sqrt{3}/4 & 1/4 \\ 1/2 & -3/4 & -\sqrt{3}/4 \\ 0 & 1/2 & -\sqrt{3}/2 \end{bmatrix}$

3. Find the inverse of the matrix in Exercise 2a.

4. Show that the product of orthogonal matrices **A** and **B** of the same order is also an orthogonal matrix, where $\mathbf{A} = \begin{bmatrix} \sqrt{2}/2 & \sqrt{2}/2 \\ -\sqrt{2}/2 & \sqrt{2}/2 \end{bmatrix}$ and $\mathbf{B} = \begin{bmatrix} 3 & 1 \\ 5 & 2 \end{bmatrix}$.

5. Show that the product of two improper orthogonal matrices **A** and **B** of the same order is a proper orthogonal matrix, where $\mathbf{A} = \begin{bmatrix} 1 & 1/2 \\ 2 & 0 \end{bmatrix}$ and $\mathbf{B} = \begin{bmatrix} -\sqrt{3}/2 & 1/2 \\ 1/2 & \sqrt{3}/2 \end{bmatrix}$.

6. Show that the inverse of the orthogonal matrix **A** is an orthogonal matrix, where $\mathbf{A} = \begin{bmatrix} 2/3 & -2/3 & 1/3 \\ 1/3 & 2/3 & 2/3 \\ 2/3 & 1/3 & -2/3 \end{bmatrix}$

3.3 Eigenvalues and Eigenvectors

Consider the homogeneous affine transformation **A** of a point, where **A** is not necessarily an isometry or similarity. Figure 3.25a shows what we might expect to happen for the general case of such a transformation that maps Q to Q'. However, there may be certain points, such as P, whose images under this transformation lie on the line OP, P' is an example of this (Figure 3.25b).

Vector Spaces

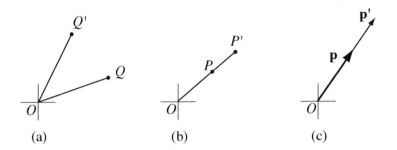

Figure 3.25 Eigenvectors: A geometric interpretation.

Since we treat points and vectors in the same way, we recognize that there are vectors such as **P** (corresponding to OP) transformed by **A** so that **p′** (corresponding to OP') is a scalar multiple, λ, of **P** (Figure 3.25c). We express this more succinctly with a few simple equations. We first write the equation of the general transformation in matrix form as

$$\mathbf{p}' = \mathbf{A}\mathbf{p} \tag{3.46}$$

and assert without qualms that under certain conditions it is possible to find

$$\mathbf{p}' = \lambda \mathbf{p} \tag{3.47}$$

if $\mathbf{p} \neq 0$ and $\lambda \neq 0$. Then it must follow that

$$\mathbf{A}\mathbf{p} = \lambda \mathbf{p} \tag{3.48}$$

Every vector **p** for which this is true is an eigenvector of **A**, and λ is the eigenvalue of **A** corresponding to **p**. (Eigenvalue is from the German "*eigenwerte*", meaning "proper values.") In the theory of vector spaces, invariant lines and eigenvectors are equivalent. Transformation **A** sends an eigenvector into a collinear vector, with the corresponding eigenvalue equal to the ratio of the two collinear vectors (this is the expansion coefficient of the eigenvector under the transformation **A**). Other names for eigenvalues are proper values, latent values, and characteristic values of a matrix; similarly for eigenvectors: proper vectors, latent vectors, and characteristic vectors of a matrix.

The Characteristic Equation

We write Equation 3.48 as

$$(\mathbf{A} - \lambda \mathbf{I})\mathbf{p} = 0 \tag{3.49}$$

This equation has nontrivial solutions if $\mathbf{p} \neq 0$ and

$$|\mathbf{A} - \lambda \mathbf{I}| = 0 \tag{3.50}$$

This last is the characteristic equation, and its solutions are the eigenvalues of **A**. Thus, if λ_i is an eigenvalue of **A**, then the nontrivial solutions of $(\mathbf{A} - \lambda_i \mathbf{I})\mathbf{p} = 0$ are the eigenvectors corresponding to λ_i.

Here is an example of a two-dimensional space with a transformation **A**, where $\mathbf{A} = \begin{bmatrix} 1 & 1 \\ 5 & -3 \end{bmatrix}$.

We set $|\mathbf{A} - \lambda \mathbf{I}| = 0$ and write $\begin{vmatrix} 1-\lambda & 1 \\ 5 & -3-\lambda \end{vmatrix} = 0$

Solving this determinant produces $\lambda^2 + 2\lambda - 8 = 0$, the characteristic equation, whose solution, in turn, produces the eigenvalues $\lambda_1 = 2, \lambda_2 = -4$.

Continuing with the example, from Equation 3.46 we have

$$\mathbf{p}' = \begin{bmatrix} 1 & 1 \\ 5 & -3 \end{bmatrix} \mathbf{p}$$

or

$$\begin{bmatrix} x' \\ y' \end{bmatrix} = \begin{bmatrix} x+y \\ 5x-3y \end{bmatrix}$$

So that for $\lambda_1 = 2$ or $\mathbf{p}' = 2\mathbf{p}$, we have

$$x + y = 2x \Rightarrow x = y$$
$$5x - 3y = 2y \Rightarrow x = y$$

These equations tell us that a vector whose x and y components are equal is an eigenvector of the given transformation. In fact, there is a family of vectors associated with the eigenvalue $\lambda_1 = 2$, so that the corresponding eigenvector has the form

$\begin{bmatrix} k \\ k \end{bmatrix}$, where $k \neq 0$.

We found two distinct eigenvalues for our example transformation matrix, so for $\lambda_2 = -4$ we have

$$x + y = -4x \Rightarrow -5x = y$$
$$5x - 3y = -4y \Rightarrow -5x = y$$

The eigenvector corresponding to this eigenvalue has the form

$$\begin{bmatrix} k \\ -5k \end{bmatrix}$$

That is, a vector whose y component is -5 times the x component is an eigenvector corresponding to the eigenvalue $\lambda_2 = -4$ of the given transformation.

Vector Spaces

We may now assert for the general case of a two-dimensional, homogeneous affine transformation that if

$$\mathbf{A} = \begin{bmatrix} a & b \\ c & d \end{bmatrix} \tag{3.51}$$

then we determine the eigenvalues from the relationship

$$\begin{vmatrix} a-\lambda & b \\ c & d-\lambda \end{vmatrix} = 0 \tag{3.52}$$

or $\quad \lambda^2 - (a+d)\lambda + (ad-bc) = 0 \tag{3.53}$

From elementary algebra we know that λ has two possible values and that these may be real and distinct, equal, or complex. When we know the values of the eigenvalues, then we can determine the values of the corresponding eigenvectors, using Equation 3.49. By doing this we find two possible forms for the ratio of the vector components, depending on the value of λ namely

$$\frac{p_x}{p_y} = -\frac{b}{a-\lambda_i} \text{ or } -\frac{d-\lambda_i}{c} \tag{3.54}$$

where $i = 1, 2$. We know these forms are equivalent from Equation 3.52:

$$(a-\lambda)(d-\lambda) - bc = 0$$

For the general case where $\mathbf{A}$ is an $n \times n$ square matrix, we develop the characteristic equation from

$$|\mathbf{A} - \lambda \mathbf{I}| = \begin{vmatrix} a_{11}-\lambda & a_{12} & \cdots & a_{1n} \\ a_{21} & a_{22}-\lambda & \cdots & a_{2n} \\ \vdots & \vdots & \ddots & \vdots \\ a_{n1} & a_{n2} & \cdots & a_{nn}-\lambda \end{vmatrix} \tag{3.55}$$

Expanding this determinant in polynomial form, we obtain the equation

$$c_0 \lambda^n + c_1 \lambda^{n-1} + \cdots + c_{n-1} \lambda + c_n = 0 \tag{3.56}$$

This equation has n roots, not necessarily distinct, and these are the eigenvalues. Each eigenvalue λ_i, where $i = 1, 2, \ldots, n$, has a corresponding eigenvector $\mathbf{p}_i$, such that. $\mathbf{A}\mathbf{p}_i = \lambda_i \mathbf{p}_i$. The coefficients of the characteristic equation are

$$c_0 = 1$$
$$c_1 = -t_1$$
$$c_2 = -\frac{1}{2}(c_1 t_1 + t_2)$$
$$c_3 = -\frac{1}{3}(c_2 t_1 + c_1 t_2 + t_3) \tag{3.57}$$
$$\vdots$$
$$c_n = -\frac{1}{n}(c_{n-1} t_1 + c_{n-2} t_2 + \cdots + c_1 t_{n-1} + t_n)$$

where t_1 is the trace of matrix $\mathbf{A}$, t_2 is the trace of $\mathbf{A}^2$,… and t_n is the trace of $\mathbf{A}^n$. (The trace of a matrix is the sum of its diagonal elements.)

For a 3×3 matrix, we express more simply the coefficients as follows:

$$c_0 = 1$$
$$c_1 = -(a_{11} + a_{22} + a_{33})$$
$$c_2 = \begin{vmatrix} a_{11} & a_{12} \\ a_{21} & a_{22} \end{vmatrix} + \begin{vmatrix} a_{11} & a_{12} \\ a_{31} & a_{33} \end{vmatrix} + \begin{vmatrix} a_{22} & a_{23} \\ a_{32} & a_{33} \end{vmatrix} \tag{3.58}$$
$$c_3 = \begin{vmatrix} a_{11} & a_{12} & a_{13} \\ a_{21} & a_{22} & a_{23} \\ a_{31} & a_{32} & a_{33} \end{vmatrix}$$

Here is an example. Find the characteristic equation of the following matrix:

$$\mathbf{A} = \begin{bmatrix} 3 & 0 & 2 \\ 0 & 1 & 2 \\ 2 & 2 & 2 \end{bmatrix}$$

First find $\mathbf{A}^2$ and $\mathbf{A}^3$:

$$\mathbf{A}^2 = \begin{bmatrix} 13 & 4 & 10 \\ 4 & 5 & 6 \\ 10 & 6 & 12 \end{bmatrix} \text{ and } \mathbf{A}^3 = \begin{bmatrix} 59 & 24 & 54 \\ 24 & 17 & 30 \\ 54 & 30 & 56 \end{bmatrix}$$

Next, we compute the traces: $t_1 = 6$, $t_2 = 30$, $t_3 = 132$. The coefficients are

$$c_0 = 1$$
$$c_1 = 6$$
$$c_2 = -(1/2)\big[(-6)(6) + 30\big] = 3$$
$$c_3 = -(1/3)\big[(3)(6) + (-6)(30) + 132\big] = 10$$

Vector Spaces

Thus, the characteristic equation is

$$\lambda^3 - 6\lambda^2 + 3\lambda + 10 = 0$$

The characteristic equation of matrix **A** is independent of the choice of basis. If Γ is the orthogonal transformation matrix from one basis to another, then the matrix **A** goes into **A'**, where

$$\mathbf{A}' = \Gamma \mathbf{A} \Gamma^{-1} \tag{3.59}$$

Substituting into $\mathbf{A} - \lambda \mathbf{I}$ and applying some simple matrix algebra, we find

$$\mathbf{A}' - \lambda \mathbf{I} = \Gamma \mathbf{A} \Gamma^{-1} - \Gamma(\lambda \mathbf{I}) \Gamma^{-1}$$
$$= \Gamma(\mathbf{A} - \lambda \mathbf{I}) \Gamma^{-1}$$

since, obviously, $\Gamma(\lambda \mathbf{I}) \Gamma^{-1} = \lambda \mathbf{I}$.

Next, from determinant theory we find that

$$|\mathbf{A}' - \lambda \mathbf{I}| = |\Gamma||\mathbf{A} - \lambda \mathbf{I}||\Gamma^{-1}|$$

and since $|\Gamma||\Gamma^{-1}| = 1$, it follows that

$$|\mathbf{A} - \lambda \mathbf{I}| = |\mathbf{A}' - \lambda \mathbf{I}| \tag{3.60}$$

This verifies our initial assertion, and we are now justified to speak of the characteristic equation of the transformation **A**. It also follows from the invariance of the characteristic equation that its coefficients are also invariant, thus the matrix of a linear transformation **A** in a space of n dimensions has n invariants.

Similarity Transformations

The matrix **A** undergoes a similarity transformation if it is pre- and post multiplied by any other matrix and its inverse. **B** is a similarity transformation of **A** if

$$\mathbf{B} = \mathbf{T} \mathbf{A} \mathbf{T}^{-1} \tag{3.61}$$

Of course the dimensions of **T** and **A** must be compatible, and **T** must have an inverse. We say that **A** and **B** are similar matrices. Similar matrices have equal determinants, the same characteristic equation and the same eigenvalues, although not necessarily the same eigenvectors (why?). This obviously leads us to conclude that similarity transformations preserve eigenvalues.

If matrix **A** is similar to a diagonal matrix **D**, then

$$\mathbf{D} = \begin{bmatrix} \lambda_1 & 0 & \cdots & 0 \\ 0 & \lambda_2 & \cdots & 0 \\ \vdots & \vdots & \ddots & \vdots \\ 0 & 0 & \cdots & \lambda_n \end{bmatrix} \tag{3.62}$$

where $\lambda_1, \lambda_2, \ldots, \lambda_n$ are the eigenvalues of **A**. This is true if there exists a nonsingular matrix **X** such that $\mathbf{XAX}^{-1} = \mathbf{D}$. Clearly, the elements in the main diagonal of a diagonal matrix **D**, are its eigenvalues. Therefore, the eigenvalues of **D** are precisely the eigenvalues of **A**. Finally, if the eigenvalues of matrix **A** are distinct, then **A** is similar to a diagonal matrix.

The virtual blizzard of assertions without rigorous proof is unavoidable. Space does not permit the full development of proofs. However, most are simple and worthwhile exercises in the algebra of matrices and determinants. The assertions here and to follow indicate the scope and power of these methods in the mathematics of transformations.

Symmetric Transformations

A real symmetric matrix, by definition, satisfies the condition $a_{ij} = a_{ji}$, or $\mathbf{A}^\mathrm{T} = \mathbf{A}$. If **A** and **B** are symmetric, so that $[\mathbf{AB}]^\mathrm{T} = \mathbf{B}^\mathrm{T}\mathbf{A}^\mathrm{T} = \mathbf{BA}$, then the product of two symmetric matrices is symmetric if and only if the two matrices commute. The eigenvalues of a real symmetric matrix are real. Furthermore, if **A** is a real symmetric matrix, then there is an orthogonal matrix **R** such that $\mathbf{R}^{-1}\mathbf{AR}$ is a diagonal matrix. Finally, if **A** is a real symmetric matrix, then the eigenvectors of **A** associated with the distinct eigenvalues are mutually orthogonal vectors.

Given a transformation **A** in the plane, if the eigenvalues of **A** are distinct (i.e., $\lambda_1 \neq \lambda_2$), then **A** has a diagonal matrix

$$\mathbf{A} = \begin{bmatrix} \lambda_1 & 0 \\ 0 & \lambda_2 \end{bmatrix} \tag{3.63}$$

in the basis defined by the eigenvectors. We interpret this geometrically as the simultaneous expansion (or contraction) along mutually perpendicular axes described by the eigenvectors corresponding to λ_1 and λ_2.

Diagonalization of Matrices

An eigenvector $\mathbf{x}_i$ such that $(\mathbf{A} - \lambda_i\mathbf{I})\mathbf{x}_i = 0$ we associate with each eigenvalue λ_i. We may write this as

$$\mathbf{A}\mathbf{x}_i = \lambda_i \mathbf{x}_i \tag{3.64}$$

We construct a square matrix **X** of order n whose columns are eigenvectors $\mathbf{x}_i$ of **A**. Now, we can write Equation 3.64 as

$$\mathbf{AX} = \mathbf{X}\Lambda \tag{3.65}$$

where Λ is a diagonal matrix whose elements are the eigenvalues of **A**:

Vector Spaces

$$\Lambda = \begin{bmatrix} \lambda_1 & 0 & \cdots & 0 \\ 0 & \lambda_2 & \cdots & 0 \\ \vdots & \vdots & \ddots & \vdots \\ 0 & 0 & \cdots & \lambda_n \end{bmatrix} \tag{3.66}$$

The matrix $\mathbf{X}$ is nonsingular if the eigenvalues are distinct, so we may multiply both sides of Equation 3.65 by $\mathbf{X}^{-1}$ to obtain

$$\mathbf{X}^{-1}\mathbf{A}\mathbf{X} = \Lambda \tag{3.67}$$

By using the matrix of eigenvectors and its inverse in this way, we find we can transform any matrix $\mathbf{A}$ with distinct eigenvalues into a diagonal matrix whose elements are the eigenvalues of $\mathbf{A}$. We call this the *diagonalization* of matrix $\mathbf{A}$.

Section 3.3 Exercises

1. Find the eigenvalues and corresponding eigenvectors of the following matrices:

   a. $\begin{bmatrix} 2 & -2 \\ -6 & 3 \end{bmatrix}$ b. $\begin{bmatrix} 3 & 5 \\ 4 & 5 \end{bmatrix}$ c. $\begin{bmatrix} 1 & 0 \\ 3 & 2 \end{bmatrix}$ d. $\begin{bmatrix} 1 & 2 \\ -2 & 5 \end{bmatrix}$

2. Find the characteristic equation, eigenvalues and corresponding eigenvectors of the following matrices;

   a. $\begin{bmatrix} 5 & 3 \\ 2 & 4 \end{bmatrix}$ b. $\begin{bmatrix} 1 & 2 \\ 4 & 3 \end{bmatrix}$ c. $\begin{bmatrix} 2 & 0 \\ 0 & 0 \end{bmatrix}$

   d. $\begin{bmatrix} 2 & 0 & 0 \\ 0 & 1 & 0 \\ 0 & 0 & 3 \end{bmatrix}$ e. $\begin{bmatrix} 2 & -2 & 3 \\ 1 & 1 & 1 \\ 1 & 3 & -1 \end{bmatrix}$ f. $\begin{bmatrix} 3 & 0 & 2 \\ 0 & 1 & 2 \\ 2 & 2 & 2 \end{bmatrix}$

3. Determine which pairs of matrices are similar.

   a. $\begin{bmatrix} 6 & 2 \\ -2 & 1 \end{bmatrix}, \begin{bmatrix} 8 & 6 \\ -3 & -1 \end{bmatrix}$ b. $\begin{bmatrix} -2 & -1 \\ 0 & 11 \end{bmatrix}, \begin{bmatrix} 0 & 1 \\ 2 & 3 \end{bmatrix}$

3.4 Tensors

Those who are not familiar with tensors are at a serious disadvantage in many fields of pure and applied mathematics, physics and engineering. They are cut off from the study of Riemannian geometry and the theory of general relativity. Even in Euclidean geometry and Newtonian mechanics, they are compelled to work in notations lacking the completeness of tensors. However, these powerful tools are gradually becoming more generally accessible to a wider range of disciplines and at a more elementary level. For example, mathematicians originally developed the notion of a tensor to describe more accurately and with greater facility the mechanics of continua. They further developed and refined it as an indispensable tool of differential geometry, ultimately incorporating it into general relativity and mathematical physics.

An important motivating force behind the invention and development of tensor calculus was the desire to express geometric forms and physical laws in a way that was not dependent on a particular coordinate system. If we express such forms and laws so that they are valid in any coordinate system, then we find it easier to understand their intrinsic invariant properties. It turns out that if a tensor equation is true in one system, then it is true in all coordinate systems. This follows from the fact that tensor transformations are linear and homogeneous.

Einstein brought tensor calculus into prominence when he presented to the world his general theory of relativity in 1916. Earlier work (1900) by Rici and Levi-Civita applied tensor methods to geometry and to mathematical physics, where it approaches the status of a universal language. Tensor calculus permits compact equations in geometry and physics.

It is interesting to note that the term "tensor" has its origin in the analysis of stress at a point of a material continuum (e.g., a structural element). If we imagine an arbitrary plane surface passing through this point, then we associate a vector with the stress transmitted across this surface. Equilibrium requires that the relationship between the stress vector and the direction of the surface normal be linear and homogeneous in the direction cosines of the normal. Thus, we may specify the state of stress at any point of the continuum by a tensor—the stress tensor.

Although our exploration here of tensors is brief, at least some discussion of their importance to more advanced geometric transformations is appropriate. We are already familiar with a particular type of tensor, the vector. Putting it the other way around, in a certain sense, we might even consider a tensor to be a very generalized kind of vector. In fact, a tensor has a unique set of components in a given coordinate system, and these components transform according to well-defined laws, as do vectors. However, more appropriately, we shall see that both scalars and vectors are special forms of tensors. Let's begin by taking another, slightly different look at vectors, and in doing this we will find that there are two different kinds.

Vector Spaces

Contravariant and Covariant Vectors

Recall that we can express the components of a vector **r** in terms of a set of basis vectors $\mathbf{e}_i$. We generalize and rewrite Equation 3.25 using the index notation

$$\mathbf{r} = r^i \mathbf{e}_i \qquad (3.68)$$

where $i = 1, 2, \ldots, n$. Here we use a superscript, or upper index, to denote the components of **r** (the superscript i does not indicate an exponentiation!). The reasons for this will soon be clear. (Remember, the repeated index implies summation—the Einstein summation convention.) For example, for a three-dimensional space $n = 3$ and $\mathbf{r} = r^1 \mathbf{e}_1 + r^2 \mathbf{e}_2 + r^3 \mathbf{e}_3$. If we want to change to another set of basis vectors $\bar{\mathbf{e}}_i$, then we generalize Equation 3.24 and write

$$\bar{\mathbf{e}}_i = a_i^j \mathbf{e}_j \qquad (3.69)$$

where the a_i^j are the elements of a linear homogeneous transformation matrix. The index notation on a_i^j denotes the element in the ith row and j^{th} column of matrix **A**.

How do the components of **r** transform to this new basis? To answer this, we proceed much as we did in describing a change of basis in Section 3.2. First, we solve Equation 3.69 for **e** in terms of $\bar{\mathbf{e}}$ to get an equation of the form

$$\mathbf{e}_i = b_i^j \bar{\mathbf{e}}_j \qquad (3.70)$$

Substituting this into Equation 3.70 produces

$$\mathbf{r} = r^i b_i^j \bar{\mathbf{e}}_j \qquad (3.71)$$

or $\qquad \mathbf{r} = \left(b_i^j r^i \right) \bar{\mathbf{e}}_j \qquad (3.72)$

Again, if our vector **r** is to have objective or invariant properties independent of the basis, then it also must be true that

$$\mathbf{r} = \bar{r}^i \bar{\mathbf{e}}_i \qquad (3.73)$$

Now compare this to Equation 3.72. As it stands alone, the choice of index letter i in Equation 3.73 is perfectly correct. When we compare this to Equation 3.72, we see that we must make them compatible, because clearly the results of an expansion of Equation 3.72 depend upon the position and denotation of the various indices. Therefore, we rewrite Equation 3.73 into the equally valid and now compatible form

$$\mathbf{r} = \bar{r}^j \bar{\mathbf{e}}_j \qquad (3.74)$$

When we compare this to Equation 3.72, we correctly identify the relationships between the vector components in the two basis systems:

$$\bar{r}^j = b_i^j r^i \tag{3.75}$$

Notice that the components $\bar{r}^j$ derive from the components r^i via the same b_i^j coefficients that give the $\mathbf{e}_i$ in terms of the $\bar{\mathbf{e}}_i$. We call $\mathbf{r}$ a contravariant vector for that very reason (i.e., the components of $\mathbf{r}$ do not transform as does the basis $\mathbf{e}$ to $\bar{\mathbf{e}}$ but contrarily as the basis $\bar{\mathbf{e}}$ to $\mathbf{e}$).

Look at the effect produced by the raised index on the components of the contravariant vector. Let's investigate an example in a two-dimensional space. First, expand Equation 3.70 for $n = 2$:

$$\begin{aligned}\mathbf{e}_1 &= b_1^1 \bar{\mathbf{e}}_1 + b_1^2 \bar{\mathbf{e}}_2 \\ \mathbf{e}_2 &= b_2^1 \bar{\mathbf{e}}_1 + b_2^2 \bar{\mathbf{e}}_2\end{aligned} \tag{3.76}$$

Next, expand Equation 3.75, also for $n = 2$,

$$\begin{aligned}\bar{r}^1 &= b_1^1 r^1 + b_2^1 r^2 \\ \bar{r}^2 &= b_1^2 r^1 + b_2^2 r^2\end{aligned}$$

Examine the relative position of the b_i^j transformation coefficients. In Equation 3.70, the summation is over the upper index of b_i^j, and in Equation 3.75 it is over the lower index. The raised index on $\bar{r}^j$ has the effect of transposing the coefficients if we interpret these expansions in terms of analogous matrix operations. This certainly conforms to what we found in Section 3.2, but we have accomplished the same thing more simply and powerfully.

We can invent another kind of vector that also has invariant properties or an objectivity that persists into any coordinate system under a homogeneous linear transformation. Let this new vector have components r_i. Again, if the basis vectors $\mathbf{e}_j$ transform to $\bar{\mathbf{e}}_i$ under a_i^j, that is $\bar{\mathbf{e}}_i = a_i^j \mathbf{e}_j$, then we also transform r_i by a_i^j, thus

$$\bar{r}_i = a_i^j r_j \tag{3.77}$$

We call $\bar{r}_i$ a *covariant* vector. Although we can visualize the contravariant vector as an arrow-headed line segment, there is no simple way to visualize a covariant vector. Here is a summary of the various transformations for basis vectors, coordinates and free vectors:

Basis vectors: $\bar{\mathbf{e}}_i = a_i^j \mathbf{e}_j$ Covariant vector: $\bar{r}_i = a_i^j r_j$

Coordinates: $\bar{x}_i = b_j^i x^j$ Contravariant vector: $\bar{r}^i = b_j^i r^j$

The relationship between a_i^j and b_i^j is easy to demonstrate. First, change the dummy index on Equation 3.70 to k (avoiding confusion in what follows), so that

Vector Spaces

$$\mathbf{e}_i = b_i^k \overline{\mathbf{e}}_k \tag{3.78}$$

Now change the index i to j, and substitute the right-hand expression of Equation 3.78 into 3.69 to obtain

$$\overline{\mathbf{e}}_i = a_i^j b_j^k \overline{\mathbf{e}}_k \tag{3.79}$$

Let $\delta_i^k = a_i^j b_j^k$, then we have

$$\overline{\mathbf{e}}_i = \delta_i^k \overline{\mathbf{e}}_k \tag{3.80}$$

We sum over k and immediately see it must be true that

$$\delta_i^k = \begin{cases} 1 \text{ if } i = k \\ 0 \text{ if } i \neq k \end{cases} \tag{3.81}$$

This is, of course, the *Kronecker delta*. There are four equivalent forms of this relationship (the first of which we just derived — try to derive the other three.). These are

$$\begin{aligned} a_i^j b_j^k &= \delta_i^k \\ b_j^k a_i^j &= \delta_i^k \\ a_j^k b_i^j &= \delta_i^k \\ b_i^j a_j^k &= \delta_i^k \end{aligned} \tag{3.82}$$

For the special case of rectangular Cartesian coordinate systems (systems of orthonormal basis vectors) and orthogonal transformations (isometries), contravariant and covariant vectors transform in the same way and are thus equivalent. The next section elaborates on this interesting state of affairs.

Contravariant and Covariant Vectors in an Orthonormal Coordinate System

Contravariant and covariant vectors transform in the same way from one rectangular Cartesian coordinate system to another. The transformation must be orthogonal to preserve the orthonormal properties of the basis vectors. In Section 3.2 we found that an orthogonal matrix has the property $\mathbf{A}\mathbf{A}^T = \mathbf{I}$ or $\mathbf{A}^T = \mathbf{A}^{-1}$, and from Equation 3.45 we have $a_i^k a_j^k = \delta_i^j$. We already know from Equation 3.82 that $a_i^k b_k^j = \delta_i^j$. So that we find $b_k^j = a_j^k$.

Comparing the components of a contravariant and covariant vector for $n = 3$, we have $\bar{r}^1 = b_1^1 r^1 + b_2^1 r^2 + b_3^1 r^3$ and $\bar{r}_1 = a_1^1 r_1 + a_1^2 r_2 + a_1^3 r_3$. Since $b_k^j = a_j^k$, we have for $\bar{r}^1$ (by substitution) $\bar{r}^1 = a_1^1 r^1 + a_1^2 r^2 + a_1^3 r^3$.

If $r^1 = r_1$, $r^2 = r_2$ and $r^3 = r_3$, then $\bar{r}^1 = \bar{r}_1$, demonstrating the equivalence of contravariant and covariant vectors in rectangular Cartesian coordinate systems.

Tensor Notation

Consider the two contravariant vectors **r** and **s** in a space of n dimensions. We express the products of their components, taken two at a time, as

$$r^1 s^1, r^1 s^2, \ldots, r^1 s^n, r^2 s^1, r^2 s^2, \ldots, r^n s^n$$

Using the index notation, we have $r^i s^j$. Counting all the variations, we find that we produce n^2 of them. How does $r^i s^j$ transform? As you might expect:

$$r^i s^j = b_k^i b_l^j r^k s^l \tag{3.83}$$

It seems obvious that we can easily generalize the notion of vectors and vector products and introduce an object of n^2 components in some arbitrary coordinate system. The components of the new object transform exactly as the vector product above. We denote this new object by T^{ab} and write

$$\bar{T}^{ab} = b_c^a b_d^b T^{cd} \tag{3.84}$$

Remember, summation occurs over the dummy indices c and d.

There is an analogous object T_{ab} that transforms as

$$\bar{T}_{ab} = a_a^c a_b^d T_{cd} \tag{3.85}$$

We can generalize even more, creating an object with a unique set of components $T_{cd\ldots}^{ab\ldots}$ in some arbitrary coordinate system. If this object transforms as

$$\bar{T}_{cd\ldots}^{ab\ldots} = a_c^i a_d^j \ldots b_k^a b_l^b \ldots T_{ij\ldots}^{kl\ldots} \tag{3.86}$$

when the coordinates transform according to $\bar{x}^i = a_j^i x^j$, then these objects are **tensors**. T^{ab} and T_{ab} are second-order contravariant and covariant tensors, respectively. The number of indices determines the order. In a three-dimensional space each has 3^2, or 9 components. In a four-dimensional space each has 4^2 or 16 components. T_{de}^{abc} is a fifth-order tensor (it has contravariant order 3 and covariant order 2, with 3^5 components in a three-dimensional space). Vectors, as we have seen, are tensors of order one. And, it turns out that scalars are tensors of order zero.

Vector Spaces

The next example illustrates an easy way to remember how more complex tensors, say T^{abc}_{de}, transform. For each covariant index write an a, and for each contravariant index write a b, all with their corresponding indices. This should look like

$$\overline{T}^{abc}_{de} = a_d a_e b^a b^b b^c T^{...?}_{...?} \tag{3.87}$$

This equation is obviously incomplete; we must add the necessary dummy indices to produce the complete transformation equation

$$\overline{T}^{abc}_{de} = a^r_d a^s_e b^a_t b^b_u b^c_v T^{tuv}_{rs} \tag{3.88}$$

There are tensors with special characteristics, such as symmetric tensors, and special tensor operations to facilitate computation and understanding, such as contraction. We will not pursue these notions here.

There are visual or graphical analogs for certain limited low-order tensors. These do not carry over well into higher order tensors. It is perhaps best to think of tensors as offering a very compact and efficient notation scheme for expressing multi-component objects and their transformations. Finally, we next consider briefly one of the most important concepts in mathematical physics and advanced geometry — the metric tensor.

The Metric Tensor

There are many ways to present and interpret the metric tensor. For the present, we will proceed on the basis that simpler is better. Consider a function g that operates on two contravariant vectors **r** and **s** so that

$$g(\mathbf{r},\mathbf{s}) = g_{ij} r^i s^j \tag{3.89}$$

g_{ij} is the metric tensor. Let's see what this expression means. If $\mathbf{r} = \mathbf{s}$, then

$$g(\mathbf{r},\mathbf{r}) = g_{ij} r^i r^j \tag{3.90}$$

and for $n = 2$ this expands to

$$g_{ij} r^i r^j = g_{11} \left(r^1\right)^2 + g_{12} r^1 r^2 + g_{21} r^2 r^1 + g_{22} \left(r^2\right)^2 \tag{3.91}$$

There is something familiar about this last equation. If $g_{12} = g_{21} = 0$ and $g_{11} = g_{22} = 1$, then $g(\mathbf{r},\mathbf{r})$ produces the square of the magnitude of the input vector. Now we assert without proof (although, try it!) that for orthonormal coordinate systems and orthogonal transformations $g_{ij} = g_{ji} = 0$ if $i \ne j$ (g is a symmetric tensor), and $g_{ij} = 1$ if $i = j$.

If $\mathbf{r} \ne \mathbf{s}$, and they are expressed in an orthonormal coordinate system, then from Equation 3.89 we have, for $n = 3$,

$$g_{ij}r^i s^j = g_{11}r^1 s^1 + g_{12}r^1 s^2 + g_{13}r^1 s^3$$
$$+ g_{21}r^2 s^1 + g_{22}r^2 s^2 + g_{23}r^2 s^3 \qquad (3.92)$$
$$+ g_{31}r^3 s^1 + g_{32}r^3 s^2 + g_{33}r^3 s^3$$

or $\quad g(\mathbf{r},\mathbf{s}) = r^1 s^1 + r^2 s^2 + r^3 s^3 \qquad (3.93)$

which is nothing more than the scalar product of **r** and **s**.

We find, of course, that g_{ij} transforms as

$$g_{ij} = a_i^k a_j^l g_{kl} \qquad (3.94)$$

The versatility of the metric tensor is not fully apparent until we introduce nonorthonormal coordinate systems and more general affine transformations. When we work in a curved space, the components of the metric tensor are sufficient to fully describe that space locally.

The Metric Tensor: Another View

How do we measure distance on a curved surface? A rectangular Cartesian coordinate system seems out of the question for the most general types of curved surfaces. It won't work for a simple spherical surface, and certainly not for more complex surfaces. However, it is always possible to construct a flexible mesh or grid over some local region of a curved surface (Figure 3.26).

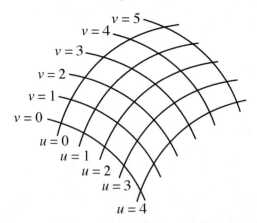

Figure 3.26 A flexible grid over a local region of a curved surface.

We use two families of curves; call them the u curves and the v curves. The u curves never cross other u curves, and the v-curves never cross other v curves. The u curves do intersect the v curves, but not necessarily at right angles. This intersecting grid of curve families allows us to locate points but not to directly measure distances between them.

Vector Spaces

If we look at a relatively small area of this grid, then the quadrilaterals formed by intersecting pairs from each family of curves closely approximate a Euclidean parallelogram (Figure 3.27). This lets us determine distances and angles, at least locally, for very complex surfaces. Here is one approach:

$OABC$ is a parallelogram of the u,v-grid, where $du = OA$ and $dv = OC$, and P is any point within the parallelogram. From the law of cosines in trigonometry we have for the distance ds from O to P

$$ds^2 = OD^2 + DP^2 + 2(OD)(DP)\cos\theta$$

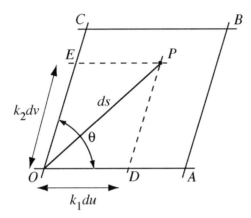

Figure 3.27 A differential area of a grid.

Since the scales in the u and v directions need not be equal, we normalize them to a common scale corresponding to ds. Thus, let $ds = k_1 du$ and $ds = k_2 dv$. Now make the appropriate substitution to obtain

$$ds^2 = k_1^2 du^2 + 2k_1 k_2 \cos\theta \, du \, dv + k_2^2 dv^2$$

In general, the coefficients of du^2, $du\,dv$ and dv^2 will vary over the surface depending on the u, v location, and so are usually considered to be functions of position, that is functions of u and v. These functions we write as $g_{11}(u,v)$, $g_{12}(u,v)$, $g_{21}(u,v)$ and $g_{22}(u,v)$, or simply as g_{11}, g_{12}, g_{21} and g_{22}. Substituting these into the equation for ds^2 gives us

$$ds^2 = g_{11} du^2 + 2 g_{12} du \, dv + g_{22} dv^2$$

where $g_{12} = g_{21}$. We call ds^2 the metric, and g_{11}, g_{12}, g_{22} the components of the metric tensor.

If we can establish the metric tensor g_{ij} for a surface, then we can determine all its metric properties. Of course, we may transform g_{ij} according to the rules of tensor transformation, and since the same distance is involved, ds^2 must be invariant. Now imagine all this generalized to geometric structures beyond two dimensions, and you have stepped well into the world of differential geometry.

Here is a simple example of the metric tensor in action, one that does not require recourse to differential geometry: In a three-dimensional space, what happens to the metric tensor when we stretch the x coordinate by a factor k? First, we express the stretching transformation as

$$\bar{x}^1 = kx^1$$
$$\bar{x}^2 = x^2$$
$$\bar{x}^3 = x^3$$

For the contravariant vector components we have

$$\bar{r}^1 = kr^1$$
$$\bar{r}^2 = r^2$$
$$\bar{r}^3 = r^3$$

This tells us that the transformation coefficients are $a_1^1 = 1/k$, $a_2^2 = 1$, $a_3^3 = 1$ and $a_i^j = 0$ for $i \neq j$. We know that $\bar{g}_{ij} = a_i^k a_j^l g_{kl}$, so that $\bar{g}_{11} = 1/k^2$ and $\bar{g}_{ij} = g_{ij}$ for $i \neq j$. Now we can compute the metric $(\bar{r})^2$

$$(\bar{r})^2 = \bar{g}_{ij}\bar{r}^i\bar{r}^j = \frac{1}{k^2}(\bar{r}^1)^2 + (\bar{r}^2)^2 + (\bar{r}^3)^2$$

We can easily verify this transformation back to the original metric form by means of the appropriate substitution for the $\bar{r}^i$'s; thus

$$(\bar{r})^2 = \frac{(kr^1)^2}{k^2} + (r^2)^2 + (r^3)^2 = (r^1)^2 + (r^2)^2 + (r^3)^2$$

Section 3.4 Exercises

1. Expand $\mathbf{r} = r^i \mathbf{e}_i$ for $n = 2, 3$ and 4.

2. How do the following tensors transform?

   a. R_b^a b. S_{bc}^a c. T_{abc} d. U_d^{abc}

3. Show that the tensor product $R_b^a S^{cd}$ is also a tensor.

4. Show that T_i^i is a scalar.

5. Show that T_{ij}^i is a covariant vector.

4 RIGID-BODY MOTION

If we limit the set of isometric transformations to just translations and rotations, then we may call the resulting set the *rigid-body motions*, or simply motions. Remember, in the theory of transformations there is an initial or original position and an image or final position. There are no intermediate positions. There is no path. However, with the motion transformations we now relax that outlook. We begin this chapter with the translation transformation and consider the Cartesian translation equations, vector-defined translations, a succession of translations, unequal translations over two points, invariants, and finally the translation group. We study the rotation transformation in two and three dimensions, equivalent rotations (where we require the services of eigenvalues and eigenvectors), products of rotations, and the rotation group. Next, we consider composite motion, a combination of translation and rotation transformations. Here we address the problem of how to mathematically express a composite motion in a computationally efficient and convenient way. Homogeneous coordinates and the homogeneous transformation matrix offer a solution. In the final section of this chapter, we briefly explore kinematic transformations. All of these transformations operate within the standard, right-hand Cartesian coordinate system, unless specified otherwise.

4.1 Translation

A translation is a mapping given by Cartesian equations of the form

$$x' = x + a$$
$$y' = y + b \qquad (4.1)$$
$$z' = z + c$$

The translation transformation of a point within the framework of a coordinate system is not very exciting and seems almost too simple mathematically to bother with. However, certain characteristics and situations make it very interesting. For example, translation complicates the linear homogeneous equations describing transformations by requiring us to add a constant term. This problem becomes more annoying when we require the product of several transformations, some of which are translations. (We'll see that homogeneous coordinates offer an effective solution.)

Given two points P_1 and P_2, there is a unique translation taking P_1 into P_2 (Figure 4.1). Let $P_1 = (x_1, y_1, z_1)$ and $P_2 = (x_2, y_2, z_2)$. There are unique numbers a, b, and c such that $x_2 = x_1 + a$, $y_2 = y_1 + b$, $z_2 = z_1 + c$ and Equations 4.1 become

$$x' = x + (x_2 - x_1)$$
$$y' = y + (y_2 - y_1)$$
$$z' = z + (z_2 - z_1)$$

This tells us that one point and its image completely determine a translation. In the preceding example, we imply that P_2 is the image of P_1.

The Equations 4.1 become more meaningful if we let $a = (x_2 - x_1) = \Delta_x$, $b = (y_2 - y_1) = \Delta_y$, and $c = (z_2 - z_1) = \Delta_z$. Then;

$$x' = x + \Delta_x$$
$$y' = y + \Delta y \qquad (4.2)$$
$$z' = z + \Delta_z$$

where Δ_x, Δ_y, and Δ_z are contributions in the x, y, and z directions to the total translation.

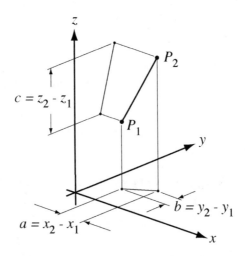

Figure 4.1 Two points determine a unique translation.

Here are some examples of translations expressed in the algebraic form given by Equations 4.2:

Straight line: Given the equation of a straight line in the plane in the slope-intercept form $y = mx + b$ and translation components Δ_x and Δ_y, find the equation for the transformed line. Since $x = x' - \Delta_x$ and $y = y' - \Delta_y$, by substitution we have

RIGID-BODY MOTION

$y' = mx' + (b - m\Delta_x + \Delta_y)$. Obviously, the slope doesn't change under a translation transformation, but the intercept now becomes $b - m\Delta_x + \Delta_y$.

Circle: The general equation of a circle in the plane with radius r and center at (x_c, y_c) is

$$(x - x_c)^2 + (y - y_c)^2 = r^2$$

If $x = x' - \Delta_x$ and $y = y' - \Delta_y$, then the translated circle is

$$[x' - (x_c + \Delta_x)]^2 + [y' - (y_c + \Delta_y)]^2 = r^2$$

where the new center coordinates are $(x'_c, y'_c) = (x_c + \Delta_x, y_c + \Delta_y)$.

Plane: The general equation of a plane in space is $Ax + By + Cz + D = 0$. If $x = x' - \Delta_x$, $y = y' - \Delta_y$, and $z = z' - \Delta_z$, then the translated plane is

$$A(x' - \Delta_x) + B(y' - \Delta_y) + C(z' - \Delta_z) + D = 0$$

or $\quad Ax' + By' + Cz' + D - A\Delta_x - B\Delta_y - C\Delta_z = 0$.

Vectors and Translation

A translation is simply the movement of a figure so that every point of it moves the same direction and distance. We express this notion of a translation operation more concisely if we use vectors. If we define a point by a vector **r** and an arbitrary translation by another vector **a**, then the transformation of **r** by **a** is **r'**, where **r' = r + a** (Figure 4.2).

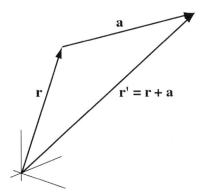

Figure 4.2 A translation transformation expressed as a vector equation.

What about translating a line? Given the vector equation for a line l, $\mathbf{r} = \mathbf{r_0} + u\mathbf{t}$ (Figure 3.10) and a translation **a**, then its image l' under **a** is simply $\mathbf{r'} = \mathbf{r}_o + u\mathbf{t} + \mathbf{a}$ (Figure 4.3). Notice that the direction of the line does not change under a translation transformation, so that l' is parallel to l.

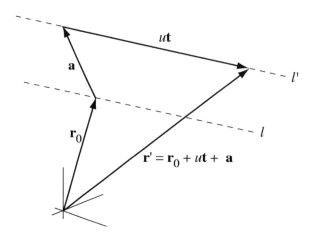

Figure 4.3 A translation of a line using vectors.

Here is one more example: The vector representation of a plane is $\mathbf{r} = \mathbf{r}_0 + u\mathbf{s} + v\mathbf{t}$ (Figure 3.11). Translating this plane a distance and direction given by the vector a produces $\mathbf{r}' = \mathbf{r}_0 + u\mathbf{s} + v\mathbf{t} + \mathbf{a}$. The image of the plane is parallel to its original. Furthermore, if $\mathbf{a} = -\mathbf{r}_0$, then $\mathbf{r}' = u\mathbf{s} + v\mathbf{t}$ and the image of the plane passes through the origin.

Finally, to translate more complex figures like Bézier or B-spline curves and surfaces, whose shapes we define by sets of control points $\mathbf{p}_i$, simply translate each control point by the same vector **a**. Thus, the transformed control points are $\mathbf{p}'_i = \mathbf{p}_i + \mathbf{a}$. This translates the figure parallel to itself a distance and direction given by **a**.

What happens if we simply want to change the placement, without rotation, of our frame of reference with respect to a figure or set of figures? Here is a vector representation of this problem and its solution (Figure 4.4):

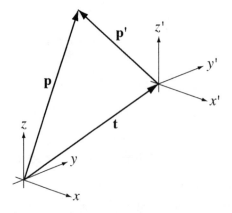

Figure 4.4 Coordinate system translation.

RIGID-BODY MOTION

Given a point represented by the vector **p** in the x, y, z coordinate system, find the vector representation of this point in the x', y', z' system, where the x', y', z' system is the mapping of the x, y, z system under a translation **t**. The solution is simple enough, as you see in the figure. We find that $\mathbf{p}' = \mathbf{p} - \mathbf{t}$ and conclude that a coordinate system translation is equivalent to an object translation of the same distance but in the opposite direction.

A Succession of Translations

A succession or product of translations $\mathbf{a}, \mathbf{b}, \mathbf{c}, \ldots$ of a point **r** produces the image point $\mathbf{r}'$, where

$$\mathbf{r}' = \big(\big(\big(\mathbf{r} + \mathbf{a}\big) + \mathbf{b}\big) + \mathbf{c}\big) + \ldots\big),$$

or more simply $\mathbf{r}' = \mathbf{r} + \mathbf{a} + \mathbf{b} + \mathbf{c} + \ldots$. As Figure 4.5 makes clear, we can, of course, sum the individual translations to produce a single equivalent translation. Thus, if $\mathbf{t} = \mathbf{a} + \mathbf{b} + \mathbf{c}$ then $\mathbf{r}' = \mathbf{r} + \mathbf{t}$.

The order in which we compute the product of two or more translations is not important. Therefore, in the above example and in the figure we find that

$$\mathbf{r}' = \mathbf{r} + \mathbf{a} + \mathbf{b} + \mathbf{c} = \mathbf{r} + \mathbf{a} + \mathbf{c} + \mathbf{b} = \mathbf{r} + \mathbf{b} + \mathbf{a} + \mathbf{c}$$

and so forth. From this we conclude that a succession of transformations is commutative.

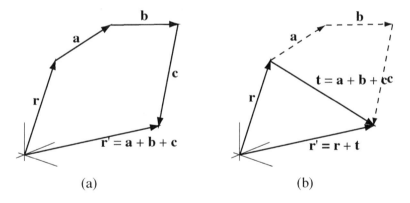

Figure 4.5 A succession of translations.

Unequal Translations of Two Points

We make things much more interesting by asking questions such as: Given two points **p** and **q** translated by **a** and **b**, producing $\mathbf{p}'$ and $\mathbf{q}'$, respectively; under what conditions are the distances $|\mathbf{p} - \mathbf{q}|$ and $|\mathbf{p}' - \mathbf{q}'|$ equal (Figure 4.6)?

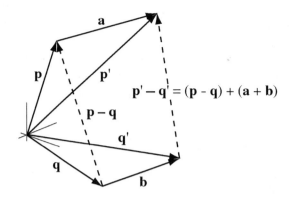

Figure 4.6 Unequal translation of two points.

In general, the distance between **p′** and **q′** will not be the same as the distance between **p** and **q** unless $\mathbf{a} = \mathbf{b}$. Under what conditions are these distances equal? In effect, we are asking for solutions to $|\mathbf{p}-\mathbf{q}| = |\mathbf{p}-\mathbf{q}+\mathbf{a}-\mathbf{b}|$. An unlimited number of solutions are possible under these conditions. However, by adding constraints we hope to reduce the number of possible solutions.

Here is an example: Start with two points **p** and **q**, where we translate **p** by **a** (Figure 4.7). To preserve the distance $|\mathbf{p}-\mathbf{q}|$, we translate **q** by **b** in such a way that **q′** lies on the surface of a sphere centered at **p′** and with radius $|\mathbf{p}-\mathbf{q}|$. We immediately see a directional constraint, since we must now direct **b** so that it intersects or is tangent to the sphere. Of course, we may impose other constraints to further simplify the solution. Notice that aspects of this problem anticipate transformations of mechanisms and linkages, and that makes this a good place to stop. Clearly, we could easily expand this study to include a greater number and more complex figures.

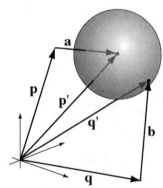

Figure 4.7 Unequal translation of two points preserving the distance between them.

Invariants under Translation

Now let's consider the invariant properties of a set of points under a translation. We begin with points in the plane and extend consequences found here quite naturally and with obvious constraints into three or higher dimensions.

A translation of a set of points preserves distances and directions between points in the set. These properties allow us to characterize translation as a rigid body motion. Because of this, the transformation fixes no points, but does fix all lines in the direction of a translation. What about other types of figures? At first, we might conclude that a translation fixes no other figures, but consider an indefinitely repeating sine wave whose axis is parallel to the direction of translation. If the magnitude of the translation is equal to the period of the sine wave, then the transformation fixes the wave. You can easily imagine other figures with similar, repetitive properties. Otherwise, a translation maps any figure onto a congruent one, and the directed distances from every point of the original figure to their corresponding image points on the translated figure are equal and parallel.

The Translation Group

Equations 4.1 define a translation. The inverses of these equations are $x = x' - a$, $y = y' - b$, $z = z' - c$. Clearly, the inverse of a translation is another translation, and the vectors describing it differ only in sign, being oppositely directed.

The product of two translations (a_1, b_1, c_1) and (a_2, b_2, c_2) begins with $x' = x + a_1$, $y' = y + b_1$, $z' = z + c_1$, and $x'' = x' + a_2$, $y'' = y' + b_2$, $z'' = z' + c_2$. Then, after some easy algebraic substitutions we find that

$$x'' = x + (a_1 + a_2)$$
$$y'' = y + (b_1 + b_2)$$
$$z'' = z + (c_1 + c_2)$$

We see that the resultant of two translations is simply another translation, where $a = a_1 + a_2$, $b = b_1 + b_2$, $c = c_1 + c_2$. The order of the product, as we also saw earlier, does not affect the result. We have just demonstrated that the set of all translations is a group, that the group is commutative, and that it is a subgroup of the group of motions.

Section 4.1 Exercises

1. Find appropriate translation component pairs Δ_x, Δ_y such that the algebraic equation of a straight line in slope-intercept form is simplified (that is, the line is translated so that it passes through the origin).

2. Find a translation that simplifies the general equation of a circle.

3. Find a translation that simplifies the general equation of a plane in space.

4. Using vectors, show that the image of a line under translation is parallel to its original.

5. How far is each point moved under the translation $x' = x+a$, $y' = y+b$, $z' = z+c$?

6. Find the equations of the translation that sends $(-4, 3)$ into $(1, 2)$.

7. Find the images of the following figures under the transformation $x' = x+2$, $y' = y-1$.

   a. $x+y-1 = 0$

   b. $x^2 + y^2 = 4$

   c. $y = 2x^2$

8. Given $T_1 \begin{cases} x' = x-1 \\ y' = y-3 \end{cases}$, $T_2 \begin{cases} x' = x+4 \\ y' = y-1 \end{cases}$, find $T_1 T_2$ and $T_2 T_1$.

9. Express each of the following translations in vector form, giving the components of each vector.

   a. $x' = x+5$, $y' = y$

   b. $x' = x-3$, $y' = y+2$

   c. $x' = x$, $y' = y$, $z' = z$

   d. $x' = x-1$, $y' = y$, $z' = z+6$

   e. $x' = x+a$, $y' = y+b$, $z' = z+c$

10. Express in scalar form a translation described by the vector $\mathbf{t} = (-2, 1, 4)$.

4.2 Rotation

A proper rotation in three dimensions is a mapping given by the familiar homogeneous equations of the form

$$\begin{aligned} x' &= r_{11}x + r_{12}y + r_{13}z \\ y' &= r_{21}x + r_{22}y + r_{23}z \\ z' &= r_{31}x + r_{32}y + r_{33}z \end{aligned} \quad (4.3)$$

where $|r_{ij}| = 1$. We immediately recognize this as a direct isometry, and we shall soon see that the transformation coefficients turn out to be trigonometric functions of the angles describing the rotation. Since all rotations have the same general form, it is the values of the coefficients of x, y, z that distinguish one rotation from another.

We develop and express the rotation transformation in terms of vectors and matrices, denoting the original point with the vector $\mathbf{p}$ and the image point with the vector $\mathbf{p}'$, and arranging their coordinates so that in matrix form $\mathbf{p} = \begin{bmatrix} x & y & z \end{bmatrix}^T$ and $\mathbf{p}' = \begin{bmatrix} x' & y' & z' \end{bmatrix}^T$. We organize the transformation coefficients into the matrix $\mathbf{R}$, the rotation matrix, where

$$\mathbf{R} = \begin{bmatrix} r_{11} & r_{12} & r_{13} \\ r_{21} & r_{22} & r_{23} \\ r_{31} & r_{32} & r_{33} \end{bmatrix}$$

Now we express Equations 4.3 with the single vector-matrix equation

$$\mathbf{p}' = \mathbf{R}\mathbf{p} \tag{4.4}$$

or in expanded matrix form

$$\begin{bmatrix} x' \\ y' \\ z' \end{bmatrix} = \begin{bmatrix} r_{11} & r_{12} & r_{13} \\ r_{21} & r_{22} & r_{23} \\ r_{13} & r_{32} & r_{33} \end{bmatrix} \begin{bmatrix} x \\ y \\ z \end{bmatrix}$$

The initial points or vectors are column matrices and are premultiplied by the rotation matrix to produce the transformed point coordinates, another column matrix.

Notice that we can interpret Equation 4.4 in two ways: as the rotation of a point or the rotation of a vector. Although we will not derive a proof for it, the row and column vectors of the rotation matrix are mutually perpendicular unit vectors. Any curve, surface, or other geometric object whose shape we derive from a set of control points is properly rotated when we rotate its control points. Examples of such objects include Bézier and B-spline curves and surfaces.

What happens if we reverse the order of the terms on the right-hand side of Equation 4.4 so that the components of $\mathbf{p}'$ = components of $\mathbf{p}$ × the transformation coefficients? Exercise 1 asks you to explore this question.

Rotation about the Origin in the Plane

In the plane, rotation about the origin is the simplest form of rotation. It is so simple, in fact, that it obscures some fascinating characteristics of rotation that we find in three dimensions. We invoke two important conventions: the right-hand rule determines the signs of the rotation angle, and the embedding coordinate system is a right-hand system (Figure 4.8).

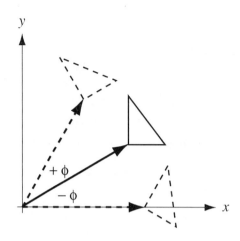

Figure 4.8 Rotation conventions in the plane.

To begin, rewrite Equations 4.3 for a plane:

$$x' = r_{11}x + r_{12}y$$
$$y' = r_{21}x + r_{22}y$$

Next, we find appropriate expressions for the coefficients r_{ij}, comprising the elements of the rotation matrix, **R**. This means we must find the image **p′** of a given point **p** rotated about the origin through an angle ϕ. Clearly, the coordinates of **p′** are functions of the coordinates of **p** and the angle ϕ. The derivation of these functions is easy and straightforward. Refer to Figure 4.9 for what follows.

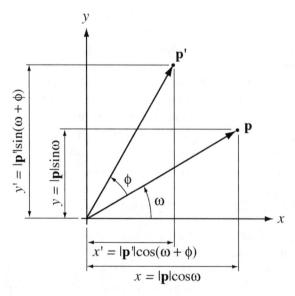

Figure 4.9 Geometry of a plane rotation about the origin.

RIGID-BODY MOTION

We know that $|\mathbf{p}'|=|\mathbf{p}|$ since these points lie on the circumference of a circle whose center is at the origin, and from elementary trigonometry, we know that

$$\cos(\omega+\phi)=\cos\omega\cos\phi-\sin\omega\sin\phi$$
$$\sin(\omega+\phi)=\sin\omega\cos\phi+\cos\omega\sin\phi$$

Using appropriate substitutions, we find that

$$x'=x\cos\phi-y\sin\phi$$
$$y'=x\sin\phi+y\cos\phi \tag{4.5}$$

and $$\mathbf{R}_\phi=\begin{bmatrix}\cos\phi & -\sin\phi\\ \sin\phi & \cos\phi\end{bmatrix} \tag{4.6}$$

The subscript on $\mathbf{R}$ denotes a specific angle of rotation.

As we should expect, $|\mathbf{R}|=1$. This is the complete, fully determined rotation matrix. It tells us how we transform the coordinates of a point, or for that matter the components of a vector, when we rotate a point or vector in the plane about the origin. There is another way to derive Equations 4.5, and that is by using polar coordinates.

With only a minor change in our point of view, we have another way to derive Equations 4.5. We begin by expressing a rotation about the origin in terms of a polar coordinate system (Figure 4.10). If the initial polar coordinates of a point are $\mathbf{p}(r,\theta)$, then the coordinates of its image after a rotation ϕ are

$$r'=r$$
$$\theta'=\theta+\phi$$

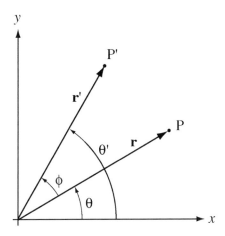

Figure 4.10 Rotations and the polar coordinate system.

Transforming to Cartesian coordinates we obtain

$$x' = r'\cos\theta'$$
$$= r\cos(\theta+\phi)$$
$$= r\cos\theta\cos\phi - r\sin\theta\sin\phi$$
$$y' = r'\sin\theta'$$
$$= r\sin(\theta+\phi)$$
$$= r\cos\theta\sin\phi + r\sin\theta\cos\phi$$

The transformation equations relating polar and Cartesian coordinates are:

$$x = r\cos\theta$$
$$y = r\sin\theta$$

After appropriate substitution, we have

$$x' = x\cos\phi - y\sin\phi$$
$$y' = x\sin\phi + y\cos\phi$$

Successive Rotations about the Origin in the Plane

We determine the product of two successive rotations ϕ_1 and ϕ_2 of a point about the origin in two steps. First, rotate the point by ϕ_1 to produce

$$\mathbf{p}' = \mathbf{R}_{\phi_1}\mathbf{p}$$

Second, rotate this result by ϕ_2 to obtain

$$\mathbf{p}'' = \mathbf{R}_{\phi_2}\mathbf{p}'$$

or

$$\mathbf{p}'' = \mathbf{R}_{\phi_2}\mathbf{R}_{\phi_1}\mathbf{p}$$

The indicated matrix multiplication, $\mathbf{R}_{\phi_2}\mathbf{R}_{\phi_1}$, produces

$$\mathbf{R}_{\phi_2}\mathbf{R}_{\phi_1} = \begin{bmatrix} \cos\phi_1\cos\phi_2 - \sin\phi_1\sin\phi_2 & -\sin\phi_1\cos\phi_2 - \cos\phi_1\sin\phi_2 \\ \cos\phi_1\sin\phi_2 + \sin\phi_1\cos\phi_2 & -\sin\phi_1\sin\phi_2 + \cos\phi_1\cos\phi_2 \end{bmatrix}$$

By employing some common trigonometric identities, we simplify this to obtain

$$\mathbf{R}_{\phi_2}\mathbf{R}_{\phi_1} = \begin{bmatrix} \cos(\phi_1+\phi_2) & -\sin(\phi_1+\phi_2) \\ \sin(\phi_1+\phi_2) & \cos(\phi_1+\phi_2) \end{bmatrix}$$

This tells us that $\mathbf{R}_{\phi_2}\mathbf{R}_{\phi_1} = \mathbf{R}_{\phi_1+\phi_2}$. In other words, merely add the angles ϕ_1 and ϕ_2 and use the resulting sum in the rotation matrix. In fact, we easily extend this to include the product of n successive rotations about the origin. Thus

RIGID-BODY MOTION

$$\mathbf{R}_{(\phi_1+\phi_2+\ldots+\phi_n)} = \begin{bmatrix} \cos(\phi_1+\phi_2+\ldots+\phi_n) & -\sin(\phi_1+\phi_2+\ldots+\phi_n) \\ \sin(\phi_1+\phi_2+\ldots+\phi_n) & \cos(\phi_1+\phi_2+\ldots+\phi_n) \end{bmatrix}$$

Notice that the order in which we perform the product of rotations in the plane about the origin is not important. Thus, we have $\mathbf{R}_{\phi_2}\mathbf{R}_{\phi_1} = \mathbf{R}_{\phi_1}\mathbf{R}_{\phi_2} = \mathbf{R}_{\phi_1+\phi_2}$, and we conclude that a sequence of rotations about the origin commutes.

Rotation of a Line Segment about the Origin

The rotation of a line segment about the origin leaves its length invariant. We demonstrate this as follows (Figure 4.11):

If two points $\mathbf{p}_1$ and $\mathbf{p}_2$ define the ends of a line segment, then its length is

$$l = \sqrt{(x_1-x_2)^2 + (y_1-y_2)^2}$$

After rotation about the origin through an angle ϕ, the length is

$$l' = \sqrt{(x_1'-x_2')^2 + (y_1'-y_2')^2}$$

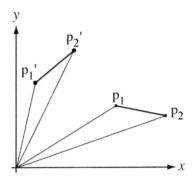

Figure 4.11 Rotating a line.

Next, we substitute from Equations 4.5 for x', x', y', and y_2' to obtain

$$l' = \sqrt{\begin{array}{l}[(x_1\cos\phi - y_1\sin\phi)-(x_2\cos\phi - y_2\sin\phi)]^2 + \\ [(x_1\sin\phi + y_1\cos\phi)-(x_2\sin\phi + y_2\cos\phi)]^2\end{array}}$$

After some rearranging, we find

$$l' = \sqrt{(x_1-x_2)^2(\sin^2\phi + \cos^2\phi) + (y_1-y_2)^2(\sin^2\phi + \cos^2\phi)}$$

but $\sin^2\phi + \cos^2\phi = 1$, so that $L' = \sqrt{(x_1-x_2)^2 + (y_1-y_2)^2}$, or $L = L'$.

Now we might ask: What happens to the algebraic equation of a line under a rotation transformation? Given the general equation for a straight line in the slope-intercept form $y = mx + b$, find the algebraic equation of its image for an arbitrary rotation ϕ about the origin in the xy plane.

Since $x' = x\cos\phi - y\sin\phi$, and $y' = x\sin\phi + y\cos\phi$, then solving for x and y we have

$$x = x'\cos\phi + y'\sin\phi$$
$$y = -x'\sin\phi + y'\cos\phi$$

Substitute these expressions into the line equation to obtain

$$y' = \left(\frac{\sin\phi + m\cos\phi}{\cos\phi - m\sin\phi}\right)x + \frac{b}{\cos\phi - m\sin\phi}$$

Of course, this also works for a parametric representation of a straight line. Thus, if $x(u) = au + b$ and $y(u) = cu + d$, then, after some algebraic aerobics, we find

$$x'(u) = (a\cos\phi - c\sin\phi)u + b\cos\phi - d\sin\phi$$
$$y'(u) = (a\sin\phi + c\cos\phi)u + b\sin\phi + d\cos\phi$$

This technique also applies to equations of higher order.

Rotation about an Arbitrary Point in the Plane

Rotation about the origin is straightforward, easy to understand and easy to do. However, we often want to perform a rotation α about some other point. What then? To achieve this, we execute a sequence of two translations interrupted by a rotation about the origin. Let $\mathbf{p}_c$ be the center point of the rotation $\mathbf{R}_\alpha$, and $\mathbf{p}_i$ one of a set of points to be rotated (Figure 4.12a-c). Here is the procedure:

Step 1. Perform a translation on $\mathbf{p}_c$ (and each $\mathbf{p}_i$) that moves $\mathbf{p}_c$ to the origin: $\mathbf{p}_i - \mathbf{p}_c$ (Figure 12.a).

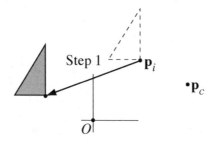

Figure 4.12a Rotation about an arbitrary point in the plane: Step 1.

Step 2. Rotate the results about the origin: $\mathbf{R}_\alpha(\mathbf{p}_i - \mathbf{p}_c)$ (Figure 4.12b).

RIGID-BODY MOTION

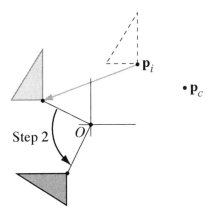

Figure 4.12b Rotation about an arbitrary point in the plane: Step 2.

Step 3. Reverse the translation of Step 1, returning the center of rotation to $\mathbf{p}_c$, thus producing (Figure 4.12c)

$$\mathbf{p}'_i = \mathbf{R}_\alpha (\mathbf{p}_i - \mathbf{p}_c) + \mathbf{p}_c \tag{4.7}$$

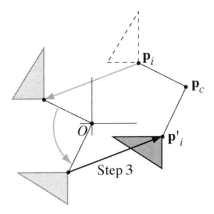

Figure 4.12c Rotation about an arbitrary point in the plane: Step 3.

Successive Rotations about Two Different Points in the Plane

Given two rotations γ_a and γ_b about $\mathbf{p}_a$ and $\mathbf{p}_b$, respectively, find the result of performing these in succession: first through γ_a about $\mathbf{p}_a$ and then through γ_b about $\mathbf{p}_b$. The first rotation produces

$$\mathbf{p}' = \mathbf{R}_{\gamma_a}(\mathbf{p} - \mathbf{p}_a) + \mathbf{p}_a$$

and the second,

$$\mathbf{p}'' = \mathbf{R}_{\gamma_b}(\mathbf{p}' - \mathbf{p}_b) + \mathbf{p}_b$$

or $\quad \mathbf{p}'' = \mathbf{R}_{\gamma_b}[\mathbf{R}_{\gamma_a}(\mathbf{p} - \mathbf{p}_a) + \mathbf{p}_a - \mathbf{p}_b] + \mathbf{p}_b \tag{4.8}$

Notice that, in general, rotations with different centers do not commute. Of course, things get even more complicated for three or more successive rotations about different points. Later in this chapter, we will see how homogeneous coordinates simplify the situation.

Rotation of the Coordinate System in the Plane

Frequently we will want to find the coordinates of a point or the components of a vector in a coordinate system rotated ϕ_c with respect to the initial system, but with a common origin (Figure 4.13).

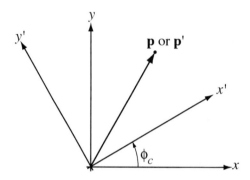

Figure 4.13 Rotation of the coordinate system in the plane.

This produces results identical to that of an equivalent but oppositely directed rotation of the point or vector in a fixed coordinate system. Thus, referring to Equation 4.6, we assert $\phi = -\phi_c$. If $\mathbf{R}_{\phi_c}$ denotes the matrix of transformation coefficients for a coordinate system rotation, then we have by substitution

$$\mathbf{R}_{\phi_c} = \begin{bmatrix} \cos(-\phi_c) & -\sin(-\phi_c) \\ \sin(-\phi_c) & \cos(-\phi_c) \end{bmatrix}$$

From elementary trigonometry we know that $\cos(-\phi_c) = \cos(\phi_c)$ and $\sin(-\phi_c) = -\sin(\phi_c)$, so that

$$\mathbf{R}_{\phi_c} = \begin{bmatrix} \cos\phi_c & \sin\phi_c \\ -\sin\phi_c & \cos\phi_c \end{bmatrix} \tag{4.9}$$

Notice that $\mathbf{R}_{\phi_c} = \mathbf{R}_\phi^T = \mathbf{R}_\phi^{-1}$, demonstrating again that geometric object and coordinate system transformations are very simply related. Finally we have

$$\mathbf{p}' = \mathbf{R}_{\phi_c} \mathbf{p}$$

Rotations in the plane produce some interesting effects on the algebraic expressions of second-degree curves. Here are some properties of second-degree curves under rotation:

Rigid-Body Motion

Given a second-degree curve represented by equations of the form

$$ax^2 + 2bxy + cy^2 + dx + ey + f = 0$$

then the determinant

$$\begin{vmatrix} a & b \\ b & c \end{vmatrix} = ac - b^2$$

called the discriminant, and the sum

$$T = a + c$$

called the trace, are invariant under a rotation of the coordinate axes about the origin. Thus

$$\begin{vmatrix} a' & b' \\ b' & c' \end{vmatrix} = \begin{vmatrix} a & b \\ b & c \end{vmatrix}$$

and $\quad a' + c' = a + c$

The proofs of these assertions, though tedious, are not difficult. Here are some consequent invariant properties under rotation transformation:

1. If the discriminant is not zero, then the curve has a center of symmetry.

2. If the discriminant is positive, then the curve is either an ellipse (or circle), a point, or null.

3. If the discriminant is negative, then the curve is either a hyperbola or two intersecting lines.

Sometimes it is possible to simplify the equation. For example, if $b \neq 0$, then there is a rotation ϕ of the coordinate axes such that $b' = 0$. This is true if

$$\cot 2\phi = \frac{a - c}{2b}$$

Rotations about the Principal Axes in Space

The rotation of bodies in three-dimensional space is far more complex and subtle than rotations in the plane. Solid objects tumble and turn in space. In doing this they do not change size, shape or handedness. How do we best describe such transformations without losing our way in a conceptual and mathematical thicket? We begin by extending our work above to include rotations in three dimensions. The simplest of these are rotations about one or more of the three principal coordinate axes. We denote the angles of rotation as ψ, θ, and ϕ about the x, y, and z axes, respectively (Figure 4.14). Remember, for now our perspective is that of rotating points or other geometric objects about and within a fixed coordinate system. Later we consider rotations of the coordinate system itself.

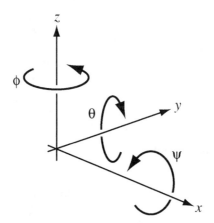

Figure 4.14 Angles of rotation about the principal axes.

Rotation about the z axis looks very much like rotation about the origin in the xy plane. To account for the z coordinate, we expand Equations 4.5 as follows:

$$x' = x\cos\phi - y\sin\phi$$
$$y' = x\sin\phi + y\cos\phi$$
$$z' = z$$

Now we must make a corresponding change in the rotation matrix; thus

$$\mathbf{R}_\phi = \begin{bmatrix} \cos\phi & -\sin\phi & 0 \\ \sin\phi & \cos\phi & 0 \\ 0 & 0 & 1 \end{bmatrix} \tag{4.10}$$

Again, notice that $|\mathbf{R}_\phi| = 1$, as is characteristic of a proper rotation.

The rotation matrices for rotation about the x and y axes follow directly, so

$$\mathbf{R}_\psi = \begin{bmatrix} 1 & 0 & 0 \\ 0 & \cos\psi & -\sin\psi \\ 0 & \sin\psi & \cos\psi \end{bmatrix} \tag{4.11}$$

and

$$\mathbf{R}_\theta = \begin{bmatrix} \cos\theta & 0 & \sin\theta \\ 0 & 1 & 0 \\ -\sin\theta & 0 & \cos\theta \end{bmatrix} \tag{4.12}$$

where $|\mathbf{R}_\psi| = |\mathbf{R}_\theta| = 1$. From the pattern established in Equations 4.10 and 4.11, it seems as though the minus sign on the element in the third row, first column of $\mathbf{R}_\theta$ is misplaced, but this is not the case, as you will discover if you carefully work through the algebra.

Successive Rotations about the Principal Axes in Space

We may specify the rotation of a point or other geometric object in space as the product of successive rotations about each of the three principal axes. It is important to establish some kind of convention describing how we might do this; here is one way (remember, order is important):

1. Rotate by $\mathbf{R}_\phi$ around the z axis, if $\phi \neq 0$.

2. Rotate by $\mathbf{R}_\theta$ around the y axis, if $\theta \neq 0$.

3. Rotate by $\mathbf{R}_\psi$ around the x axis, if $\psi \neq 0$.

Here is another way of looking at these rotations. In the plane or in space we produce a so-called improper rotation if $|\mathbf{R}| = -1$. This is a combination rotation and reflection. Let's investigate improper rotations in the plane first. Two different arrays exhibit the distinguishing characteristic. We derive them from the proper rotation matrix as follows:

$$\begin{bmatrix} \cos\phi & -\sin\phi \\ \sin\phi & \cos\phi \end{bmatrix} \begin{bmatrix} 1 & 0 \\ 0 & -1 \end{bmatrix} = \begin{bmatrix} \cos\phi & \sin\phi \\ \sin\phi & -\cos\phi \end{bmatrix}$$

and

$$\begin{bmatrix} \cos\phi & -\sin\phi \\ \sin\phi & \cos\phi \end{bmatrix} \begin{bmatrix} -1 & 0 \\ 0 & 1 \end{bmatrix} = \begin{bmatrix} -\cos\phi & -\sin\phi \\ -\sin\phi & \cos\phi \end{bmatrix}$$

These produce the following pairs of transformation equations:

$$x' = x\cos\phi + y\sin\phi$$
$$y' = x\sin\phi - y\cos\phi$$

and

$$x' = -x\cos\phi - y\sin\phi$$
$$y' = -x\sin\phi + y\cos\phi$$

Compare these to the proper rotation transformation equations:

$$x' = x\cos\phi - y\sin\phi$$
$$y' = x\sin\phi + y\cos\phi$$

We see at once that the first pair produces a reflection across the x axis of the proper rotation given by $-\phi$. The second pair of transformation equations produces a reflection across the y axis of the proper rotation given by $-\phi$. Thus, an improper rotation in the plane changes the sign of the angle and induces a reflection about the x or y axis.

Improper rotations in space are mathematically more complex, although easy to describe in words. An improper rotation through ϕ about some axis $\mathbf{a}$ is the equivalent of a proper rotation through $-\phi$ and a simultaneous reflection through a plane perpendicular to **a**. This transformation is easier for us to express and understand in terms of rotations in space about an arbitrary axis through the origin.

Proper rotations are the only orthogonal transformations that we can generate by a continuous variation of the coefficients, corresponding to a continuous movement of a rigid body from some initial position to a final one. However, this is not true for improper rotations where a discontinuity must arise so that the sign of $|\mathbf{R}|$ might change.

For the rotation of a point or vector, then, we have

$$\mathbf{p}' = \mathbf{R}_\psi \mathbf{R}_\theta \mathbf{R}_\phi \mathbf{p}$$

Let $\mathbf{R}_{\psi\theta\phi} = \mathbf{R}_\psi \mathbf{R}_\theta \mathbf{R}_\phi$, so that after appropriately multiplying the matrices of Equations 4.10-4.12, we obtain

$$\mathbf{R}_{\psi\theta\phi} = \begin{bmatrix} \cos\theta\cos\phi & -\cos\theta\sin\phi & \sin\theta \\ \cos\psi\sin\phi + \sin\psi\sin\theta\cos\phi & \cos\psi\cos\phi - \sin\psi\sin\theta\sin\phi & -\sin\psi\cos\theta \\ \sin\psi\sin\phi - \cos\psi\sin\theta\cos\phi & \sin\psi\cos\phi + \cos\psi\sin\theta\sin\phi & \cos\psi\cos\theta \end{bmatrix} \quad (4.13)$$

Since $|\mathbf{R}_\psi| = |\mathbf{R}_\theta| = |\mathbf{R}_\phi| = 1$, then from matrix algebra we know that

$$|\mathbf{R}_{\psi\theta\phi}| = |\mathbf{R}_\psi||\mathbf{R}_\theta||\mathbf{R}_\phi| = 1$$

Throughout this section, we have asserted that the condition $|\mathbf{R}| = +1$ characterizes a proper rotation. However, if $|\mathbf{R}| = -1$, then the transformation describes an improper rotation, comprised of a simultaneous rotation and reflection.

Given R, Find ϕ, θ, ψ

After a long series of rotations, sometimes we find that we must solve the inverse problem: Given any proper rotation matrix **R**, find a set of angles that generates an equivalent matrix. Right away, we realize that this problem is too loosely stated. There are many conventions we can choose that help us to limit the problem. For example, in Equation 4.13 we assume a specific sequence of angles to describe the rotation of a point about the principal axes. However, we are free to choose other ordering sequences (we'll see other sequences later in this section). Let's work with Equation 4.13, for now. To begin, we equate the arbitrary, proper rotation matrix **R** to $\mathbf{R}_{\psi\theta\phi}$, requiring, of course, that $|\mathbf{R}| = +1$. This produces

$$\begin{bmatrix} r_{11} & r_{12} & r_{13} \\ r_{21} & r_{22} & r_{23} \\ r_{31} & r_{32} & r_{33} \end{bmatrix} = \begin{bmatrix} \cos\theta\cos\phi & -\cos\theta\sin\phi & \sin\theta \\ \cos\psi\sin\phi + \sin\psi\sin\theta\cos\phi & \cos\psi\cos\phi - \sin\psi\sin\theta\sin\phi & -\sin\psi\cos\theta \\ \sin\psi\sin\phi - \cos\psi\sin\theta\cos\phi & \sin\psi\cos\phi + \cos\psi\sin\theta\sin\phi & \cos\varphi\cos\theta \end{bmatrix} \quad (4.14)$$

Matrix equality implies element-by-element equality. So we may immediately write

$$r_{11} = \cos\theta\cos\phi \quad (4.15)$$

$$r_{21} = \cos\psi\sin\phi + \sin\psi\sin\theta\cos\phi$$

$$r_{31} = \sin\psi\sin\phi - \cos\psi\sin\theta\cos\phi$$

$$r_{12} = -\cos\theta\sin\phi$$

$$r_{22} = \cos\psi\cos\phi - \sin\psi\sin\theta\sin\phi$$

$$r_{32} = \sin\psi\cos\phi + \cos\psi\sin\theta\sin\phi$$

$$r_{13} = \sin\theta \quad (4.16)$$

$$r_{23} = -\sin\psi\cos\theta$$

$$r_{33} = \cos\psi\cos\theta \quad (4.17)$$

At first glance, we are tempted to solve these equations for ϕ, θ, ψ as follows. From Equation 4.16:

$$\theta = \sin^{-1} r_{13} \quad (4.18)$$

From Equations 4.17 and 4.18:

$$\psi = \cos^{-1}(r_{33}/\cos\theta) \quad (4.19)$$

From Equations 4.15 and 4.18

$$\phi = \cos^{-1}(r_{11}/\cos\theta) \quad (4.20)$$

However, these equations present us with the potential for serious computational problems. For instance, computing an angle using the arc sine function makes the distinction between the angles θ and $\pi - \theta$ ambiguous; and at $\theta = 0, \pi$, $\sin\theta = 0$. Equations 4.19 and 4.20 are undefined when $\theta = \pm\pi/2$, since $\cos(\pm\pi/2) = 0$. We must take a slightly different, more cautious approach. A safer way for us is to use tangent and arc tangent functions. Here is one way to extract such functions: Premultiply both sides of Equation 4.14 by $\mathbf{R}_\psi^{-1}$ from Equation 4.11 to produce

$$\begin{bmatrix} r_{11} & r_{12} & r_{13} \\ r_{21}\cos\psi + r_{31}\sin\psi & r_{22}\cos\psi + r_{32}\sin\psi & r_{23}\cos\psi + r_{33}\sin\psi \\ -r_{21}\sin\psi + r_{31}\cos\psi & -r_{22}\sin\psi + r_{32}\cos\psi & -r_{23}\sin\psi + r_{33}\cos\psi \end{bmatrix}$$

(4.21)

$$= \begin{bmatrix} \cos\theta\cos\phi & -\cos\theta\sin\phi & \sin\theta \\ \sin\phi & \cos\phi & 0 \\ -\sin\theta\cos\phi & \sin\theta\sin\phi & \cos\theta \end{bmatrix}$$

We see that the element in the second row and third column is zero. This means that equating this element in the matrix on the right to that on the left yields

$$r_{23}\cos\psi + r_{33}\sin\psi = 0$$

$$\tan\psi = (-r_{23}/r_{33})$$

$$\psi = \tan^{-1}(-r_{23}/r_{33})$$

There are two possible solutions, 180° apart. It is also possible that $r_{23} = r_{33} = 0$; then the angle is undefined, and we arbitrarily set $\psi = 0$. Once we have computed a value for ψ, then we know all the elements of the left side of Equation 4.21, and we easily compute θ and ϕ. Thus

$$\sin\theta = r_{13}$$

$$\cos\theta = -r_{23}\sin\psi + r_{33}\cos\psi$$

Thus $\quad \theta = \tan^{-1}\left(\dfrac{r_{13}}{-r_{23}\sin\psi + r_{33}\cos\psi}\right)$

Since both the sine and cosine are defined, we can compute a unique value for the angle θ. Finally, for ϕ we have

$$\sin\phi = r_{21}\cos\psi + r_{31}\sin\psi$$

$$\cos\phi = r_{23}\cos\psi + r_{32}\sin\psi$$

$$\phi = \tan^{-1}\left(\dfrac{r_{21}\cos\psi + r_{31}\sin\psi}{r_{23}\cos\psi + r_{32}\sin\psi}\right)$$

Rotation of the Coordinate System in Space

As with rotations in the plane, there are many occasions when we must find point coordinates or vector components in a coordinate system rotated with respect to the initial system, but with a common origin. Now, however, we must deal with three angles of rotational displacement, ϕ_c, θ_c, and ψ_c, some of which may be equal to zero (Figure 4.15).

RIGID-BODY MOTION

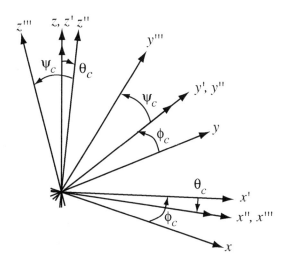

Figure 4.15 Rotation of the coordinate system in space.

Notice that ϕ_c, θ_c, and ψ_c are not so strictly tied to the x, y, and z axes as are ϕ, θ, and ψ. Under a rotation of a point or vector, the axes remain fixed; however, under the coordinate system rotation there is a succession of original and primed axes. In fact as we develop other coordinate system rotations, we'll find that ϕ_c, θ_c, and ψ_c lose any direct relationship to particular coordinate axes, but instead indicate the order of rotation: where ϕ_c is the first rotation, θ_c the second, and ψ_c the third. For each of these rotations, we must specify the axis.

Figure 4.15 may seem confusing at first glance, but stepping through the sequence of rotations should clarify its meaning. The figure shows what happens to the coordinate axes when we subject them to rotations ϕ_c, θ_c, and ψ_c, in that order, and when none of them is zero. Rotation ϕ_c produces a displacement of the x, y axes to the x', y' axes, but the z' axis is coincident with the initial z axis, since it is the axis of the rotation. Next, rotation θ_c displaces the x' and z' axes to x'' and z'', and leaves the y'' axis coincident with the y' axis, since the y' axis is the axis of rotation. Finally, rotation ψ_c displaces the y'' and z'' axes to y''' and z''', leaving the x''' axis fixed and coincident with the x'' axis. So let $\mathbf{R}_{\phi_c \theta_c \psi_c} = \mathbf{R}_{xyz}$, where we understand the x, y, z notation to indicate the sequential axes of rotation.

Now, since $\phi = -\phi_c$, $\theta = -\theta_c$, $\psi = -\psi_c$, using Equation 4.13 we have

$$\mathbf{R}_{xyz} = \begin{bmatrix} 1 & 0 & 0 \\ 0 & \cos\psi_c & \sin\psi_c \\ 0 & -\sin\psi_c & \cos\psi_c \end{bmatrix} \begin{bmatrix} \cos\theta_c & 0 & -\sin\theta_c \\ 0 & 1 & 0 \\ \sin\theta_c & 0 & \cos\theta_c \end{bmatrix} \begin{bmatrix} \cos\phi_c & \sin\phi_c & 0 \\ -\sin\phi_c & \cos\phi_c & 0 \\ 0 & 0 & 1 \end{bmatrix}$$

and the product is

$$\mathbf{R}_{xyz} = \begin{bmatrix} \cos\theta_c \cos\phi_c \\ -\cos\psi_c \sin\phi_c + \sin\psi_c \sin\theta_c \cos\phi_c \\ \sin\psi_c \sin\phi_c + \cos\psi_c \sin\theta_c \cos\phi_c \end{bmatrix}$$

$$\begin{matrix} \cos\theta_c \sin\phi_c & -\sin\theta_c \\ \cos\psi_c \cos\phi_c + \sin\psi_c \sin\theta_c \sin\psi_c & \sin\psi_c \cos\theta_c \\ -\sin\psi_c \cos\phi_c + \cos\psi_c \sin\theta_c \sin\phi_c & \cos\psi_c \cos\theta_c \end{matrix}$$

(4.22)

Compare this to Equation 4.13. Some of the terms change sign, but the forms are identical. But wait ... there is something interesting going on here. Look again at Equations 4.13 and 4.22. We clearly see that $\mathbf{R}_{\psi_c \theta_c \phi_c}$ is neither the transpose nor the inverse of $\mathbf{R}_{\psi\theta\phi}$, as we found to be the case earlier for a coordinate system rotation in the plane. We demonstrate this using some simple matrix algebra to show that $\mathbf{R}_{\psi_c \theta_c \phi_c} \neq \mathbf{R}_{\psi\theta\phi}^T$, thus

$$\mathbf{R}_{\psi_c \theta_c \phi_c} = \mathbf{R}_\psi^{-1} \mathbf{R}_\theta^{-1} \mathbf{R}_\phi^{-1} = \mathbf{R}_\psi^T \mathbf{R}_\theta^T \mathbf{R}_\phi^T = \left[\mathbf{R}_\phi \mathbf{R}_\theta \mathbf{R}_\psi\right]^T$$

and $\quad \mathbf{R}_{\psi\theta\phi}^T = \left[\mathbf{R}_\psi \mathbf{R}_\theta \mathbf{R}_\phi\right]^T$

In general

$$\left[\mathbf{R}_\phi \mathbf{R}_\theta \mathbf{R}_\psi\right]^T \neq \left[\mathbf{R}_\psi \mathbf{R}_\theta \mathbf{R}_\phi\right]^T$$

(See the development of Equation 4.42 in the discussion of rotation groups.)

Euler Angle Rotations

When we perform a rotation transformation in or on the x, y, z coordinate system, as a specified sequence of rotations about a succession of coordinate axes, we call the corresponding set of angles the *Euler angles*. The sequence we have just developed above is often called the x, y, z convention. This seems to be the convention preferred by many aerospace engineers and scientists, at least in the U. S. and Britain. This particularly applies to engineering applications involving the orientation of moving air- or spacecraft. Here the first rotation is the yaw angle ϕ about the z axis, the second is the pitch angle θ about the y' axis, and the third is the roll angle ψ about the x'' axis.

Here is another interpretation of the roll, pitch, yaw convention. Roll, pitch, and yaw are terms originating in naval architecture and refer to the motion of ships. They eventually came to denote analogous motions in airplanes and spacecraft. The roll axis, usually the x axis, is in the direction of the ship's forward motion (Figure 4.16).

RIGID-BODY MOTION

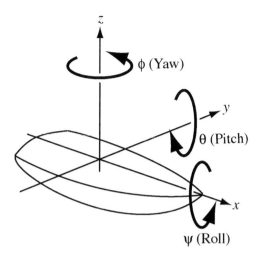

Figure 4.16 Roll, pitch, and yaw.

When wind or waves cause a ship to heel alternately to port and starboard passengers and crew experience a rolling motion. An airplane banks or rolls to the left or right when turning. We define pitch as the rotation around the axis perpendicular to the roll axis and in the horizontal plane. A ship exhibits a pitching motion when its bow rises and falls in a seaway. Yaw is rotation about a vertical axis. In our Cartesian system, then, roll corresponds to a rotation ψ about the x axis, pitch corresponds to a rotation θ about the y-axis, and yaw corresponds to a rotation ϕ about the z axis. Let $\mathbf{R}_R, \mathbf{R}_P, \mathbf{R}_Y$ denote the rotation matrices for roll, pitch, and yaw through the angles ψ, θ, and ϕ, respectively. If we fix the coordinate frame and move the ship within it, in the order roll, pitch, and yaw, then we transform points of the ship according to $\mathbf{p}' = \mathbf{R}_Y \mathbf{R}_P \mathbf{R}_R \mathbf{p}$, where $\mathbf{R}_Y \mathbf{R}_P \mathbf{R}_R = \mathbf{R}_\phi \mathbf{R}_\theta \mathbf{R}_\psi$ and

$$\mathbf{R}_\phi \mathbf{R}_\theta \mathbf{R}_\psi = \begin{bmatrix} \cos\phi & -\sin\phi & 0 \\ \sin\phi & \cos\phi & 0 \\ 0 & 0 & 1 \end{bmatrix} \begin{bmatrix} \cos\theta & 0 & \sin\theta \\ 0 & 1 & 0 \\ -\sin\theta & 0 & \cos\theta \end{bmatrix} \begin{bmatrix} 1 & 0 & 0 \\ 0 & \cos\psi & -\sin\psi \\ 0 & \sin\psi & \cos\psi \end{bmatrix}$$

or

$$\mathbf{R}_\phi \mathbf{R}_\theta \mathbf{R}_\psi = \begin{bmatrix} \cos\phi\cos\theta & \cos\phi\sin\theta\sin\psi - \sin\phi\cos\psi & \cos\phi\sin\theta\cos\psi + \sin\phi\sin\psi \\ \sin\phi\cos\theta & \sin\phi\sin\theta\sin\psi + \cos\phi\cos\psi & \sin\phi\sin\theta\cos\psi - \cos\phi\sin\psi \\ -\sin\theta & \cos\theta\sin\psi & \cos\theta\cos\psi \end{bmatrix}$$

This, of course, looks something like Equation 4.3 and 4.22, but the order, signs, and function sequences are different because the order of matrix multiplication is different.

The x convention, which we'll now briefly introduce, is used in celestial mechanics, applied mechanics, as well as in molecular and solid state physics. Here we begin the sequence by rotating the initial coordinate system axes through the angle ϕ_c counterclockwise about the z axis. Next, we rotate the resulting x', y', z' system through θ_c about the x' axis. Finally, we rotate the x'', y'', z'' system through ψ_c about the z'' axis, producing the x''', y''', z''' axes. The final product matrix is

$$\mathbf{R}_{zxz} = \begin{bmatrix} \cos\psi_c & \sin\psi_c & 0 \\ -\sin\psi_c & \cos\psi_c & 0 \\ 0 & 0 & 1 \end{bmatrix} \times$$

$$\begin{bmatrix} 1 & 0 & 0 \\ 0 & \cos\theta_c & \sin\theta_c \\ 0 & -\sin\theta_c & \cos\theta_c \end{bmatrix} \begin{bmatrix} \cos\phi_c & \sin\phi_c & 0 \\ -\sin\phi_c & \cos\phi_c & 0 \\ 0 & 0 & 1 \end{bmatrix}$$

or

$$\mathbf{R}_{zxz} = \begin{bmatrix} \cos\psi_c \cos\phi_c - \cos\theta_c \sin\phi_c \sin\psi_c & \cos\psi_c \sin\phi_c + \cos\theta_c \cos\phi_c \sin\psi_c & \sin\psi_c \sin\theta_c \\ -\sin\psi_c \cos\phi_c - \cos\theta_c \sin\phi_c \cos\psi_c & -\sin\psi_c \sin\phi_c + \cos\theta_c \cos\phi_c \cos\psi_c & \cos\psi_c \sin\theta_c \\ \sin\theta_c \sin\psi_c & -\sin\theta_c \cos\phi_c & \cos\theta_c \end{bmatrix}$$

Yet another sequence, the y convention, is often used in quantum mechanics, nuclear physics, and particle physics. Here the rotation sequence is $\mathbf{R}_{xyz}$. The details are easy to work our, and the final matrix product is

$$\mathbf{R}_{zyz} = \begin{bmatrix} -\sin\psi_c \sin\phi_c + \cos\theta_c \cos\phi_c \cos\psi_c & \sin\psi_c \cos\phi_c + \cos\theta_c \sin\phi_c \cos\psi_c & -\cos\psi_c \sin\theta_c \\ -\cos\psi_c \sin\phi_c - \cos\theta_c \cos\phi_c \sin\psi_c & \cos\psi_c \cos\phi_c - \cos\theta_c \sin\phi_c \sin\psi_c & \sin\psi_c \sin\theta_c \\ \sin\theta_c \cos\psi_c & \sin\theta_c \sin\psi_c & \cos\theta_c \end{bmatrix}$$

There are twelve possible conventions for Euler angle rotations. We can take the initial rotation about any of the three coordinate axes. For the subsequent two rotations, the only limitation we face is that no two successive rotations can be about

RIGID-BODY MOTION

the same axis. So, in general, we note that the sequence of rotations we might use is somewhat arbitrary. The *x* and *y* conventions become problematical when the original and transformed systems are only minimally displaced from one another. The angles ϕ and ψ become indistinguishable and their respective axes are nearly coincident. To avoid this problem we take each rotation around a different axis.

Rotation about an Arbitrary Axis in Space

Now let's look at a more general rotation transformation in three-dimensional space, the rotation of points or other geometric objects about an arbitrary axis. The process is not as difficult as it may seem, for we do it one step at a time using simple translations and rotations about the principal axes. There are a number of ways to describe the axis. At a minimum, we must know the direction or orientation of the axis and some point on it, as well as an angle of rotation.

In Figure 4.17, the unit vector **a** establishes the direction of the axis, $\mathbf{p}_a$ is a point on it, and α is the angle through which points or objects rotate. The right-hand rule applies to the sign of α. If the thumb of the right hand points in the direction of **a**, then the fingers of the right hand curl around **a** in the positive sense of the angle.

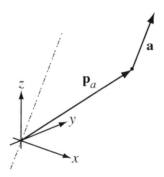

Figure 4.17 Vector description of an arbitrary axis of rotation.

Our computational strategy begins by translating the arbitrary axis so that it passes through the origin, and then rotating it so that it is coincident with one of the principal axes. We transform all points and other objects to be rotated about this arbitrary axis in the same way, and then rotate them through the angle α about the coincident principal axis. Finally, we reverse the two rotations and single translation to return the arbitrary axis to its original orientation and position. Again, these movements define the transformations we apply to points and other geometric objects that we want to rotate about this arbitrary axis. Translating and rotating the arbitrary axis until it is collinear with a principal axis is a convenient way to generate the appropriate transformation matrix for the geometric objects in question. It is, of course, not necessary to apply these transformations to the arbitrary axis itself. What follows is a more mathematical presentation of this algorithm (Figure 4.18).

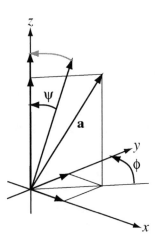

Figure 4.18 A computation strategy for rotation about an arbitrary axis.

Step 1. If $\mathbf{p}_a$ is a point on the arbitrary axis, say the axis defined by the vector $\mathbf{a}$, and $\mathbf{p}_i$ are points we want to rotate, then translate the $\mathbf{p}_i$ so that $\mathbf{p}'_i = \mathbf{p}_i - \mathbf{p}_a$. This is the translation that moves the $\mathbf{a}$ axis so that it passes through the origin.

Step 2. Rotate through ϕ about the z axis so that $\mathbf{p}'_i = \mathbf{R}_\phi(\mathbf{p}_i - \mathbf{p}_a)$, where $\phi = \tan^{-1}(a_x / a_y)$. This places the $\mathbf{a}$ axis in the y, z plane.

Step 3. Rotate through ψ about the x axis so that $\mathbf{p}'_i = \mathbf{R}_\psi \mathbf{R}_\phi(\mathbf{p}_i - \mathbf{p}_a)$, where $\psi = \sin^{-1}(a_x^2 + a_y^2)^{1/2}$, making the $\mathbf{a}$ axis collinear with the z axis.

Step 4. Rotate through α about the z axis so that $\mathbf{p}'_i = \mathbf{R}_\alpha \mathbf{R}_\psi \mathbf{R}_\phi(\mathbf{p}_i - \mathbf{p}_a)$.

Step 5. Reverse the rotation about the x axis so that
$\mathbf{p}'_i = \mathbf{R}_\psi^{-1} \mathbf{R}_\alpha \mathbf{R}_\psi \mathbf{R}_\phi(\mathbf{p}_i - \mathbf{p}_a)$.

Step 6. Reverse the rotation about the z axis so that now
$\mathbf{p}'_i = \mathbf{R}_\phi^{-1} \mathbf{R}_\psi^{-1} \mathbf{R}_a \mathbf{R}_\psi \mathbf{R}_\phi(\mathbf{p}_i - \mathbf{p}_a)$

Step 7. Reverse the translation, finally producing

$$\mathbf{p}'_i = \mathbf{R}_\phi^{-1} \mathbf{R}_\psi^{-1} \mathbf{R}_\alpha \mathbf{R}_\psi \mathbf{R}_\phi(\mathbf{p}_i - \mathbf{p}_a) + \mathbf{p}_a \quad (4.23)$$

If the axis already passes through the origin, then Equation 4.23 simplifies to

$$\mathbf{p}'_i = \mathbf{R}_\phi^{-1} \mathbf{R}_\psi^{-1} \mathbf{R}_\alpha \mathbf{R}_\psi \mathbf{R}_\phi \mathbf{p}_i \quad (4.24)$$

Rigid-Body Motion

Equivalent Rotations in Space

Let's begin this topic by visualizing a child's building block on whose faces appear the first six letters of the alphabet. Leonhard Euler (1752) demonstrated that no matter how many times we twist and turn the block relative to some reference position, we can reach every possible new position via a single equivalent rotation.

Here is a very simple example (Figure 4.19). First turn the block 90° about an axis through the center of the cube and perpendicular to the F-face, placing it in Position 1. Then, rotate it 180° about an axis through the center of the cube and perpendicular to the B-face to place it in Position 2.

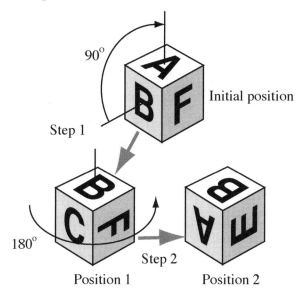

Figure 4.19 A two-step rotation.

However, if we rotate the block from its reference position 180° about the axis passing through the centers of opposite edges *AB* and *CD*, then we find that this single rotation also places the block in Position 2 (Figure 4.20).

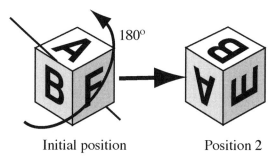

Figure 4.20 An equivalent rotation in one step.

So, now let's restrict our studies to rotations about any arbitrary axis through the origin in the direction of the unit vector **a** and through the angle α. We pose this problem: Given a proper rotation matrix **R** determined perhaps by a sequence of rotations about the principal axes, find the equivalent rotation about a single arbitrary axis **a**, and find the angle of rotation α about this axis. Our strategy is this: Execute the multiplication of the five rotation matrices as indicated in Equation 4.24 to produce a resultant matrix, then, by performing some simple arithmetical operations on equivalent elements of the resultant matrix and **R**, we determine α and the components of **a** in terms of the elements of **R**. (Note, this is the second type of "given **R**, find ..." problem. Earlier we solved "given **R**, find ϕ, θ, and ψ.")

From Equation 4.24, let $\mathbf{R} = \mathbf{R}_\phi^{-1}\mathbf{R}_\psi^{-1}\mathbf{R}_\alpha\mathbf{R}_\psi\mathbf{R}_\phi$

$$\mathbf{R} = \begin{bmatrix} \cos\phi & \sin\phi & 0 \\ -\sin\phi & \cos\phi & 0 \\ 0 & 0 & 1 \end{bmatrix} \begin{bmatrix} 1 & 0 & 0 \\ 0 & \cos\psi & \sin\psi \\ 0 & -\sin\psi & \cos\psi \end{bmatrix} \times$$

$$\begin{bmatrix} \cos\alpha & -\sin\alpha & 0 \\ \sin\alpha & \cos\alpha & 0 \\ 0 & 0 & 1 \end{bmatrix} \begin{bmatrix} 1 & 0 & 0 \\ 0 & \cos\psi & -\sin\psi \\ 0 & \sin\psi & \cos\psi \end{bmatrix} \begin{bmatrix} \cos\phi & -\sin\phi & 0 \\ \sin\phi & \cos\phi & 0 \\ 0 & 0 & 1 \end{bmatrix}$$

or

$$\mathbf{R} = \begin{bmatrix} \cos\alpha\left(\cos^2\phi + \sin^2\phi\,\cos^2\psi\right) + \sin^2\phi\,\sin^2\psi \\ (1-\cos\alpha)\left(\cos\phi\sin\phi\sin^2\psi\right) + \sin\alpha\cos\psi \\ (1-\cos\alpha)(\sin\phi\cos\psi\sin\psi) - \sin\alpha\cos\phi\sin\psi \end{bmatrix}$$

$$\begin{matrix} (1-\cos\alpha)\left(\cos\phi\sin\phi\sin^2\psi\right) - \sin\alpha\cos\psi \\ \cos\alpha\left(\cos^2\phi\,\cos^2\psi + \sin^2\phi\right) + \cos^2\phi\,\sin^2\psi \\ (1-\cos\alpha)(\cos\phi\cos\psi\sin\psi) + \sin\alpha\sin\phi\sin\psi \end{matrix} \quad (4.25)$$

$$\begin{matrix} (1-\cos\alpha)(\sin\phi\cos\psi\sin\psi) + \sin\alpha\cos\phi\sin\psi \\ (1-\cos\alpha)(\cos\phi\cos\psi\sin\psi) - \sin\alpha\sin\phi\sin\psi \\ \cos\alpha\sin^2\psi + \cos^2\psi \end{matrix}$$

Rewrite this is a more concise form as

RIGID-BODY MOTION

$$\mathbf{R} = \begin{bmatrix} r_{11} & r_{12} & r_{13} \\ r_{21} & r_{22} & r_{23} \\ r_{31} & r_{32} & r_{33} \end{bmatrix} \quad (4.26)$$

and we are given the elements of $\mathbf{R}$, where

$r_{11} = \cos\alpha(\cos^2\phi + \sin^2\phi \cos^2\psi) + \sin^2\phi \sin^2\psi$

$r_{12} = (1-\cos\alpha)(\cos\phi \sin\phi \sin^2\psi) - \sin\alpha \cos\psi$

$r_{13} = (1-\cos\alpha)(\sin\phi \cos\psi \sin\psi) + \sin\alpha \cos\phi \sin\psi$

$r_{21} = (1-\cos\alpha)(\cos\phi \sin\phi \sin^2\psi) + \sin\alpha \cos\psi$

$r_{22} = \cos\alpha(\cos^2\phi \cos^2\psi + \sin^2\phi) + \cos^2\phi \sin^2\psi$

$r_{23} = (1-\cos\alpha)(\cos\phi \cos\psi \sin\psi) - \sin\alpha \sin\phi \sin\psi$

$r_{31} = (1-\cos\alpha)(\sin\phi \cos\psi \sin\psi) - \sin\alpha \cos\phi \sin\psi$

$r_{32} = (1-\cos\alpha)(\cos\phi \cos\psi \sin\psi) + \sin\alpha \sin\phi \sin\psi$

$r_{33} = \cos\alpha \sin^2\psi + \cos^2\psi$

From our work in the previous section and with reference to Figure 4.18, we also know

$$\cos\phi = a_y \Big/ \sqrt{(a_x^2 + a_y^2)} \quad (4.27)$$

$$\sin\phi = a_x \Big/ \sqrt{(a_x^2 + a_y^2)} \quad (4.28)$$

$$\cos\psi = a_z \quad (4.29)$$

$$\sin\psi = \sqrt{(a_x^2 + a_y^2)} \quad (4.30)$$

To begin, we sum the diagonal elements of $\mathbf{R}$ and set this equal to the sum of the diagonal elements of the resultant matrix in Equation 4.25.

$r_{11} + r_{22} + r_{33} = 2\cos\alpha + 1$

From this relationship, we find that the cosine of the angle of rotation is

$$\cos\alpha = \frac{1}{2}(r_{11} + r_{22} + r_{33} - 1) \quad (4.31)$$

Next, we take the difference of the off-diagonal elements for each matrix.

$$r_{32} - r_{23} = 2\sin\alpha \sin\phi \sin\psi$$
$$r_{13} - r_{31} = 2\sin\alpha \cos\phi \sin\psi$$
$$r_{21} - r_{12} = 2\sin\alpha \cos\psi$$

Substituting appropriately from Equations 4.27-4.30 into these three equations, we obtain

$$r_{32} - r_{23} = 2a_x \sin\alpha \qquad (4.32)$$

$$r_{13} - r_{31} = 2a_y \sin\alpha \qquad (4.33)$$

$$r_{21} - r_{12} = 2a_z \sin\alpha \qquad (4.34)$$

Now we would like to isolate $\sin\alpha$ from the components of the as yet undetermined vector **a**. To do this, we square and add Equations 4.32-4.34:

$$(r_{32} - r_{23})^2 + (r_{13} - r_{31})^2 + (r_{21} - r_{12})^2 = 4(a_x^2 + a_y^2 + a_z^2)\sin^2\alpha$$

Since **a** is a unit vector, $a_x^2 + a_y^2 + a_z^2 = 1$, and therefore

$$(r_{32} - r_{23})^2 + (r_{13} - r_{31})^2 + (r_{21} - r_{12})^2 = 4\sin^2\alpha$$

or
$$\sin\alpha = \pm\frac{1}{2}\sqrt{(r_{32} - r_{23})^2 + (r_{13} - r_{31})^2 + (r_{21} - r_{12})^2} \qquad (4.35)$$

If we define a positive rotation about **a** as satisfying $0 \le \alpha \le \pi$, then we use the positive root of Equation 4.35, and the angle of rotation is uniquely defined by $\tan\alpha = \sin\alpha/\cos\alpha$, or

$$\alpha = \tan^{-1}\left(\sqrt{(r_{32} - r_{23})^2 + (r_{13} - r_{31})^2 + (r_{21} - r_{12})^2} \Big/ r_{11} + r_{22} + r_{33} - 1\right)$$

This allows us to compute the components of **a** from Equations 4.32-4.34.

$$a_x = \frac{r_{32} - r_{23}}{2\sin\alpha} \qquad (4.36)$$

$$a_y = \frac{r_{13} - r_{31}}{2\sin\alpha} \qquad (4.37)$$

$$a_z = \frac{r_{21} - r_{12}}{2\sin\alpha} \qquad (4.38)$$

We use these solutions with caution, however. As we saw earlier, there are computational pitfalls lurking behind those trigonometric functions. For example, when α approaches 0° or 180°, the magnitude of the numerator and denominator become very small. When $\alpha = 180°$ Equations 4.36-4.38 become 0/0, and we obtain no information whatsoever about the vector **a**. So, as α approaches 180° we must try another line of attack. Using the diagonal elements of Equations 4.25 and 4.26 and substituting appropriately from Equations 4.27 – 4.30, we obtain

RIGID-BODY MOTION

$$a_x = \pm\sqrt{\frac{r_{11} - \cos\alpha}{1 - \cos\alpha}} \tag{4.39}$$

$$a_y = \pm\sqrt{\frac{r_{22} - \cos\alpha}{1 - \cos\alpha}} \tag{4.40}$$

$$a_z = \pm\sqrt{\frac{r_{33} - \cos\alpha}{1 - \cos\alpha}} \tag{4.41}$$

According to conditions we established (i.e., $\alpha < 180°$, and approaching this value), the sine of α must be positive. Therefore, the signs determined by the expressions on the left side of Equations 4.32-4.34 determine the signs for Equations 4.39-4.41. We will use Equations 4.39-4.41 to compute only the largest component of **a** (which will correspond to the equation containing the most positive of the terms r_{11}, r_{22}, r_{33}), and more accurately compute the remaining two components using equations derived by summing pairs of off-diagonal elements of the matrices in Equations 4.25 and 4.26.

$$r_{21} + r_{12} = 2a_x a_y (1 - \cos\alpha)$$
$$r_{32} + r_{23} = 2a_y a_z (1 - \cos\alpha)$$
$$r_{31} + r_{13} = 2a_x a_z (1 - \cos\alpha)$$

Here then is a summary of this last step in determining the components of **a** when α approaches 180°.

If a_x is the largest component, then

$$a_y = \frac{r_{21} + r_{12}}{2a_x(1 - \cos\alpha)} \quad \text{and} \quad a_z = \frac{r_{31} + r_{13}}{2a_x(1 - \cos\alpha)}$$

If a_y is the largest component, then

$$a_x = \frac{r_{21} + r_{12}}{2a_y(1 - \cos\alpha)} \quad \text{and} \quad a_z = \frac{r_{32} + r_{23}}{2a_y(1 - \cos\alpha)}$$

If a_z is the largest component, then

$$a_x = \frac{r_{31} + r_{13}}{2a_z(1 - \cos\alpha)} \quad \text{and} \quad a_y = \frac{r_{32} + r_{23}}{2a_z(1 - \cos\alpha)}$$

There is another way to find the equivalent angle and axis of rotation given any arbitrary proper rotation matrix **R**. It uses eigenvalues and eigenvectors in a mathematically more elegant way, yet still intuitive, and more computationally efficient. The following section discusses this approach.

Eigenvectors and Equivalent Rotations

We know that a rotation in the plane fixes one point. A rotation or a series of rotations in space (say about the origin) fixes a line, the equivalent axis of rotation. We represent this axis by some vector **a**. Given a rotation matrix **R** describing some arbitrary rotation, perhaps the net result of successive rotations about the principal axes, it follows that

$$\mathbf{Ra} = \lambda \mathbf{a}$$

In other words, **a** must be an eigenvector of **R**, so we apply the mathematics of eigenvalues and eigenvectors to find **a** given **R**.

We derive and restate this in terms of classical mechanics, where we consider the physical motions of a rigid body. At any time t we can specify the orientation of a body by an orthogonal transformation. Since the orientation changes with time, the transformation matrix must be a function of time, say $\mathbf{R}(t)$. If we impose a coordinate system fixed with respect to the body (the body system) and make this coincident with the world coordinate system (defining the space in which the body moves), then initially the transformation is simply $\mathbf{R}(0) = \mathbf{I}$. At some later time $\mathbf{R}(t)$ will in general be different from the identity transformation. However, since the rigid motion of a physical body must be continuous, $\mathbf{R}(t)$ must be a continuous function of time. We conclude that the transformation must evolve continuously from the identity transformation.

Now we introduce Euler's theorem, which states: the general displacement of a rigid body with one fixed point is a rotation about some axis. If the fixed point is the origin, then the displacement produces no translation of the body system and only rotational motion is present. Furthermore, we know that it is a characteristic of any rotation in space that one line is fixed, the axis of rotation. any vector along the axis of rotation must have the same components in both the initial and final position of the body, where the orthogonality property preserves the magnitude of the vector. We prove Euler's theorem by showing that there exists a vector **a** having the same components in both systems. This is called the eigenvalue problem and the solution is the eigenvector (s) of **R**. In Chapter 3, we saw the general solution to this. Now we restate Euler's theorem as: The real orthogonal matrix specifying the physical motion of a rigid body with one point fixed has an eigenvalue equal to one.

Given that an eigenvector of **R** for the eigenvalue $\lambda = 1$ is in the direction of the axis, we then compute it as any column vector of the matrix of cofactors of $\mathbf{R} - \mathbf{I}$. We compute the angle of rotation α from the relationship

$$\operatorname{tr}\mathbf{R} = 1 + 2\cos\alpha$$

as we have already seen (Equation 4.31). The sign of α is the same as the sign of the evaluated 3×3 determinant $\det(\mathbf{p}, \mathbf{Rp}, \mathbf{a})$ formed from the column vectors **p**, **Rp**, and **a**, where **p** is any vector not parallel to **a**. The right-hand rule applies. Notice that the sign of α changes if we replace **a** with $-\mathbf{a}$. (We will not develop a general proof here.) We can also compute the eigenvector somewhat more efficiently as any column vector of the symmetric matrix

$$\left(\mathbf{R}+\mathbf{R}^T\right)-\left(\mathrm{tr}\mathbf{R}-1\right)\mathbf{I}$$

Try to work out the proof of this assertion.

Rotation Groups

Given the set of all rotations in the plane about the origin, we immediately see that it forms a group, since the product of any two rotations is obviously another rotation. The same is true for the inverse of any rotation. These conditions are easy to verify. Let

$$\mathbf{A} = \begin{bmatrix} \cos\alpha & -\sin\alpha \\ \sin\alpha & \cos\alpha \end{bmatrix}, \quad \mathbf{B} = \begin{bmatrix} \cos\beta & -\sin\beta \\ \sin\beta & \cos\beta \end{bmatrix}$$

Then the matrix product

$$\mathbf{AB} = \begin{bmatrix} \cos\alpha\cos\beta - \sin\alpha\sin\beta & -\cos\alpha\sin\beta + \sin\alpha\sin\beta \\ \sin\alpha\cos\beta + \cos\alpha\sin\beta & -\sin\alpha\sin\beta + \cos\alpha\cos\beta \end{bmatrix}$$

$$= \begin{bmatrix} \cos(\alpha+\beta) & -\sin(\alpha+\beta) \\ \sin(\alpha+\beta) & \cos(\alpha+\beta) \end{bmatrix}$$

clearly corresponds to a rotation through the angle $\alpha+\beta$, and the inverse of a rotation, say

$$\mathbf{A}^{-1} = \begin{bmatrix} \cos\alpha & \sin\alpha \\ -\sin\alpha & \cos\alpha \end{bmatrix}$$

corresponds to a rotation through $-\alpha$, and satisfies the relation $\mathbf{A}^{-1}\mathbf{A} = \mathbf{A}\mathbf{A}^{-1} = \mathbf{I}$. This, too, we easily verify:

$$\mathbf{AA}^{-1} = \begin{bmatrix} \cos\alpha & -\sin\alpha \\ \sin\alpha & \cos\alpha \end{bmatrix} \begin{bmatrix} \cos\alpha & \sin\alpha \\ -\sin\alpha & \cos\alpha \end{bmatrix}$$

$$= \begin{bmatrix} \cos^2\alpha + \sin^2\alpha & \cos\alpha\sin\alpha - \cos\alpha\sin\alpha \\ \cos\alpha\sin\alpha - \cos\alpha\sin\alpha & \cos^2\alpha + \sin^2\alpha \end{bmatrix}$$

$$= \begin{bmatrix} 1 & 0 \\ 0 & 1 \end{bmatrix}$$

$$= \mathbf{I}$$

Rotations in three dimensions also form a group. We merely need to consider for a moment some of the preceding work. For example, we have demonstrated that any sequence of rotations about the principal axes is equivalent to a rotation about some arbitrary axis, and vice versa. In fact, we can find an equivalent single rotation for any sequence of rotations. Furthermore, Equation 4.13 is the transformation matrix for the rotation sequence ϕ about the z axis, θ about the y axis, and ψ about the x axis. If we reverse this transformation sequence to rotate $-\psi$ about the x axis, $-\theta$ about the y axis, and $-\phi$ about the z axis, then we produce the matrix $\mathbf{R}^{-1}_{\phi\theta\psi}$, where

$$\mathbf{R}^{-1}_{\phi\theta\psi} = \begin{bmatrix} \cos\phi & \sin\phi & 0 \\ -\sin\phi & \cos\phi & 0 \\ 0 & 0 & 1 \end{bmatrix} \begin{bmatrix} \cos\theta & 0 & -\sin\theta \\ 0 & 1 & 0 \\ \sin\theta & 0 & \cos\theta \end{bmatrix} \begin{bmatrix} 1 & 0 & 0 \\ 0 & \cos\psi & \sin\psi \\ 0 & -\sin\psi & \cos\psi \end{bmatrix}$$

or

$$\mathbf{R}^{-1}_{\phi\theta\psi} = \begin{bmatrix} \cos\phi\cos\theta & \cos\phi\sin\theta\sin\psi + \sin\phi\cos\psi & -\cos\phi\sin\theta\cos\psi + \sin\phi\sin\psi \\ -\sin\phi\cos\theta & -\sin\phi\sin\theta\sin\psi + \cos\phi\cos\psi & \sin\phi\sin\theta\cos\psi + \cos\phi\sin\psi \\ \sin\theta & -\cos\theta\sin\psi & \cos\theta\cos\psi \end{bmatrix} \quad (4.42)$$

and we find that

$$\mathbf{R}_{\psi\theta\phi}\mathbf{R}^{-1}_{\phi\theta\psi} = \mathbf{I}$$

This demonstrates the inverse property of rotations in three dimensions. There are rotation subgroups, important to many areas of physics and engineering that we will investigate in the next chapter.

Finally, because of the orthogonality of the rotation matrix ($|\mathbf{R}| = 1$ and $\mathbf{R}^{-1} = \mathbf{R}^T$), it is very easy to find $\mathbf{p}$ given $\mathbf{p}' = \mathbf{R}\mathbf{p}$; indeed, we immediately see that

$$\mathbf{p} = \mathbf{R}^T\mathbf{p}'$$

This simply means that if we have for a given rotation transformation
$x' = r_{11}x + r_{12}y + r_{13}z$
$y' = r_{21}x + r_{22}y + r_{23}z$
$z' = r_{31}x + r_{32}y + r_{33}z$

then by interchanging the rows and columns of the system of coefficients we obtain
$x = r_{11}x' + r_{21}y' + r_{31}z'$
$y = r_{12}x' + r_{22}y' + r_{32}z'$
$z = r_{13}x' + r_{23}y' + r_{33}z'$

Rotations and Quaternions

Given a matrix that defines a general rotation in three dimensions, we have seen that it is somewhat difficult to extract from it an equivalent rotation axis and angle. Quaternions, invented and described by W. R. Hamilton in 1843, suggest an alternative that is simpler and more intuitive. Hamilton worked to create and investigate "hyper numbers" ... the algebra of number pairs, triplets, and quadruples. His efforts resulted in quaternions. Gibbs and Heaviside subsequently refined and extended Hamilton's work, producing what we now know as scalars and vectors.

Let's look at quaternions a little more closely, and then develop their connection with rotations. A real quaternion, Q, is represented by the expression $a + b\mathbf{i} + c\mathbf{j} + d\mathbf{k}$, where a, b, c, and d are real coefficients. The word *quaternion* (from the Latin word *quaternion* meaning set of four) sometimes refers to the set $\{\pm 1, \pm \mathbf{i}, \pm \mathbf{j}, \pm \mathbf{k}\}$. The following properties apply

$$\mathbf{i}^2 = \mathbf{j}^2 = \mathbf{k}^2 = \mathbf{ijk} = -1 \quad \text{and} \quad \begin{array}{ll} \mathbf{ij} = \mathbf{k} & \mathbf{ji} = -\mathbf{k} \\ \mathbf{jk} = \mathbf{i} & \mathbf{kj} = -\mathbf{i} \\ \mathbf{ki} = \mathbf{j} & \mathbf{ik} = -\mathbf{j} \end{array}$$

The multiplication of two quaternions is not commutative. Quaternion addition proceeds term-by-term, and quaternions preserve the other arithmetic properties associated with real and complex numbers. The set of quaternions for which $\mathbf{i}$, $\mathbf{j}$, and $\mathbf{k} = 0$ is equivalent to the set of real numbers. The set of quaternions for which only one of $\mathbf{i}$, $\mathbf{j}$, and $\mathbf{k}$ is non-zero is equivalent to the set of complex numbers.

A quaternion whose first term equals zero is equivalent to a vector. Given the quaternion $Q = a + b\mathbf{i} + c\mathbf{j} + d\mathbf{k}$: if $a = 0$, then we may associate Q with the three-dimensional vector $\mathbf{q}$, where $\mathbf{i}, \mathbf{j}$, and $\mathbf{k}$ are interpreted as unit vectors along the coordinate axes. Thus, the quaternion $Q = 0 + x\mathbf{i} + y\mathbf{j} + z\mathbf{k}$ represents the vector $\mathbf{q} = x\mathbf{i} + y\mathbf{j} + z\mathbf{k}$. This means that $a + k\mathbf{q}$ is the quaternion $Q = a + kx\mathbf{i} + ky\mathbf{j} + kz\mathbf{k}$, or $Q = a + k\mathbf{q}$.

Let Q and R be two quaternions that represent the vectors $\mathbf{q}$ and $\mathbf{r}$, respectively. Then the product of Q and R is

$$QR = -\mathbf{q} \bullet \mathbf{r} + \mathbf{q} \times \mathbf{r}$$

where $\mathbf{q} \bullet \mathbf{r}$ and $\mathbf{q} \times \mathbf{r}$ are the scalar and vector products, respectively.

It is useful to think of a quaternion as the formal sum $a + \mathbf{r}$ of a real number, a, and a vector, $\mathbf{r}$. The sum of two quaternions $A = a + \mathbf{r}$ and $B = b + \mathbf{s}$ is another quaternion $C = c + \mathbf{t}$, whose scalar part c is simply $a + b$, and whose vector part $\mathbf{t}$ is $\mathbf{r} + \mathbf{s}$. Thus

$$A + B = (a + \mathbf{r}) + (b + \mathbf{s}) = (a + b) + (\mathbf{r} + \mathbf{s})$$

We define the product of two quaternions as follows

$$AB = (a+\mathbf{r})(b+\mathbf{s})$$
$$= (ab - \mathbf{r} \bullet \mathbf{s}) + (a\mathbf{s} + b\mathbf{r} + \mathbf{r} \times \mathbf{s})$$

Here, the presence of the term $\mathbf{r} \times \mathbf{s}$ tells us that the product AB is not commutative. That is $AB \neq BA$.

Unlike vectors, quaternion algebra permits inverses. We generate the inverse Q^{-1} of Q by subtracting the vector part of Q from the scalar part and dividing this by the square of the modulus of Q. Subtracting the vector part of Q from the scalar part produces $C(Q)$, the conjugate of Q. This is analogous to the conjugate of a complex number. If m is the modulus of Q, then

$$Q^{-1} = \frac{C(Q)}{m^2}$$

where $m^2 = a^2 + b^2 + c^2 + d^2$.

We can use a quaternion to define a rotation in space about an axis through the origin. The nature of quaternion algebra allows us to define sequences of rotations as products of quaternions, permitting the direct extraction of the equivalent or resultant angle and axis of rotation. We may also use quaternions to define rotations whose axes do not pass through the origin, although the procedure here involves translating the axis so that it passes through the origin, just as we did above for the typical rotation matrix (the extension to quaternions is obvious, so we won't discuss it here).

Let the unit vector $\mathbf{a}$ represent the axis of rotation passing through the origin, and let α represent the angle of rotation about this axis. We assert that this rotation is, in turn, represented by the quaternion

$$R = \cos(\alpha/2) + \mathbf{a}\sin(\alpha/2)$$

To rotate a point $\mathbf{p}$ through α about $\mathbf{a}$, perform the quaternion multiplication

$$P' = RPR^{-1}$$

where P is the quaternion representing $\mathbf{p}$ (i. e., the scalar part of P is zero and the vector part is simply $\mathbf{p}$), and P' is the rotated quaternion from which we extract $\mathbf{p}'$.

Although we will not formally prove the validity of the two expressions above, it is easy to show that when P represents a vector and R a rotation, that RPR^{-1} also represents a vector. Simply carry out the indicated quaternion multiplication and inspect the form of P'. Also, compare RPR^{-1} to the standard rotation product $\mathbf{Rp}$ that we developed earlier in this chapter.

We easily extend the above process to a sequence of rotations about axes all of which pass through the origin. Let the quaternions R and S define two distinct rotations. Under R point $\mathbf{p}$ rotates to RPR^{-1}, followed by rotation S that takes $\mathbf{p}$ to $S(RPR^{-1})S^{-1}$. Quaternion multiplication is associative, so that

$$S(RPR^{-1})S^{-1} = (SR)P(R^{-1}S^{-1})$$

We note that $R^{-1}S^{-1} = (SR)^{-1}$, since it is obviously true that

$$(SR)(R^{-1}S^{-1}) = S(RR^{-1})S^{-1} = S(1)S^{-1} = SS^{-1} = 1$$

We now assert that $S(RPR^{-1})S^{-1} = TPT^{-1}$, where $T = SR$ and T represents the rotation equivalent to the sequence stated as: rotate by R first then by S. From this we conclude that any sequence of rotations about axes passing through the origin is equivalent to a single rotation.

The components of the equivalent rotation given by T we extract, as you might expect, as follows: The scalar component of T is, of course, the cosine of one-half the rotation angle. The vector component of T defines the axis of rotation (although not the unit vector along the axis). If we divide each of the vector components of T by the sine of one-half the angle of rotation, then we obtain the components of the unit vector lying along the axis of rotation.

Section 4.2 Exercises

1. Reverse the order of the terms on the right side of Equation 4.4. Give the appropriate matrix equation.

2. Find the rotation matrix for rotations 30°, 45°, 90° and 180° about origin.

3. Compute the determinant of each of the rotation matrices found for Exercise 2, above.

4. If a line is defined by a vector equation $\mathbf{p} = \mathbf{p}_0 + u\mathbf{p}_1$, where $\mathbf{p}_0$ is any point on the line and $\mathbf{p}_1$ is a vector parallel to the line, then find the vector equation of this line after it has been rotated by $\mathbf{R}$.

5. Find the general algebraic (non-vector) equations for rotating a point x, y through an angle θ about x_c, y_c.

6. Consider successive rotations about two different points. Is the outcome independent of the order in which we perform the rotations? Explain your answer.

7. Verify Equation 4.13 by setting
   a. $\alpha = \beta = 0$
   b. $\alpha = \gamma = 0$
   c. $\beta = \gamma = 0$

8. Rewrite Equation 4.23 for an arbitrary axis passing through the origin.

9. Rewrite Equation 4.23 for the case where the arbitrary axis passes through the origin and lies in the *xy* plane.

10. Prove that $|AB|=|BA|=|A||B|=|B||A|$ for any two square matrices **A** and **B**. (Note that this is true for square matrices of any order.)

11. Find the equation of the image of each of the following curves under the rotation transformation $x'=(x-y)/\sqrt{2}$, $y'=(x+y)/\sqrt{2}$. Find the angle of rotation, ϕ.

    a. $y=0$

    b. $y=x$

    c. $y=x+3$

    d. $x=3$

    e. $x^2+y^2=1$

    f. $xy=1$

    g. $(x-1)^2+y^2=2$

    h. $y=x^2$

12. Show that for a series of points p_i and a given rotation **R**, Equation 4.7 looks like a rotation of each p_i plus a constant translation.

13. Show that the sum of the squares of any row or column of a proper rotation matrix is equal to one. This is the distinguishing characteristic of a normalized matrix.

14. Show that the scalar product of any pair of rows (or columns) of a proper rotation matrix is zero. A matrix with this property is orthogonal.

15. Show that for the convention representing points or vectors as column matrices that the columns of **R** represent the components in the initial coordinate system of unit vectors along the coordinate axes of the rotated system.

16. Show that for the convention representing points or vectors as column matrices that the rows of **R** represent the components in the rotated coordinate system of unit vectors along the axes of the original system.

17. Find the rotation matrix for very small angles of rotation. Do this for two- and three-dimensional rotations.

4.3 Composite Motion

When a geometric object undergoes a rigid-body motion that is the result of a sequence of rotations and translations, the resulting net transformation is a composite motion. We generate the simpler swept objects this way and execute rotation about an arbitrary point in the plane or about an arbitrary axis in space this way also. In fact, we can represent almost all the affine transformations we have encountered, including rotation, reflection, scaling, and shear, as linear homogeneous equations of the form $\mathbf{p}' = \mathbf{A}\mathbf{p}$.

We define each of these transformations relative to the origin of the coordinate system, and combine a sequence of such transformations simply as a series of matrix multiplication. Translation is the exception. To rotate an object about an axis not through the origin, then we combine it with appropriate translations. This destroys the homogeneous character of the transformation and requires us to alter the equation to a nonhomogeneous form; for example, $\mathbf{p}' = \mathbf{A}\mathbf{p} + \mathbf{t}$. A sequence of transformations of this type quickly becomes a nested mess of matrix additions and multiplication. If we can find a way to incorporate translations into a form such as that of the homogeneous affine matrix $\mathbf{A}$, then we can express a mixed sequence of transformations of all types, including translations, as a product of the matrices describing these transformations, producing a single equivalent matrix.

Homogeneous Coordinates

Since it is not possible to express a general linear transformation of the plane or space in matrix form where the transformation matrix is a square matrix of order two or three, respectively, we must use homogeneous coordinates. We begin to solve this problem by introducing an extended point or position vector:

$$\mathbf{r} = \begin{bmatrix} x \\ y \\ z \\ 1 \end{bmatrix} \tag{4.43}$$

This allows us to express a translation transformation $\mathbf{r}' = \mathbf{r} + \mathbf{t}$ in matrix form as $\mathbf{r}' = \mathbf{T}\mathbf{r}$, where $\mathbf{T}$ must be

$$\mathbf{T} = \left[\begin{array}{ccc|c} 1 & 0 & 0 & t_x \\ 0 & 1 & 0 & t_y \\ 0 & 0 & 1 & t_z \\ \hline 0 & 0 & 0 & 1 \end{array} \right] \tag{4.44}$$

To make this approach even more useful we introduce yet another modification to the position vector. Let $\mathbf{p}_h = h\mathbf{r}$. We call the components (hx, hy, hz, h) of the vector $\mathbf{p}_h$ the homogeneous coordinates of the point $\mathbf{p}$ and extract the values of the Cartesian coordinates x, y, and z from the vector $\mathbf{p}_h$ by division, so that

$$x = \frac{hx}{h}, \quad y = \frac{hy}{h}, \quad z = \frac{hz}{h}$$

Introducing the scalar h allows us to use this format to describe certain projective transformations as well. The ramifications of this are explored in Chapter 7. For now we assume $h = 1$ unless indicated otherwise.

Notice that if we use homogeneous coordinates, then the representation of a point in Cartesian space is no longer unique. For example, the homogeneous coordinates $(6, 10, 2, 2)$ and $(9, 15, 3, 3)$ both represent the same point $(3, 5, 1)$ in three-dimensional Cartesian space. Furthermore, we assert that any coordinate system representing a point in three-dimensional space by 4 coordinates, where any scalar multiple $\lambda \mathbf{r}$ of a given vector $\mathbf{r}$ represents the same point in 3-space, is a homogeneous coordinate system. Clearly, we easily generalize this with the following observation: The homogeneous coordinates of a point in n-dimensional space consist of $n+1$ numbers.

The Homogeneous Transformation Matrix

We use a 4×4 matrix, $\mathbf{H}$, to describe all the three-dimensional transformations we have encountered thus far. This is the homogeneous transformation matrix, and it applies to all affine transformations. Let's drop the subscript on the vector $\mathbf{p}_h$; and, with $h = 1$, assume that the vector $\mathbf{p}$ represents either the matrix of homogeneous coordinates $\begin{bmatrix} x & y & z & 1 \end{bmatrix}^T$ or the matrix of Cartesian coordinates $\begin{bmatrix} x & y & z \end{bmatrix}^T$, depending on the algebraic or conceptual context. $\mathbf{H}$ becomes a 3×3 matrix for transformations in the plane (the analogy is direct and obvious). Now we write

$$\mathbf{p}' = \mathbf{H}\mathbf{p}$$

For the homogeneous affine transformations, $\mathbf{A}_h$, we have

$$\mathbf{A}_h = \left[\begin{array}{ccc|c} a_{11} & a_{12} & a_{13} & 0 \\ a_{21} & a_{22} & a_{23} & 0 \\ a_{31} & a_{32} & a_{33} & 0 \\ \hline 0 & 0 & 0 & 1 \end{array}\right] \tag{4.45}$$

where $\mathbf{A} = \begin{bmatrix} a_{11} & a_{12} & a_{13} \\ a_{21} & a_{22} & a_{23} \\ a_{31} & a_{32} & a_{33} \end{bmatrix}$ from the equations $\begin{cases} x' = a_{11}x + a_{12}y + a_{13}z \\ y' = a_{21}x + a_{22}y + a_{23}z \\ z' = a_{31}x + a_{32}y + a_{33}z \end{cases}$

RIGID-BODY MOTION

For rotations,

$$\mathbf{R} = \left[\begin{array}{ccc|c} r_{11} & r_{12} & r_{13} & 0 \\ r_{21} & r_{22} & r_{23} & 0 \\ r_{31} & r_{32} & r_{33} & 0 \\ \hline 0 & 0 & 0 & 1 \end{array}\right] \tag{4.46}$$

Here is an example: Find the resultant homogeneous transformation matrix describing the rotation ϕ of any point $\mathbf{p}$ in the plane about the point $\mathbf{p}_c$. First, translate the point $\mathbf{p}$ (the entire coordinate plane goes with it) so that $\mathbf{p}_c$ is at the origin, producing

$$\mathbf{p}' = \begin{bmatrix} 1 & 0 & -x_c \\ 0 & 1 & -y_c \\ 0 & 0 & 1 \end{bmatrix} \mathbf{p}$$

where $\mathbf{p} = \begin{bmatrix} x & y & 1 \end{bmatrix}^T$ and $\mathbf{p}' = \begin{bmatrix} x' & y' & 1 \end{bmatrix}^T$.

Next, rotate the result through ϕ about the origin, producing

$$\mathbf{p}' = \begin{bmatrix} \cos\phi & -\sin\phi & 0 \\ \sin\phi & \cos\phi & 0 \\ 0 & 0 & 1 \end{bmatrix} \begin{bmatrix} 1 & 0 & -x_c \\ 0 & 1 & -y_c \\ 0 & 0 & 1 \end{bmatrix} \mathbf{p}$$

Finally, translate the rotated plane back to its original position:

$$\mathbf{p}' = \begin{bmatrix} 1 & 0 & x_c \\ 0 & 1 & y_c \\ 0 & 0 & 1 \end{bmatrix} \begin{bmatrix} \cos\phi & -\sin\phi & 0 \\ \sin\phi & \cos\phi & 0 \\ 0 & 0 & 1 \end{bmatrix} \begin{bmatrix} 1 & 0 & -x_c \\ 0 & 1 & -y_c \\ 0 & 0 & 1 \end{bmatrix} \mathbf{p}$$

If we carry out the indicated matrix multiplication, then we should find that

$$\mathbf{p}' = \begin{bmatrix} \cos\phi & -\sin\phi & -x_c\cos\phi + y_c\sin\phi + x_c \\ \sin\phi & \cos\phi & -x_c\sin\phi - y_c\cos\phi + y_c \\ 0 & 0 & 1 \end{bmatrix} \mathbf{p}$$

In terms of the coordinates,

$$x' = (x - x_c)\cos\phi - (y - y_c)\sin\phi + x_c$$
$$y' = (x - x_c)\sin\phi + (y - y_c)\cos\phi + y_c$$

If $x_c = y_c = 0$, then the equations describe a simple rotation about the origin.

This technique allows us to concatenate a sequence of transformations, each expressed as a homogeneous transformation matrix, into a single matrix product. Thus, for a sequence of n transformations we have

$$\begin{aligned} \mathbf{p}' &= \mathbf{H}_n \mathbf{H}_{n-1} \ldots \mathbf{H}_2 \mathbf{H}_1 \mathbf{p} \\ &= \mathbf{H}_c \mathbf{p} \end{aligned} \qquad (4.47)$$

where $\mathbf{H}_c$ is the resulting product of the sequence of matrix multiplications, representing a composite motion. Again, the order of matrix multiplication is important in this process.

The Screw Transformation

The screw transformation is an interesting special case in three dimensions, consisting of a translation and a rotation about an axis parallel to the direction of translation. There is nothing comparable to it in the plane. Its transformation matrix looks like this:

$$\mathbf{H}_\phi = \begin{bmatrix} \cos\phi & -\sin\phi & 0 & 0 \\ \sin\phi & \cos\phi & 0 & 0 \\ 0 & 0 & 1 & k\phi \\ 0 & 0 & 0 & 1 \end{bmatrix}$$

and it produces these transformation equations:

$$\begin{aligned} x' &= x\cos\phi - y\sin\phi \\ y' &= x\sin\phi + y\cos\phi \\ z' &= z + k\phi \end{aligned}$$

We interpret this as being analogous to turning a screw through an angle ϕ, where k corresponds to the pitch of the screw. If we increase k, then the screw advances farther for a given angular turn, ϕ. Of course, the screw axis can lie along any line, but we must adjust the transformation matrix accordingly. "Twist" or "glide rotation" are other terms we use to describe this transformation.

Composite motion is an essential ingredient of kinematics, our next subject of study. Beyond that, composite motion is useful to us for generating some of the simpler swept shapes (see the next section).

Sweeps

Now we can begin to think of some powerful applications of composite motion transformations and matrix methods. Geometric sweep transformations are a good example. Here we execute a continuous sequence of composite transformations on some generator shape to produce a geometric sweep. A surface of revolution is a sweep, or swept surface, that we create by rotating a line or curve about some axis.

RIGID-BODY MOTION

The following examples are quite simple, but more complex situations are direct and easy extensions.

Given a generator shape defined by a set of control points $\mathbf{p}_i$ and a transformation expressed as a continuous function $\mathbf{T}(u)$ of some independent parametric variable, u, then the set of points $\mathbf{p}'_i$ define the swept shape, where $\mathbf{p}'_i = \mathbf{T}(u)\mathbf{p}_i$. In the figure (Figure 4.21),

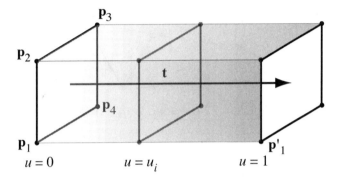

Figure 4.21 A sweep.

the four points $\mathbf{p}_i = \mathbf{p}_1, \mathbf{p}_2, \mathbf{p}_3, \mathbf{p}_4$ are sufficient to define a rectangular generator shape. The variable vector $u\mathbf{t}$, where $u \in [0,1]$, defines a sequence of translation transformations that in turn produce the swept surface of a rectangular tube. We then have

$$\mathbf{p}'_i = \mathbf{T}(u)\mathbf{p}_i = \begin{bmatrix} 1 & 0 & 0 & ut_x \\ 0 & 1 & 0 & ut_y \\ 0 & 0 & 1 & ut_z \\ 0 & 0 & 0 & 1 \end{bmatrix} \mathbf{p}_i \qquad i = 1, 4$$

Of course, we can extend these ideas to include rotations and combinations of translations and rotations. Look at what happens when we uniformly rotate the rectangle about the x axis at the same time we translate it along that axis (Figure 4.22). We have $t_y = t_z = 0$ and

$$\mathbf{p}'_i = \begin{bmatrix} 1 & 0 & 0 & ut_x \\ 0 & \cos(u\psi) & -\sin(u\psi) & 0 \\ 0 & \sin(u\psi) & \cos(u\psi) & 0 \\ 0 & 0 & 0 & 1 \end{bmatrix} \mathbf{p}_i$$

Note that nothing prevents $\mathbf{T}(u)$ from being a matrix of nonlinear functions of u.

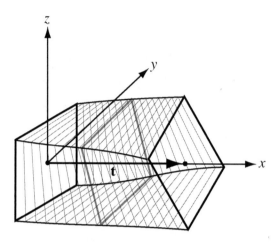

Figure 4.22 A sweep with a twist.

Section 4.3 Exercises

1. Given a resultant homogeneous transformation **H**, partition this matrix as in Equation 4.50. Show that the 3×3 submatrix in the upper left partition is always orthogonal.

2. Show that the determinant of every translation transformation in homogeneous coordinates is equal to one when expressed as Equation 4.49.

3. Show that the product of a translation **H** and its inverse $\mathbf{H}^{-1}$ produces the identity matrix **I** (use homogeneous coordinate matrices).

4.4 Kinematics

Visualize a collection of rigid solid bodies joined so that only certain constrained motions are possible. Such an arrangement is a mechanism. Kinematics is the study of the motion of a mechanism. A serial link mechanism (this could be the manipulator arm of a robot, for example) is a sequence of links connected by actuated joints. We will examine only a simple, highly idealized serial link mechanism. However, keep in mind that mechanisms with extremely complex joint geometries and motions are possible.

A serial link mechanism with n links and n joints has n degrees of freedom. Usually we define a reference base with Link 1 connected to it through Joint 1. There is no joint at the end of the last link. The rigid links maintain a fixed geometric relationship between the joints at each end of the link. We face two problems: first, given a set of joint motions (expressed as relative transformations), find the resultant position of the end of the last link; second, given an end position, find an appropriate corresponding set of joint motions.

RIGID-BODY MOTION

Here is a simple example, consisting of a planar assembly of two links and two joints constrained to joint rotations in the plane (Figure 4.23).

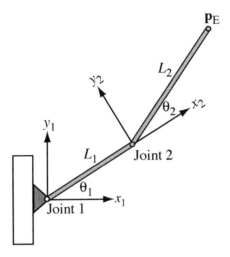

Figure 4.23 An articulated mechanism.

Our first job is to express the relationships between the links. To do this we establish a coordinate frame for each link. The frame for Link 1 we fix relative to the base with its origin at Joint 1. The frame for Link 2 we fix relative to Link 1 with its origin at Joint 2 and its x axis collinear with the length of Link 1. The relationship between Link 1 and Link 2 we express as a product of homogeneous transformations, $\mathbf{H}_1$, relating the coordinate frame of Link 2 to the coordinate frame of Link 1. For our example, this is simply the product of a rotation and translation. Thus,

$$\mathbf{H}_1 = [\text{Rotation } \theta_1][\text{Translation } L_1]$$

$$= \begin{bmatrix} \cos\theta_1 & -\sin\theta_1 & 0 \\ \sin\theta_1 & \cos\theta_1 & 0 \\ 0 & 0 & 1 \end{bmatrix} \begin{bmatrix} 1 & 0 & L_1 \\ 0 & 1 & 0 \\ 0 & 0 & 1 \end{bmatrix}$$

Finally, we must relate the position of the end of Link 2, E, to the coordinate frame of Link 2. This is nothing more than the product, $\mathbf{H}_2$, of rotation and translation matrices describing motion at Joint 2.

$$\mathbf{H}_2 = [\text{Rotation } \theta_2][\text{Translation } L_2] = \begin{bmatrix} \cos\theta_2 & -\sin\theta_2 & 0 \\ \sin\theta_2 & \cos\theta_2 & 0 \\ 0 & 0 & 1 \end{bmatrix} \begin{bmatrix} 1 & 0 & L_2 \\ 0 & 1 & 0 \\ 0 & 0 & 1 \end{bmatrix}$$

It follows that the position of E relative to Frame 1, or the base, is

$$\mathbf{p}_E = \mathbf{H}_1 \mathbf{H}_2 \mathbf{p}_0 \tag{4.48}$$

where we set $\mathbf{p}_0 = \begin{bmatrix} 0 & 0 & 1 \end{bmatrix}^T$. (Do you see why?)

In three dimensions, we might encounter links whose relative joint rotation axes are skew lines in space, perhaps with telescoping lengths and joints that twist. The number of potential motion variables is almost unlimited. We express the position of the end as a generalization of Equation 4.48:

$$\mathbf{p}_E = \mathbf{H}_1 \mathbf{H}_2 \ldots \mathbf{H}_{n-1} \mathbf{H}_n \mathbf{p}_0 \text{ or } \mathbf{p}_E = \prod \mathbf{H}_i \mathbf{p}_0 \qquad (4.49)$$

Here, however, the $\mathbf{H}_i$ are certainly more complex than a rotation and translation.

Given the end position, E, there are usually many sets of joint motions possible to achieve it. In Figure 4.24, we see that there are two possibilities for the 2-link mechanism described above. As you would imagine, the number of sets of joint motions that will produce a given end position increases rather rapidly as the number and complexity of the individual joints increase.

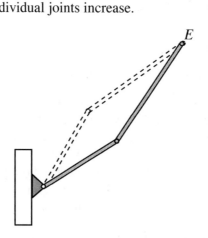

Figure 4.24 Multiple joint positions for a given end position.

A central problem in mechanism control is of obtaining a solution for the joint coordinates given any net transformation $\mathbf{H}$. We usually know where we want to position the end of the mechanism, but we must find the appropriate joint coordinates describing the movement to achieve this position. Thus, given any transformation $\mathbf{H}$, find the Euler angles ϕ, θ, and ψ. In other words, given the numerical values of the elements of $\mathbf{H}$, what are the corresponding values of ϕ, θ, and ψ? Recall that we have already solved a similar problem: Given $\mathbf{R}$, find ϕ, θ, and ψ. The presence of translations is only a minor complication.

5 REFLECTION AND SYMMETRY

Reflections and the more complex symmetry transformations differ from rigid body motions only because parity or handedness may be reversed, as in a mirror reflection. However, the size and shape of an object or figure do not change under these transformations, except in special cases such as oblique reflections. In fact, we saw in Chapter 2 that an appropriate set of reflections will generate any rigid motion.

There are many reasons for studying the reflection transformations. For example, only by supplementing the direct isometries (rigid motions) with reflections can we generate all the crystal structures that we find in nature. Reflection transformations also play a significant role in our study of symmetry, where centers of symmetry and lines of symmetry are a natural result. Consequently, the mathematics of reflections has considerable aesthetic appeal. As we shall see, reflections are orientation reversing; therefore, we must exclude them from the family of allowable topological deformations. The distinguishing characteristic of a reflection is that the determinant of its transformation matrix, $\mathbf{R}_f$, is equal to minus one:

$$|\mathbf{R}_f| = -1 \tag{5.1}$$

and this is what tells us that every reflection is an orientation reversing one.

Reflection transformations fixing a plane in space produce what we call a mirror image. Reflection in the plane and in space can also fix a point or a line. If it fixes a point, it is a central inversion or, more simply, an inversion. We began our study of reflections in Chapter 2, using synthetic definitions and Cartesian equations. Now we will extend this work, showing how vectors, matrices, and homogeneous coordinates greatly simplify the representation and analysis of these transformations.

The concept of symmetry evokes images of decorative patterns, tilings, balance in design, and repetitive shapes. These images are part of our everyday intuitive, qualitative experience of symmetry. They are some of the important ingredients with which we form aesthetic judgments. Symmetry as an idea to study and contemplate was once thought to be unrelated to the more quantitative and concrete forms of mathematical and scientific thinking; but, of course, one of the great achievements of nineteenth century mathematics was the discovery of the intimate relationship between mathematical geometry and symmetry. For example, the concept of a group arises in the study of symmetries of mathematical and scientific objects. In addition, with the growth of quantum mechanics and elementary particle physics, we find that the conservation laws of physics are closely related to the concept of the symmetry of space and time, and in certain dynamic processes where symmetry is "broken." Regarding Einstein's theory of relativity, H. Weyl has observed that it is the inherent symmetry of the four-dimensional spacetime continuum that relativity elaborates.

The dodecahedron has many axes of symmetry. For example, the figure shows one of its five-fold axes of rotational symmetry. Rotation of the dodecahedron around this axis in 72-degree increments leaves it unchanged. We will explore this and other forms of polyhedral symmetry in detail in this chapter.

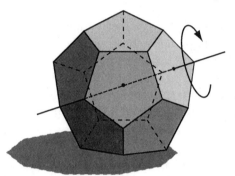

The history of science, and particularly that of physics, has even more to tell us about the importance of symmetry to physics. Einstein's 1905 paper on relativity theory is titled *On the Electrodynamics of Moving Bodies*. In the very first sentence of this paper Einstein states, "It is known that Maxwell's electrodynamics–as usually understood at the present time–when applied to moving bodies, leads to asymmetries which do not appear to be inherent in the phenomena." In other words, Maxwell's theory does not reflect the symmetry demonstrably inherent in the phenomena. It was this incorrect or incomplete understanding as revealed by a lack of symmetry that led Einstein to an alternative interpretation of the phenomena and to what we now call the theory of relativity.

It is important to note that in his correspondence during the first two years after his revolutionary 1905 paper, Einstein always referred not to his "theory of relativity" but to an "*invariantentheorie*" (invariation theory). The theory of relativity is the logical result of Einstein's insistence that the laws of physics should remain invariant for all observers under uniform motion. In fact, symmetry and invariance are complementary notions. We now recognize that an object or physical law, and so forth, is symmetrical to the extent that it is invariant under certain kinds of transformations. It turns out that symmetry-breaking is also important in almost all the sciences, including physics, chemistry, and biology. Here we discover dynamic processes wherein symmetry is either created or destroyed...a fascinating subject in its own right. We go beyond the idea of symmetry as a property of geometric figures to the idea that symmetry is, of course, properly thought of as a transformation. That certain transformations can be applied to certain figures without changing them is a property of the figure. These properties are often easily detected visually.

Reflection and Symmetry

We will consider reflection transformations and variations of them under the following major categories: central inversion, reflection in the plane, and reflection in space. Then, after a brief review of the general mathematics of symmetry transformations and symmetry groups, we'll study several special topics, including the ornamental groups, polygonal symmetry and tilings, and polyhedral symmetry.

5.1 Central Inversion

Recall the earlier synthetic definition of an inversion: Given some point M, the inversion of P through M produces P' such that M is the midpoint of the line PP'. Conversely, given two arbitrary points A and B, there is exactly one reflection σ_M such that $\sigma_M(A) = B$ or $\sigma_M(B) = A$. The axis of reflection (the line fixed by the reflection) is the perpendicular bisector of line segment AB. The term inversion has other meanings, as well, so we use the terms central inversion or point reflection to distinguish this transformation from others, when necessary.

A central inversion fixes the point of inversion and all lines through it, but not point-wise (Exercise 1). This transformation also fixes certain types of repeating figures. For example, if the point of inflection of a sine wave is coincident with the point of inversion then the transformation fixes the sine wave; that is, it maps it onto itself.

Inversion Fixing the Origin

These equations describe the inversion of a point through the origin:

$$x' = -x$$
$$y' = -y \qquad (5.2)$$

Clearly, this transformation does fix the origin (Figure 5.1a).

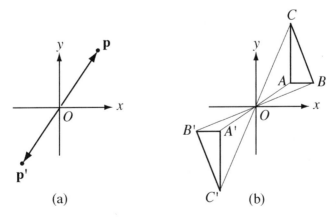

Figure 5.1 Inversion of a point in the plane through the origin.

It is also very easy to rewrite this set of equations as a single matrix equation:

$$\mathbf{p}' = \mathbf{R}_f \mathbf{p} \tag{5.3}$$

where $\quad \mathbf{R}_f = \begin{bmatrix} -1 & 0 & 0 \\ 0 & -1 & 0 \\ 0 & 0 & 1 \end{bmatrix}$ (5.4)

and $\quad \mathbf{p}' = \begin{bmatrix} x' & y' & 1 \end{bmatrix}^T, \mathbf{p} = \begin{bmatrix} x & y & 1 \end{bmatrix}^T$.

We immediately notice that $|\mathbf{R}_f| = +1$ and that this transformation preserves the orientation of triangle ABC in Figure 5.1b. What is going on here? Before answering this we first look at an inversion in space, which the following equations describe (Figure 5.2):

$$\begin{aligned} x' &= -x \\ y' &= -y \\ z' &= -z \end{aligned} \tag{5.5}$$

or in matrix form

$$\mathbf{p}' = \mathbf{R}_f \mathbf{p} \tag{5.6}$$

where $\quad \mathbf{R}_f = \begin{bmatrix} -1 & 0 & 0 & 0 \\ 0 & -1 & 0 & 0 \\ 0 & 0 & -1 & 0 \\ 0 & 0 & 0 & 1 \end{bmatrix}$ (5.7)

and $\quad \mathbf{p}' = \begin{bmatrix} x' & y' & z' & 1 \end{bmatrix}^T, \mathbf{p} = \begin{bmatrix} x & y & z & 1 \end{bmatrix}^T$.

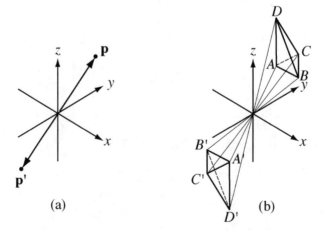

Figure 5.2 Inversion of a point in space through the origin.

REFLECTION AND SYMMETRY

The matrix forms of the inversion equations in two and three dimensions look the same (Equations 5.3 and 5.6). This demonstrates the power of this notation system. However, look at what happens to the determinant of $\mathbf{R}_f$.

$$|\mathbf{R}_f| = -1 \tag{5.8}$$

Obviously, this time the transformation does reverse the orientation of a figure (Figure 5.2b). If we investigate inversions in higher dimensions, then we will soon conclude that central inversion is a reflection only in spaces of an odd number of dimensions. If the number of dimensions is even, then the transformation produces a rotation through 180°, or a halfturn. Is central inversion an isometry? Yes, and here is a brief proof of this:

Central inversion preserves the distances between corresponding pairs of points (Figure 5.3). If this is true, then $|\mathbf{p}'_2 - \mathbf{p}'_1| = |\mathbf{p}_2 - \mathbf{p}_1|$ or

$$\sqrt{(x'_2 - x'_1)^2 + (y'_2 - y'_1)^2 + (z'_2 - z'_1)^2} = \sqrt{(x_2 - x_1)^2 + (y_2 - y_1)^2 + (z_2 - z_1)^2}$$

Substitute $x'_1 = -x_1$, $x'_2 = -x_2, \ldots, z'_2 = -z_2$ into the equation above so that

$$\sqrt{(-x_2 + x_1)^2 + (-y_2 + y_1)^2 + (-z_2 + z_1)^2} = \sqrt{(x_2 - x_1)^2 + (y_2 - y_1)^2 + (z_2 - z_1)^2}$$

Obviously, $(-x_2 + x_1)^2 = (x_2 - x_1)^2$, and similarly for y and z. This means that the initial assertion is true, and a central inversion is indeed an isometry.

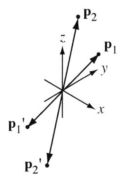

Figure 5.3 Central inversion is an isometry.

Inversion Fixing an Arbitrary Point in Space

The idea of an inversion that fixes an arbitrary point, not necessarily the origin, is easy to understand when we express this transformation in terms of vectors (Figure 5.4). Let the vector $\mathbf{p}_I$ denote the arbitrary point of inversion, then

$$\mathbf{p}' = \mathbf{p} + 2(\mathbf{p}_I - \mathbf{p}), \text{ or } \mathbf{p}' = -\mathbf{p} + 2\mathbf{p}_I \tag{5.9}$$

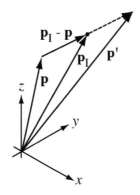

Figure 5.4 Inversion through an arbitrary point.

The product of three transformation matrices yields the same result. First, translate $\mathbf{p}$ by $-\mathbf{p}_I$, in effect placing $\mathbf{p}_I$ at the origin. Second, perform the central inversion fixing the origin. Finally, reverse the first translation, producing $\mathbf{p}'$, which is the image of $\mathbf{p}$ under inversion in $\mathbf{p}_I$. Let $\mathbf{T}_I$ and $\mathbf{T}_I^{-1}$ denote the translation matrix and its inverse, respectively, and let $\mathbf{R}_f$ denote the inversion matrix. Then

$$\mathbf{p}' = \mathbf{T}_I \mathbf{R}_f \mathbf{T}_I^{-1} \mathbf{p} \tag{5.10}$$

where
$$\mathbf{T}_I \mathbf{R}_f \mathbf{T}_I^{-1} = \begin{bmatrix} 1 & 0 & 0 & p_{I_x} \\ 0 & 1 & 0 & p_{I_y} \\ 0 & 0 & 1 & p_{I_z} \\ 0 & 0 & 0 & 1 \end{bmatrix} \begin{bmatrix} -1 & 0 & 0 & 0 \\ 0 & -1 & 0 & 0 \\ 0 & 0 & -1 & 0 \\ 0 & 0 & 0 & 1 \end{bmatrix} \begin{bmatrix} 1 & 0 & 0 & -p_{I_x} \\ 0 & 1 & 0 & -p_{I_y} \\ 0 & 0 & 1 & -p_{I_z} \\ 0 & 0 & 0 & 1 \end{bmatrix} \tag{5.11}$$

or
$$\mathbf{p}' = \begin{bmatrix} -1 & 0 & 0 & 2p_{I_x} \\ 0 & -1 & 0 & 2p_{I_y} \\ 0 & 0 & -1 & 2p_{I_z} \\ 0 & 0 & 0 & 1 \end{bmatrix} \tag{5.12}$$

The matrix equation produces the same result as the vector equation. (Compare Equations 5.9 and 5.12.)

The Product of Two Inversions

The product of two successive inversions in the origin is trivial. If there is an even number of such transformations, then the outcome is the identity transformation. If there is an odd number, then the outcome is the equivalent of a single inversion. It is an entirely different matter if we take the product of successive inversions through two different points. Let's look at a vector interpretation first (Figure 5.4). Let

REFLECTION AND SYMMETRY

$\mathbf{p}_I$ and $\mathbf{p}_J$ denote the inversion points. We make use of the obvious vector relationships to write

$$\mathbf{p}'' = -\mathbf{p}' + 2\mathbf{p}_J \tag{5.13}$$

Substitute from Equation 5.9 for $\mathbf{p}'$ to produce

$$\mathbf{p}'' = \mathbf{p} + 2(\mathbf{p}_J - \mathbf{p}_I) \tag{5.14}$$

Figure 5.5 and Equation 5.14 tell us that the product of two inversions in two distinct arbitrary points is equal to a translation, namely $2(\mathbf{p}_J - \mathbf{p}_I)$. The translation is obviously parallel to the line between $\mathbf{p}_I$ and $\mathbf{p}_J$. The distance is twice that between these points, and the order of the successive inversions determines the translation direction.

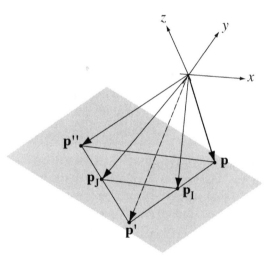

Figure 5.5 The product of two inversions.

We can also express this transformation as a product of matrices, and we require no less than six to do it. The meaning of the notation in the following matrix equation should be self-evident.

$$\mathbf{p}' = \mathbf{T}_J \mathbf{R}_f \mathbf{T}_J^{-1} \mathbf{T}_I \mathbf{R}_f \mathbf{T}_I^{-1} \mathbf{p} \tag{5.15}$$

However, the matrix product on the right side of this equation would benefit from some notational compression, in that it would be convenient to refer to this and similar products yet to be presented with a shorter form. So now we'll introduce $\mathbf{H}$ (see Section 4.3) and reserve its use for just such multi-term matrix products. We'll rewrite the equation as

$$\mathbf{p}' = \mathbf{H}\mathbf{p} \tag{5.16}$$

where $\quad \mathbf{H} = \mathbf{T}_J \mathbf{R}_f \mathbf{T}_J^{-1} \mathbf{T}_I \mathbf{R}_f \mathbf{T}_I^{-1} \tag{5.17}$

Now let's compute **H**. We already have $\mathbf{T_I R}_f \mathbf{T_I^{-1}}$ from Equation 5.12, so

$$\mathbf{H} = \begin{bmatrix} 1 & 0 & 0 & p_{J_x} \\ 0 & 1 & 0 & p_{J_y} \\ 0 & 0 & 1 & p_{J_z} \\ 0 & 0 & 0 & 1 \end{bmatrix} \begin{bmatrix} -1 & 0 & 0 & 0 \\ 0 & -1 & 0 & 0 \\ 0 & 0 & -1 & 0 \\ 0 & 0 & 0 & 1 \end{bmatrix} \begin{bmatrix} 1 & 0 & 0 & -p_{J_x} \\ 0 & 1 & 0 & -p_{J_y} \\ 0 & 0 & 1 & -p_{J_z} \\ 0 & 0 & 0 & 1 \end{bmatrix} \begin{bmatrix} -1 & 0 & 0 & 2p_{I_x} \\ 0 & -1 & 0 & 2p_{I_y} \\ 0 & 0 & -1 & 2p_{I_z} \\ 0 & 0 & 0 & 1 \end{bmatrix} \quad (5.18)$$

Complete the indicated matrix multiplication to find

$$\mathbf{H} = \begin{bmatrix} 1 & 0 & 0 & 2(p_{J_x} - p_{I_x}) \\ 0 & 1 & 0 & 2(p_{J_y} - p_{I_y}) \\ 0 & 0 & 1 & 2(p_{J_z} - p_{I_z}) \\ 0 & 0 & 0 & 1 \end{bmatrix} \quad (5.19)$$

We see that the matrix **H** in this case describes a translation. The first three elements in the fourth column tell us that the translation is equal to twice the distance between $\mathbf{p_I}$ and $\mathbf{p_J}$, and in the direction of the vector $\mathbf{p_J} - \mathbf{p_I}$. This is in complete agreement with the conclusions we made after deriving Equation 5.14.

The Product of Three Inversions

A succession of three inversions produces some interesting results. Let the vectors $\mathbf{p_I}$, $\mathbf{p_J}$, and $\mathbf{p_K}$ denote the inversion points (Figure 5.6). After some considerable, but quite ordinary, vector algebra we find that

$$\mathbf{p}''' = -\mathbf{p} + 2(\mathbf{p_I} - \mathbf{p_J} + \mathbf{p_K}) \quad (5.20)$$

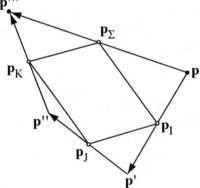

Figure 5.6 The product of three inversions in the plane.

REFLECTION AND SYMMETRY   201

The product of three inversions is the equivalent of an inversion through a fourth point, $\mathbf{p}_\Sigma$, where $\mathbf{p}_\Sigma = \mathbf{p}_I - \mathbf{p}_J + \mathbf{p}_K$. From vector algebra, we find that $\mathbf{p}_\Sigma - \mathbf{p}_I = -\mathbf{p}_J + \mathbf{p}_K$ and $\mathbf{p}_K - \mathbf{p}_\Sigma = -\mathbf{p}_I + \mathbf{p}_J$, demonstrating that opposite sides of the quadrilateral are of equal length and direction. This tells us that $\mathbf{p}_I$, $\mathbf{p}_J$, $\mathbf{p}_K$, and $\mathbf{p}_\Sigma$ are the vertices of a parallelogram. We see that this is true from the geometry of Figure 5.7, where if $\mathbf{p}$ is not coplanar with $\mathbf{p}_I$, $\mathbf{p}_J$, and $\mathbf{p}_K$, then $\mathbf{p}'$ is as distant from this plane as $\mathbf{p}$ but on the opposite side, similarly for $\mathbf{p}''$ and $\mathbf{p}'''$.

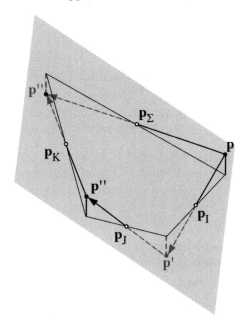

Figure 5.7 The $\mathbf{p}_I$, $\mathbf{p}_J$, $\mathbf{p}_K$ and $\mathbf{p}_\Sigma$ parallelogram.

The matrix expression for this transformation is

$$\mathbf{H} = \mathbf{T}_K \mathbf{R}_f \mathbf{T}_K^{-1} \mathbf{T}_J \mathbf{R}_f \mathbf{T}_J^{-1} \mathbf{T}_I \mathbf{R}_f \mathbf{T}_I^{-1} \tag{5.21}$$

where, of course, $\mathbf{p}' = \mathbf{H}\mathbf{p}$. We determine $\mathbf{H}$ by building upon Equation 5.19. Thus

$$\mathbf{H} = \mathbf{T}_K \mathbf{R}_f \mathbf{T}_K^{-1} \begin{bmatrix} 1 & 0 & 0 & 2(p_{J_x} - p_{I_x}) \\ 0 & 1 & 0 & 2(p_{J_y} - p_{I_y}) \\ 0 & 0 & 1 & 2(p_{J_z} - p_{I_z}) \\ 0 & 0 & 0 & 1 \end{bmatrix} \tag{5.22}$$

or

$$H = \begin{bmatrix} 1 & 0 & 0 & p_{K_x} \\ 0 & 1 & 0 & p_{K_y} \\ 0 & 0 & 1 & p_{K_z} \\ 0 & 0 & 0 & 1 \end{bmatrix} \begin{bmatrix} -1 & 0 & 0 & 0 \\ 0 & -1 & 0 & 0 \\ 0 & 0 & -1 & 0 \\ 0 & 0 & 0 & 1 \end{bmatrix} \begin{bmatrix} 1 & 0 & 0 & -p_{K_x} \\ 0 & 1 & 0 & -p_{K_y} \\ 0 & 0 & 1 & -p_{K_z} \\ 0 & 0 & 0 & 1 \end{bmatrix} \begin{bmatrix} 1 & 0 & 0 & 2(p_{J_x} - p_{I_x}) \\ 0 & 1 & 0 & 2(p_{J_y} - p_{I_y}) \\ 0 & 0 & 1 & 2(p_{J_z} - p_{I_z}) \\ 0 & 0 & 0 & 1 \end{bmatrix} \quad (5.23)$$

So that finally we have

$$H = \begin{bmatrix} -1 & 0 & 0 & 2(p_{I_x} - p_{J_x} + p_{K_x}) \\ 0 & -1 & 0 & 2(p_{I_y} - p_{J_y} + p_{K_y}) \\ 0 & 0 & -1 & 2(p_{I_z} - p_{J_z} + p_{K_z}) \\ 0 & 0 & 0 & 1 \end{bmatrix} \quad (5.24)$$

Let $p_{\Sigma_x} = p_{I_x} - p_{J_x} + p_{K_x}$ and similarly for p_{Σ_y}, p_{Σ_z}; then

$$H = \begin{bmatrix} -1 & 0 & 0 & 2p_{\Sigma_x} \\ 0 & -1 & 0 & 2p_{\Sigma_y} \\ 0 & 0 & -1 & 2p_{\Sigma_z} \\ 0 & 0 & 0 & 1 \end{bmatrix} \quad (5.25)$$

Equations 5.24 and 5.25 confirm our initial assertion.

Three distinct equivalent points of inversion, $\mathbf{p}_\Sigma$, are possible. The order in which we traverse the three given points of inversion to produce the final transformation determines the equivalent $\mathbf{p}_\Sigma$ (Figure 5.8). The succession of inversions in the order $\mathbf{p}_I \to \mathbf{p}_J \to \mathbf{p}_K$ or $\mathbf{p}_K \to \mathbf{p}_J \to \mathbf{p}_I$ produces the equivalent point of inversion $\mathbf{p}_{\Sigma_1}$; $\mathbf{p}_J \to \mathbf{p}_K \to \mathbf{p}_I$ or $\mathbf{p}_I \to \mathbf{p}_K \to \mathbf{p}_J$ produces $\mathbf{p}_{\Sigma_2}$; and $\mathbf{p}_K \to \mathbf{p}_I \to \mathbf{p}_J$ or $\mathbf{p}_J \to \mathbf{p}_I \to \mathbf{p}_K$ produces $\mathbf{p}_{\Sigma_3}$. The six points $\mathbf{p}_I$, $\mathbf{p}_J$, $\mathbf{p}_K$, $\mathbf{p}_{\Sigma_1}$, $\mathbf{p}_{\Sigma_2}$, and $\mathbf{p}_{\Sigma_3}$ delineate four congruent triangles.

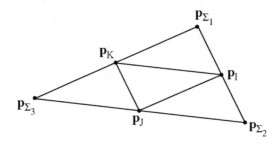

Figured 5.8 Three possible distinct equivalent points of inversion.

Reflection and Symmetry

If we continue this line of investigation to include the products of four or more inversions, then we would no doubt discover more properties that are interesting. We already know some characteristics; for example, an even number of inversions always produces a translation, whereas an odd number produces an equivalent inversion. We can also vary the dimensions of the space we work in, although the odd dimension spaces seem to produce the most interesting result. Do you suppose this is always true, even for much higher-dimension spaces?

Center of Symmetry

Point C is a center of symmetry to set S if the inversion of S in C maps S onto itself. We can easily create a center of symmetry: The point of inversion of a set of points P_i is the center of symmetry P_c of the combined set $P_i \cup P_i'$, where P_i' is the image of P_i under the inversion (Figure 5.9). This applies to any geometric object. Thus, the point of inversion of a geometric object G is the center of symmetry of $G \cup G'$. That the center of a circle is also its center of symmetry is an example.

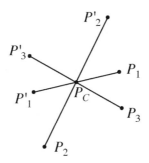

Figure 5.9 Center of symmetry.

Section 5.1 Exercises

1. Show synthetically and analytically that a central inversion in the plane fixes all lines through the point of inversion.

2. Sketch the central inversion of a rectangle in the plane, and discuss the orientation of the image.

3. Sketch the central inversion of a tetrahedron in space, numbering each vertex. Is there a rigid-body motion that superimposes the original onto the image so that corresponding vertices are coincident? Explain your answer.

4. Sketch the geometry describing the product of two inversions for each of the two possible orders. Describe anything about the results you find interesting.

5.2 Reflections in the Plane

Here we study reflections fixing a line in the *xy plane*. Recall the earlier synthetic definition describing the reflection of points: If P' is the reflection of P fixing a line m, then m is the perpendicular bisector of the line segment PP' (Figure 5.10a). Furthermore, reflection fixing a line is orientation-reversing (Figure 5.10b).

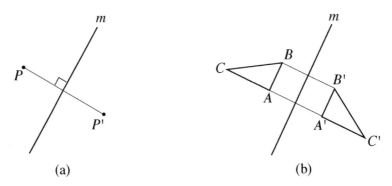

Figure 5.10 Reflection in a line.

Reflection Fixing a Line Parallel to a Principal Axis

The simplest reflection fixing a line parallel to a principal axis is reflection in the axis itself. We describe a reflection in the x and y axes as (Figure 5.11):

$$\begin{aligned} x' &= x \\ y' &= -y \end{aligned} \quad \text{and} \quad \begin{aligned} x' &= -x \\ y' &= y \end{aligned} \tag{5.26}$$

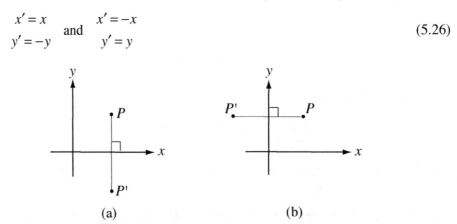

Figure 5.11 Reflection in the *x* and *y* axes.

The transformation matrix for reflection in the *x*-axis is

$$\mathbf{R}_f = \begin{bmatrix} 1 & 0 & 0 \\ 0 & -1 & 0 \\ 0 & 0 & 1 \end{bmatrix} \tag{5.27}$$

REFLECTION AND SYMMETRY

and for reflection in the y axis is

$$\mathbf{R}_f = \begin{bmatrix} -1 & 0 & 0 \\ 0 & 1 & 0 \\ 0 & 0 & 1 \end{bmatrix} \tag{5.28}$$

Reflection fixing the line $y = d$ parallel to the x axis is only slightly more complicated (Figure 5.12). We must first impose a translation so that the line $y = d$ coincides with the x axis, then reflect points about this axis, and finally translate so that the axis is back to its original position. In matrix form we have

$$\mathbf{H} = \mathbf{T}_d \mathbf{R}_f \mathbf{T}_d^{-1} \tag{5.29}$$

or

$$\mathbf{H} = \begin{bmatrix} 1 & 0 & 0 \\ 0 & 1 & d \\ 0 & 0 & 1 \end{bmatrix} \begin{bmatrix} 1 & 0 & 0 \\ 0 & -1 & 0 \\ 0 & 0 & 1 \end{bmatrix} \begin{bmatrix} 1 & 0 & 0 \\ 0 & 1 & -d \\ 0 & 0 & 1 \end{bmatrix} = \begin{bmatrix} 1 & 0 & 0 \\ 0 & -1 & 2d \\ 0 & 0 & 1 \end{bmatrix} \tag{5.30}$$

The complete transformation matrix is

$$\mathbf{p}' = \begin{bmatrix} 1 & 0 & 0 \\ 0 & -1 & 2d \\ 0 & 0 & 1 \end{bmatrix} \mathbf{p} \tag{5.31}$$

A similar equation applies to reflection in a line parallel to the y axis.

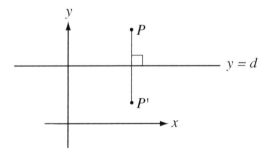

Figure 5.12 Reflection in a line parallel to the x axis.

Reflection Fixing an Arbitrary Line through the Origin

Given line m passing through the origin at angle ϕ with the x axis (Figure 5.13), then the transformation matrix producing the reflected image of $\mathbf{p}$ in m is

$$\mathbf{H} = \mathbf{R}_\phi \mathbf{R}_f \mathbf{R}_\phi^{-1} \tag{5.32}$$

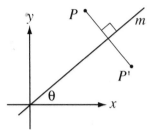

Figure 5.13 Reflection in an arbitrary line through the origin.

This tells us we must first perform a rotation so that m is coincident with the x axis. Next, we perform a reflection in this axis. Finally, we perform a rotation that returns m to its original position. Thus

$$\mathbf{H} = \begin{bmatrix} \cos\phi & -\sin\phi & 0 \\ \sin\phi & \cos\phi & 0 \\ 0 & 0 & 1 \end{bmatrix} \begin{bmatrix} 1 & 0 & 0 \\ 0 & -1 & 0 \\ 0 & 0 & 1 \end{bmatrix} \begin{bmatrix} \cos\phi & \sin\phi & 0 \\ -\sin\phi & \cos\phi & 0 \\ 0 & 0 & 1 \end{bmatrix} \quad (5.33)$$

so that $\quad \mathbf{H} = \begin{bmatrix} \cos 2\phi & \sin 2\phi & 0 \\ \sin 2\phi & -\cos 2\phi & 0 \\ 0 & 0 & 1 \end{bmatrix} \quad (5.34)$

This matrix is very similar to a rotation matrix, but we see that this cannot be since $|\mathbf{H}| = -1$. The following section discusses reflection in the principal diagonals.

Reflection Fixing the Principal Diagonals in the *xy* plane

There are two principal diagonals in the xy plane: the lines $y = x$ and $y = -x$ (Figures 5.15 and 5.16). The matrix equations that map points on the plane onto their reflected images across these two lines are

$$\mathbf{p}' = \begin{bmatrix} 0 & 1 & 0 \\ 1 & 0 & 0 \\ 0 & 0 & 1 \end{bmatrix} \mathbf{p} \quad \text{and} \quad \mathbf{p}' = \begin{bmatrix} 0 & -1 & 0 \\ -1 & 0 & 0 \\ 0 & 0 & 1 \end{bmatrix} \mathbf{p}$$

Notice that in both cases the determinant of the transformation matrix is equal to -1.

Reflection Fixing an Arbitrary Line

Now let's use vectors to describe reflection fixing an arbitrary line in the xy plane (Figure 5.16). Given any point $\mathbf{p}$ and a reflection fixing the line $\mathbf{r}_0 + u\mathbf{t}$, then the reflected image $\mathbf{p}'$ of $\mathbf{p}$ is

$$\mathbf{p}' = \mathbf{p} + 2\mathbf{s} \quad (5.35)$$

where $2\mathbf{s}$ is the vector corresponding to the line segment joining $\mathbf{p}$ and $\mathbf{p}'$.

REFLECTION AND SYMMETRY

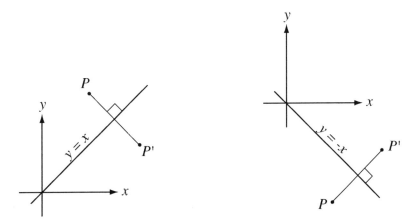

Figure 5.14 Reflection in line *y* = *x*.   **Figure 5.15** Reflection in line *y* = −*x*.

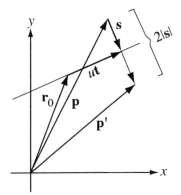

Figure 5.16 Reflection in an arbitrary line.

We apply some straightforward vector algebra to determine **s**. First, we know that $\mathbf{s} = \mathbf{r}_0 + u\mathbf{t} - \mathbf{p}$ and $\mathbf{s} \bullet \mathbf{t} = 0$, where **t** is the unit vector in the direction of the line. This leads to

$$\mathbf{s} = \mathbf{r}_0 - \mathbf{p} - \left[(\mathbf{r}_0 - \mathbf{p}) \bullet \mathbf{t}\right]\mathbf{t} \tag{5.36}$$

and $\quad \mathbf{p}' = -\mathbf{p} + 2\left\{\mathbf{r}_0 - \left[(\mathbf{r}_0 - \mathbf{p}) \bullet \mathbf{t}\right]\mathbf{t}\right\} \tag{5.37}$

Equation 5.37 does not lend itself to ready interpretation in matrix form. So, we will derive the transformation matrix independent of it. We begin with the translation $-\mathbf{r}_0$, then the rotation $-\phi$ (where $\phi = \tan^{-1} t_y/t_x$), then a reflection in the *x* axis, the rotation ϕ, and finally the translation $\mathbf{r}_0$. This product looks like

$$\mathbf{H} = \mathbf{T}_{\mathbf{r}_0} \mathbf{R}_\phi \mathbf{R}_f \mathbf{R}_\phi^{-1} \mathbf{T}_{\mathbf{r}_0}^{-1} \tag{5.38}$$

which works out to be

$$\mathbf{H} = \begin{bmatrix} \cos 2\phi & \sin 2\phi & -r_{0_x}\cos 2\phi - r_{0_y}\sin 2\phi + r_{0_x} \\ \sin 2\phi & -\cos 2\phi & -r_{0_x}\sin 2\phi + r_{0_y}\cos 2\phi + r_{0_y} \\ 0 & 0 & 1 \end{bmatrix} \quad (5.39)$$

Reflection in Two Parallel Lines

Reflection in two parallel lines is equivalent to a translation and therefore is orientation-preserving. We will investigate the simplest arrangement: that of two lines parallel to one of the principal axes (Figure 5.17).

From the figure, we obtain the following

$$p''_x = p_x \quad (5.40)$$

and $\quad p''_y = p_y + 2(d_2 - d_1) \quad (5.41)$

The matrix describing this transformation is the product

$$\mathbf{H} = \mathbf{T}_{d_2} \mathbf{R}_f \mathbf{T}_{d_2}^{-1} \mathbf{T}_{d_1} \mathbf{R}_f \mathbf{T}_{d_1}^{-1} \quad (5.42)$$

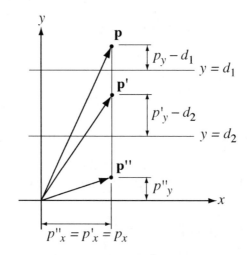

Figure 5.17 Reflection in two parallel lines.

Using the results of Equation 5.31, we find

$$\mathbf{H} = \begin{bmatrix} 1 & 0 & 0 \\ 0 & 1 & d_2 \\ 0 & 0 & 1 \end{bmatrix} \begin{bmatrix} 1 & 0 & 0 \\ 0 & -1 & 0 \\ 0 & 0 & 1 \end{bmatrix} \begin{bmatrix} 1 & 0 & 0 \\ 0 & 1 & -d_2 \\ 0 & 0 & 1 \end{bmatrix} \begin{bmatrix} 1 & 0 & 0 \\ 0 & -1 & 2d_1 \\ 0 & 0 & 1 \end{bmatrix} \quad (5.43)$$

Multiplying this out produces

REFLECTION AND SYMMETRY

$$\mathbf{H} = \begin{bmatrix} 1 & 0 & 0 \\ 0 & 1 & 2(d_2 - d_1) \\ 0 & 0 & 1 \end{bmatrix} \quad (5.44)$$

This matrix describes a translation of $2(d_2 - d_1)$ parallel to the y axis. It is twice the distance between the two parallel lines. This confirms Equations 5.40 and 5.41, as well as the assertions made earlier in the course of our synthetic derivation of this transformation.

Reflection in Three Parallel Lines

The transformation matrix describing the triple reflection in lines parallel to the y axis is (Figure 5.18)

$$\mathbf{H} = \begin{bmatrix} -1 & 0 & 2x_3 \\ 0 & 1 & 0 \\ 0 & 0 & 1 \end{bmatrix} \begin{bmatrix} -1 & 0 & 2x_2 \\ 0 & 1 & 0 \\ 0 & 0 & 1 \end{bmatrix} \begin{bmatrix} -1 & 0 & 2x_1 \\ 0 & 1 & 0 \\ 0 & 0 & 1 \end{bmatrix}$$

or $\quad \mathbf{H} = \begin{bmatrix} -1 & 0 & 2(x_1 - x_2 + x_3) \\ 0 & 1 & 0 \\ 0 & 0 & 1 \end{bmatrix}$

This has the curious property $\mathbf{H}^2 = \mathbf{I}$. Or is it really all that curious? No, because a reflection in three parallel lines is the equivalent of a single reflection. For the example above, it is a reflection in the line $x = x_1 - x_2 + x_3$, and the product of a reflection with itself is the identity transformation.

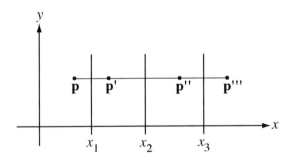

Figure 5.18 Reflection in three parallel lines.

Reflection in Intersecting Lines

To determine the transformation matrix for the product of reflections in intersecting lines m_1 and m_2, we begin with the simple arrangement where their point of intersection is the origin (Figure 5.19).

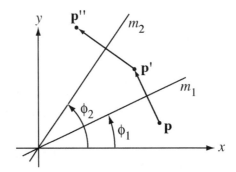

Figure 5.19 Reflection in intersecting lines that pass through the origin.

Denote the angles the lines make with the x axis as ϕ_1 and ϕ_2, where the subscripts indicate the order of reflection. We make use of Equation 5.34 (it serves to describe reflection about m_1), so that the product is

$$\mathbf{H} = \begin{bmatrix} \cos\phi_2 & -\sin\phi_2 & 0 \\ \sin\phi_2 & \cos\phi_2 & 0 \\ 0 & 0 & 1 \end{bmatrix} \begin{bmatrix} 1 & 0 & 0 \\ 0 & -1 & 0 \\ 0 & 0 & 1 \end{bmatrix} \begin{bmatrix} \cos\phi_2 & \sin\phi_2 & 0 \\ -\sin\phi_2 & \cos\phi_2 & 0 \\ 0 & 0 & 1 \end{bmatrix} \begin{bmatrix} \cos 2\phi_1 & \sin 2\phi_1 & 0 \\ \sin 2\phi_1 & -\cos 2\phi_1 & 0 \\ 0 & 0 & 1 \end{bmatrix} \quad (5.45)$$

or $\quad \mathbf{H} = \begin{bmatrix} \cos 2(\phi_2 - \phi_1) & -\sin 2(\phi_2 - \phi_1) & 0 \\ \sin 2(\phi_2 - \phi_1) & \cos 2(\phi_2 - \phi_1) & 0 \\ 0 & 0 & 1 \end{bmatrix} \quad (5.46)$

This transformation is the equivalent of a rotation of $2(\phi_2 - \phi_1)$ about the origin, that is, a rotation through twice the angle between m_1 and m_2. The more general case is that in which the point of intersection of m_1 and m_2 does not coincide with the origin (Figure 5.20). Here we must properly introduce translations that first bring the point of intersection $\mathbf{r}_1$ into coincidence with the origin, and later in the sequence return the lines to their original positions.

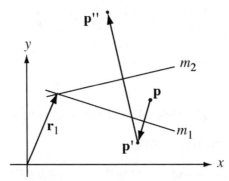

Figure 5.20 Reflection in intersecting lines.

Oblique Reflection

Before we leave the subject of reflections in the plane, consider this following interesting variation on the reflection:

In all the reflections we've discussed thus far, the line segments joining corresponding points of a geometric object and its image are perpendicular to the reflection axis. Mathematicians call these "orthogonal line reflections". We define another type of reflection if we relax the requirement for orthogonality. If we reflect a geometric object in a line m and construct the line segments joining corresponding points of the object and its image so that they are parallel and have their midpoints on the line m and at an angle ϕ to it, then we produce an oblique reflection of the object (Figure 5.21). Notice that an oblique reflection preserves area.

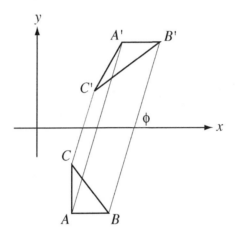

Figure 5.21 Oblique reflection.

If the oblique reflection fixes the x axis and ϕ is the angle between the axis and the parallel line segments joining corresponding points on object and image, then the Cartesian equations describing this transformation are

$$x' = x - 2y \tan \phi$$
$$y' = -y$$

In addition to area, invariant properties include parallelism and distances measured between points parallel to the axis of reflection.

Lines of Symmetry

If a reflection in line m maps a figure onto itself, then the line m is a line of symmetry of the figure. The line of reflection is a line of symmetry of the set $P_i \cup P_i'$ if the set of points P_i' is the reflected image of the set P_i (Figure 5.22). Many letters of the alphabet, in certain fonts, have one or more lines of symmetry.

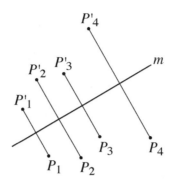

Figure 5.22 Line of symmetry.

Section 5.2 Exercises

1. Find the matrix and Cartesian equations that describe a reflection fixing the line $x = d$ parallel to the y axis.

2. Find the reflection matrix that maps any point on the plane onto its mirror image across the line $y = x/\sqrt{3}$.

3. Show that the product of the reflection matrices

   $$\begin{bmatrix} 0 & 1 & 0 \\ 1 & 0 & 0 \\ 0 & 0 & 1 \end{bmatrix} \text{ and } \begin{bmatrix} 0 & -1 & 0 \\ -1 & 0 & 0 \\ 0 & 0 & 1 \end{bmatrix}$$ is a rotation. Describe the rotation.

4. Prove that the set of reflections is not a group.

5.3 Reflections in Space

In three-dimensional space, we can reflect points, objects, or the space itself in a point (central inversion), in a line, or in a plane. We have looked at central inversion in space, and now we'll investigate reflection fixing lines and planes in space.

Reflection Fixing a Line in Space

Reflection fixing a line in space is simply an extension of the work that led to the development of Equations 5.37-5.39 in the previous section. However, reflection fixing a line in space is orientation preserving and is the equivalent of a halfturn about the fixed line, the axis of reflection. This means it is not a true reflection. Let's take a brief look at a reflection fixing a principal axis, say the z axis (Figure 5.23).

REFLECTION AND SYMMETRY 213

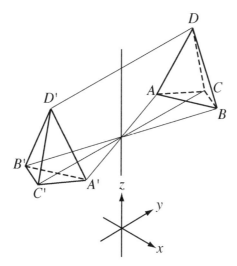

Figure 5.23 Reflection in a principal axis in space.

Clearly, the orientation of the tetrahedron $ABCD$ and its image $A'B'C'D'$ are the same. Although the line segments joining corresponding points on object and image are not parallel, they are perpendicular to the z axis and bisected by it. The matrix equation for this transformation is not hard to deduce. It is

$$\mathbf{p}' = \begin{bmatrix} -1 & 0 & 0 & 0 \\ 0 & -1 & 0 & 0 \\ 0 & 0 & 1 & 0 \\ 0 & 0 & 0 & 1 \end{bmatrix} \mathbf{p} \quad (5.47)$$

with similar equations for reflections in the x and y axis.

Reflection Fixing a Plane Parallel to a Principal Plane

The Cartesian equations for a reflection fixing the xy plane are

$$\begin{aligned} x' &= x \\ y' &= y \\ z' &= -z \end{aligned} \quad (5.48)$$

or in matrix form

$$\mathbf{p}' = \begin{bmatrix} 1 & 0 & 0 & 0 \\ 0 & 1 & 0 & 0 \\ 0 & 0 & -1 & 0 \\ 0 & 0 & 0 & 1 \end{bmatrix} \mathbf{p} \quad (5.49)$$

If the reflection plane is the $z = z_f$ plane (Figure 5.24), then two translations are required, and the transformation matrix is

$$\mathbf{H} = \begin{bmatrix} 1 & 0 & 0 & 0 \\ 0 & 1 & 0 & 0 \\ 0 & 0 & 1 & z_f \\ 0 & 0 & 0 & 1 \end{bmatrix} \begin{bmatrix} 1 & 0 & 0 & 0 \\ 0 & 1 & 0 & 0 \\ 0 & 0 & -1 & 0 \\ 0 & 0 & 0 & 1 \end{bmatrix} \begin{bmatrix} 1 & 0 & 0 & 0 \\ 0 & 1 & 0 & 0 \\ 0 & 0 & 1 & -z_f \\ 0 & 0 & 0 & 1 \end{bmatrix} \qquad (5.50)$$

After performing the indicated matrix multiplication we find

$$\mathbf{H} = \begin{bmatrix} 1 & 0 & 0 & 0 \\ 0 & 1 & 0 & 0 \\ 0 & 0 & -1 & 2z_f \\ 0 & 0 & 0 & 1 \end{bmatrix} \qquad (5.51)$$

and, of course, $\mathbf{p}' = \mathbf{H}\mathbf{p}$.

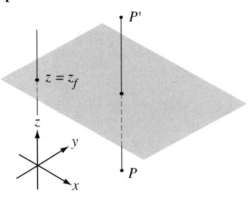

Figure 5.24 Reflection in a plane parallel to a principal plane.

Reflection Fixing an Arbitrary Plane

Consider reflection fixing an arbitrary plane that passes through the origin (Figure 5.25a). Note, for a review of various equations defining a plane, see the end of this section.

First, rotate the point set or geometric object in such a way that the unit normal **n** to the reflection plane is in the direction of the unit normal of a principal plane, say the z plane. This makes the reflection plane coincident with a principal plane, the xy plane (Figure 5.25b). Next, reflect the objects in question through this plane. Finally, rotate the intermediate image of the point set so that the unit normal, **n**, points in its original direction. Notice that to bring **n** into alignment with the z axis usually requires two rotations: ϕ about the z axis to bring **n** into the yz plane, followed by ψ about the x axis to bring **n** into alignment with the z axis.

REFLECTION AND SYMMETRY

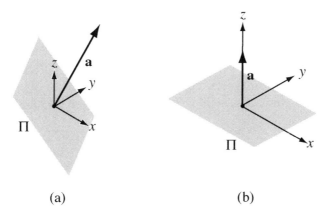

(a)   (b)

Figure 5.25 Reflection in a plane through the origin.

The matrix product that generates the final transformation matrix is

$$\mathbf{H} = \mathbf{R}_\phi \mathbf{R}_\psi \mathbf{R}_f \mathbf{R}_\psi^{-1} \mathbf{R}_\phi^{-1} \tag{5.52}$$

where $\mathbf{R}_f$ is a reflection in the xy plane, $\mathbf{R}_\phi$ is a rotation through $\phi = \tan^{-1} n_x/n_y$ about the z axis, and $\mathbf{R}_\varphi$ is a rotation through $\psi = \cos^{-1} n_z$.

If the reflection plane does not pass through the origin (Figure 5.26), then we must add a pair of translations. These are easy to determine using the Cartesian equation for a plane. For example, a translation in the x direction of $-a$ or D/A puts the plane through the origin and Equation 5.52 becomes

$$\mathbf{H} = \mathbf{T}_a \mathbf{R}_\phi \mathbf{R}_\psi \mathbf{R}_f \mathbf{R}_\psi^{-1} \mathbf{R}_\phi^{-1} \mathbf{T}_a^{-1} \tag{5.53}$$

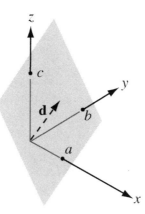

Figure 5.26 Reflection in an arbitrary plane.

There are many other possibilities for generating the reflection transformation in question. We might, for example, translate $-b$ along the y axis and then align the normal **d** with the x axis. Sometimes other aspects of a transformation problem will dictate a preferred sequence. There is also a very simple vector solution to this problem, presented in the next section.

Here is the review of several ways to define a plane (Figure 5.27):

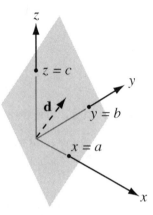

Figure 5.27 Plane intercepts with the principal axes.

Cartesian form: $Ax + By + Cz + D = 0$ or $\dfrac{x}{a} + \dfrac{y}{b} + \dfrac{z}{c} = 1$

If $D = 0$, then the plane passes through the origin.

Normal distance from origin to the plane: $d = |\mathbf{d}| = D\big/\sqrt{A^2 + B^2 + C^2}$

Unit normal components: $\mathbf{n} = \begin{bmatrix} n_x & n_y & n_z \end{bmatrix}$

where

$n_x = A\big/\sqrt{A^2 + B^2 + C^2}$

$n_y = B\big/\sqrt{A^2 + B^2 + C^2}$

$n_z = C\big/\sqrt{A^2 + B^2 + C^2}$

Intercepts with the principal axes:

x-axis: $x = -D/A$, $y = z = 0$ (Let $a = -D/A$)

y-axis: $y = -D/B$, $x = z = 0$ (Let $b = -D/B$)

z-axis: $z = -D/C$, $x = y = 0$ (Let $b = -D/C$)

Reflection Fixing an Arbitrary Plane: A Vector Solution

Given a plane of reflection $Ax + By + Cz + D = 0$, we compute the components of its normal, **n**, from

$$n_x = A\Big/\sqrt{A^2 + B^2 + C^2}, \quad n_y = B\Big/\sqrt{A^2 + B^2 + C^2}, \quad n_z = C\Big/\sqrt{A^2 + B^2 + C^2}$$

We find the reflected image **p′** of **p** as follows. Since the plane is the perpendicular bisector of the line segment (represented by) **p′ − p**, then we assert that (Figure 5.28)

$$\mathbf{p}' = \mathbf{p} + u\mathbf{n}$$

and $\quad \mathbf{p}_I = \mathbf{p} + \dfrac{1}{2}u\mathbf{n}$

where $\mathbf{p}_I$ is the point of intersection of the plane with the line segment joining **p** and **p′**.

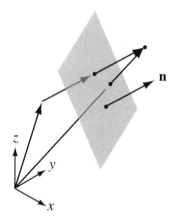

Figure 5.28 Reflection fixing a arbitrary plane: a vector solution.

These conditions lead us to the following four equations:

$$Ax_I + Bx_I + Cx_I + D = 0$$
$$x_I = p_x + un_x/2$$
$$y_I = p_y + un_y/2$$
$$z_I = p_z + un_z/2$$

From this system of equations we readily calculate u and then **p′**.

Product of Reflection in Parallel Planes

Find the product of reflection in two parallel planes Π and Δ. If the planes are parallel to a principal axis, then insert the appropriate translation matrices into the sequence. For example, given planes Π and Δ parallel to the xy plane (Figure 5.29), the matrix describing the product of reflection in these planes is (for $z_\Delta > z_\Pi$)

$$\mathbf{H} = \begin{bmatrix} 1 & 0 & 0 & 0 \\ 0 & 1 & 0 & 0 \\ 0 & 0 & 1 & z_\Delta \\ 0 & 0 & 0 & 1 \end{bmatrix} \begin{bmatrix} 1 & 0 & 0 & 0 \\ 0 & 1 & 0 & 0 \\ 0 & 0 & -1 & 0 \\ 0 & 0 & 0 & 1 \end{bmatrix} \begin{bmatrix} 1 & 0 & 0 & 0 \\ 0 & 1 & 0 & 0 \\ 0 & 0 & 1 & -z_\Delta \\ 0 & 0 & 0 & 1 \end{bmatrix} \begin{bmatrix} 1 & 0 & 0 & 0 \\ 0 & 1 & 0 & 0 \\ 0 & 0 & 1 & z_\Pi \\ 0 & 0 & 0 & 1 \end{bmatrix} \times$$

$$\begin{bmatrix} 1 & 0 & 0 & 0 \\ 0 & 1 & 0 & 0 \\ 0 & 0 & -1 & 0 \\ 0 & 0 & 0 & 1 \end{bmatrix} \begin{bmatrix} 1 & 0 & 0 & 0 \\ 0 & 1 & 0 & 0 \\ 0 & 0 & 1 & -z_\Pi \\ 0 & 0 & 0 & 1 \end{bmatrix}$$

(5.54)

or

$$\mathbf{H} = \begin{bmatrix} 1 & 0 & 0 & 0 \\ 0 & 1 & 0 & 0 \\ 0 & 0 & 1 & 2(z_\Delta - z_\Pi) \\ 0 & 0 & 0 & 1 \end{bmatrix}$$

(5.55)

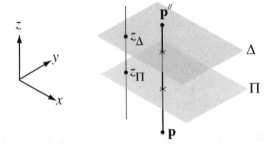

Figure 5.29 Product of reflection in two planes parallel to a principal plane.

This is not a true reflection since $|\mathbf{H}| = +1$, which, of course, means it is orientation preserving. In fact, it is a translation in the z direction a distance $2(z_\Delta - z_\Pi)$. If planes Π and Δ are not parallel to a principal plane, then we must introduce rotations that align the normal to these planes with a principal axis. This is the same procedure we used to derive Equation 5.52.

Reflection in Intersecting or Perpendicular Planes

Finding the product of reflection in two intersecting planes is, as in so many other transformations we've already investigated, largely a matter of developing a

Reflection and Symmetry

sequence of rotations and translations that successively bring each plane into a position coincident to a principal plane so that we can perform a simple reflection. This transformation is the equivalent of a rotation through some angle about the line of intersection of the two planes, and is orientation preserving.

The product of three reflections in any three mutually perpendicular planes with a common point M is the equivalent of an inversion in M. The line PP''' is the body diagonal of a rectangular parallelepiped centered at M whose faces are parallel to the reflection planes (Figure 5.30).

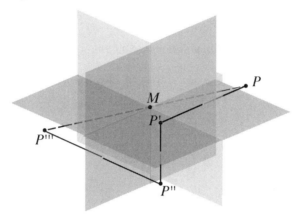

Figure 5.30 Reflection in three perpendicular planes.

Section 5.3 Exercises

1. Find the general expression for **H** (similar to Equation 5.53) for reflection in an arbitrary plane. Translate along the z axis to put the plane through the origin, and align **n**, the normal to the plane, with the y axis. Give the translation distance and the rotation angles, as well as $\mathbf{R}_f$.

5.4 Summary of Reflection Matrices

We have used $\mathbf{R}_f$ to denote a variety of reflection and pseudo-reflection transformation matrices (let's call the transformation where $|\mathbf{R}_f| = +1$ a pseudo-reflection). These matrices are special instances of the following general expressions:

$$\mathbf{R}_f = \begin{bmatrix} \pm 1 & 0 & 0 \\ 0 & \pm 1 & 0 \\ 0 & 0 & 1 \end{bmatrix}, \text{ in the plane} \tag{5.56}$$

$$\mathbf{R}_f = \begin{bmatrix} \pm 1 & 0 & 0 & 0 \\ 0 & \pm 1 & 0 & 0 \\ 0 & 0 & \pm 1 & 0 \\ 0 & 0 & 0 & 1 \end{bmatrix}, \text{ in space} \tag{5.57}$$

If an odd number of minus signs appear, then $|\mathbf{R}_f| = -1$ and the transformation is orientation reversing; otherwise $|\mathbf{R}_f| = +1$ and it is orientation preserving. Here is a summary of the special instances of $\mathbf{R}_f$ in the plane and in space. In each case, both the matrix formulation and Cartesian equations are given.

Reflection in the *y* axis:

$$\mathbf{R}_f = \begin{bmatrix} -1 & 0 & 0 \\ 0 & 1 & 0 \\ 0 & 0 & 1 \end{bmatrix}, \begin{matrix} x' = -x \\ y' = y \end{matrix}$$

Reflection in the *x* axis:

$$\mathbf{R}_f = \begin{bmatrix} 1 & 0 & 0 \\ 0 & -1 & 0 \\ 0 & 0 & 1 \end{bmatrix}, \begin{matrix} x' = x \\ y' = -y \end{matrix}$$

Central inversion:

$$\mathbf{R}_f = \begin{bmatrix} -1 & 0 & 0 \\ 0 & -1 & 0 \\ 0 & 0 & 1 \end{bmatrix}, \begin{matrix} x' = -x \\ y' = -y \end{matrix}$$

Reflection in the *yz* plane:

$$\mathbf{R}_f = \begin{bmatrix} -1 & 0 & 0 & 0 \\ 0 & 1 & 0 & 0 \\ 0 & 0 & 1 & 0 \\ 0 & 0 & 0 & 1 \end{bmatrix}, \begin{matrix} x' = -x \\ y' = y \\ z' = z \end{matrix}$$

Reflection in the *xz* plane:

$$\mathbf{R}_f = \begin{bmatrix} 1 & 0 & 0 & 0 \\ 0 & -1 & 0 & 0 \\ 0 & 0 & 1 & 0 \\ 0 & 0 & 0 & 1 \end{bmatrix}, \begin{matrix} x' = x \\ y' = -y \\ z' = z \end{matrix}$$

Halfturn about the *x* axis:

$$\mathbf{R}_f = \begin{bmatrix} 1 & 0 & 0 & 0 \\ 0 & -1 & 0 & 0 \\ 0 & 0 & -1 & 0 \\ 0 & 0 & 0 & 1 \end{bmatrix}, \begin{matrix} x' = x \\ y' = -y \\ z' = -z \end{matrix}$$

Halfturn about the *y* axis:

$$\mathbf{R}_f = \begin{bmatrix} -1 & 0 & 0 & 0 \\ 0 & 1 & 0 & 0 \\ 0 & 0 & -1 & 0 \\ 0 & 0 & 0 & 1 \end{bmatrix}, \begin{matrix} x' = -x \\ y' = y \\ z' = -z \end{matrix}$$

Halfturn about the *z* axis:

$$\mathbf{R}_f = \begin{bmatrix} -1 & 0 & 0 & 0 \\ 0 & -1 & 0 & 0 \\ 0 & 0 & 1 & 0 \\ 0 & 0 & 0 & 1 \end{bmatrix}, \begin{matrix} x' = -x \\ y' = -y \\ z' = z \end{matrix}$$

Reflection and Symmetry

Central inversion (reflection in the origin):

$$\mathbf{R}_f = \begin{bmatrix} -1 & 0 & 0 & 0 \\ 0 & -1 & 0 & 0 \\ 0 & 0 & -1 & 0 \\ 0 & 0 & 0 & 1 \end{bmatrix}, \begin{array}{l} x' = -x \\ y' = -y \\ z' = -z \end{array}$$

5.5 Symmetry Basics

Before venturing into more rigorous mathematical interpretations of symmetry, let's investigate the symmetry of a simple geometric object, the equilateral triangle (Figure 5.31).

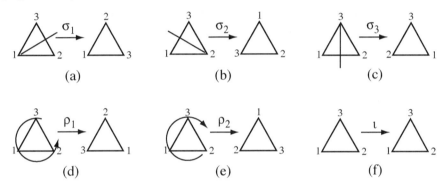

Figure 5.31 Six possible symmetries of an equilateral triangle.

We discover symmetries of this triangle by finding ways of transforming it or, more simply, ways of moving it that do not change its general appearance and metrical properties. A symmetry preserves distances and angles between the vertices and edges. There are six possible symmetries of an equilateral triangle. We can flip the triangle about a line bisecting the angle at vertex 1 leaving it unchanged while interchanging vertices 2 and 3 (Figure 5.31a). This, of course, is a reflection transformation. The image triangle is identical metrically to the original and occupies the same position and orientation. If we ignore the vertex labels, then the original and image are in all ways identical. Next, we can flip or reflect the triangle about a line bisecting the angle at vertex 2, interchanging vertices 1 and 3 (Figure 5.31b); and we can flip it about a line bisecting the angle at vertex 3, interchanging vertices 1 and 2 (Figure 5.31c). We can rotate the triangle 120° counterclockwise about its center (Figure 5.31d), moving vertex 1 to 2, 2 to 3, and 3 to 1; or, we can rotate it 120° clockwise (Figure 5.31e), moving vertex 1 to 3, 3 to 2, and 2 to 1. Finally, we can rotate the equilateral triangle 360° clockwise or counterclockwise, which is the same as the identity transformation, which is the equivalent of doing nothing to it (Figure 5.31f).

Inversion, it turns out, is not a symmetry transformation of an equilateral triangle (Figure 5.32). It is easy to demonstrate that the six symmetries of the equilateral triangle form a group, with all the requisite group properties. For example, label each symmetry transformation $\sigma_1, \sigma_2, \sigma_3, \rho_1, \rho_2$ and ι, as in Figure 5.31. Now consider the various products of these symmetries. The product $\sigma_2 \sigma_1$ (reflection 1 followed by reflection 2) first interchanges vertices 2 and 3 (i.e., interchange the vertex at the top with the vertex at the lower right), then interchange the vertex at the top with the vertex at the lower left. We immediately see that $\sigma_2 \sigma_1 = \rho_2$. In fact, we can construct a 6×6 Cayley table showing the results of the various possible products taken two at a time. The product of any two symmetries in this group is the equivalent of another symmetry of this group. All symmetry transformations of a figure form a group, the symmetry group of the figure. Reflection, rotation and inversion, subject to special conditions, are capable of producing symmetry transformations.

When we assert that a figure is symmetrical, we mean that there is an isometry or congruent transformation that leaves it invariant or unchanged as a whole, merely interchanging (permuting) its component elements. This is true in the plane, in space, or in higher dimensions. For example, the inversion of a square in its center leaves it invariant, and we call it, therefore, a symmetry transformation.

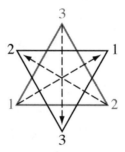

Figure 5.32 Inversion is not a symmetry transformation of an equilateral triangle.

We can express this idea of symmetry more precisely through recourse to the mathematics of Chapter 2. Recall that for any isometry σ and any set of points P we write σP for the image of P under σ. By a symmetry of a set P we mean an isometry that maps P onto itself, that is $\sigma P = P$. Thus, any rotation of a circle about its center, or any reflection about a diameter, is a symmetry of the circle (or, more strictly, a symmetry of the set of points whose locus defines a circle). Clearly, a symmetry is always its own inverse.

We must make two additional observations: First, there is an isometry that maps every point onto itself. This we know as the identity transformation, and it is a symmetry of every set. Second, we do not distinguish between a counterclockwise rotation of θ and a clockwise rotation of $2\pi - \theta$, nor between a rotation of θ and a rotation of $\theta + 2k\pi$, for any integer k. Only the final result of a mapping is important, not the way we arrive at that result.

Reflection and Symmetry

Centers, Lines, and Planes of Symmetry

We can identify special points or sets of points that are fixed under certain symmetry transformations. They are easy to identify and visualize for simple geometric objects (Figure 5.33). The center of a circle is a point of symmetry under inversion in the center, which of course fixes the center. If a line bisects a plane figure into mirror images, then it is a line of symmetry. A cylinder in space has an axis of symmetry: a three-dimensional line of symmetry. A cube has many planes of symmetry (try to find them all), as well as a point and several lines of symmetry.

This is not the end of it. We can imagine or invent many higher forms of invariant point sets under more exotic symmetry transformations. Think of a torus, for example. Later in this chapter we will see that "wallpaper" patterns, especially friezes, offer many examples of figures with infinitely many centers and lines of symmetry.

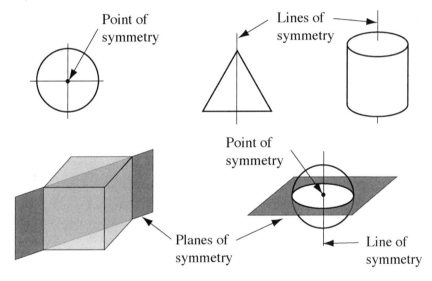

Figure 5.33 Centers, lines, and planes of symmetry.

Analysis of Symmetry

Although in a strict sense the concept of symmetry is intimately bound to reflection transformations, it is possible to elaborate the presence or condition of symmetry as it applies to a distinct geometric object or figure. In a symmetrical figure, the points comprising its points, lines or planes of symmetry are subsets of the total set of points defining the figure itself. See for example, the equilateral triangle of Figure 5.34a. Compare this with reflecting an arbitrary figure G through some line l to produce its image G'. The line l is a line of symmetry of the combined figure $G + G'$ (Figure 5.34b). This latter process is a way of producing and interpreting symmetric patterns.

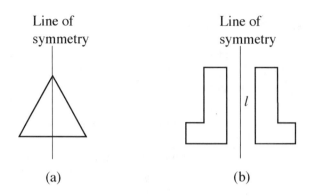

Figure 5.34 Symmetrical figures.

All the synthetic techniques we used in Chapter 2 to reveal and understand the properties of reflection transformations we can equally apply to symmetries. For example, if a symmetry in the plane fixes line l (i.e., l is a line of symmetry) and maps point P onto P', then l is the perpendicular bisector of the line joining P and P'.

Analytical techniques are also available to us. Given a plane curve $f(x,y)=0$, we can make the following assertions about possible symmetries:

1. If $f(x,y)=\pm f(-x,-y)$, then the curve is symmetric about the origin. A circle with its center at the origin is a good example of this (Figure 5.35a).

2. If $f(x,y)=\pm f(x,-y)$, then the curve is symmetric about the x axis. The parabola $x=y^2$ is an example (Figure 5.35b).

3. If $f(x,y)=\pm f(-x,y)$, then the curve is symmetric about the y axis. Here the parabola $y=x^2$ serves as an example (Figure 5.35c).

4. If $f(x,y)=\pm f(y,x)$, then the curve is symmetric about the line $x=y$. See the hyperbola in Figure 5.35d.

5. If $f(x,y)=\pm f(-y,-x)$, then the curve is symmetric about the line $x=-y$. See the hyperbola in Figure 5.35e.

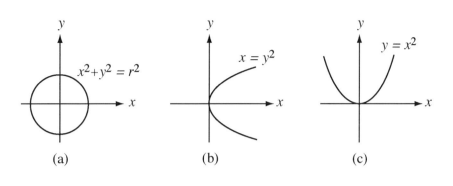

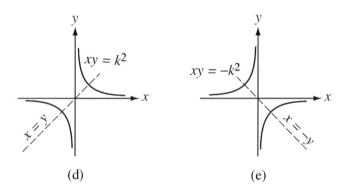

Figure 5.35 Symmetry of a curve.

Section 5.5 Exercises

1. Given the circle $(x-2)^2 + (y+3)^2 = 25$, show that the lines $x = 2$ or $y = -3$ are lines of symmetry.
2. What points and lines are invariant under a reflection of the x, y-plane in a line ℓ?
3. What are the lines of symmetry for a circle?
4. Does an angle have a point of symmetry? A line of symmetry?
5. Does a plane have a line of symmetry?
6. Is the origin a center of symmetry for the curve $x = y^3$?
7. Is either axis a line of symmetry for the curve $x = y^4$?
8. Determine how many planes of symmetry each of the following solids has.
   a. A rectangular solid, no square faces.
   b. A rectangular pyramid with isosceles triangles for lateral faces and a rectangular, not square, base.
   c. A rectangular pentagonal prism, lateral faces not square.

d. A right circular cone.
   e. A triangular pyramid with all faces scalene.
   f. A cube.
   g. A regular square-base pyramid.
   h. A right square prism, not a cube.
   i. A regular hexagonal prism, lateral faces not square.
   j. A non-right circular cone.
   k. A sphere.
   l. A triangular pyramid, all faces equilateral.

9. Describe any symmetries of the curve $y = x^3$.

10. Describe any symmetries of the curve $x^2 + xy + y^2 = 1$.

11. Describe any symmetries of the curve $2x^2 + 3y^2 = 1$.

12. Describe any symmetries of the curve $x^4 + y^4 = 1$.

13. Describe any symmetries of the curve $xy = 3$.

14. Describe any symmetries of the curve $3x^2 - 4y^2 = -5$.

15. Describe any symmetries of the curve $x^3 + x^2y + xy^2 + y^3 = 7$.

16. Given the curve $x^a y^b + x^a - y^b = 0$, where a and b are positive integers:

    a. If a and b are both even, is the curve symmetric about the origin?

    b. If a and b are both odd, is the curve symmetric about the origin?

    c. If $a = b$ and both are odd, is the curve symmetric about the line $x = y$? About $x = -y$?

    d. If a is even and b is odd, is the curve symmetric about the x axis? About the y axis?

5.6 Symmetry Groups

If there is one overarching principle that unifies and organizes symmetry, it is group theory. For it is the mathematics of group theory that allows us to formally define and classify symmetry. Two properties characterize the set of all symmetry transformations on a system of points. This greatly simplifies the study of these systems. First, the product of two successive symmetry transformations is always another symmetry transformation. Second, a transformation that we see undo or reverse a prior symmetry transformation is itself always another symmetry transformation. These conditions, corresponding to the closure and inverse properties of groups, lead us to suspect that the symmetry transformations themselves form a group (as defined in Chapter 2), as indeed they do.

REFLECTION AND SYMMETRY

Recall that the study of geometric transformations is intrinsically bound up with the study of mathematical structures we know as groups. The simplest geometric objects are easily understood in terms of groups. This motivates us to study groups as part of our study of geometric transformations. We know that a group is a finite set of elements; in our case these are geometric transformations and an operation that connects or combines them. Again, in our case, the operation generates the product of two geometric transformations, which must itself be another geometric transformation in the initial set.

Consider a set of symmetry transformations using translations to map a figure onto itself: for example, the translations that map the pattern shown in Figure 5.36 onto itself. Assume that this pattern extends indefinitely far in both directions. The most obvious symmetry transformation is translation by the vector **t**. But notice that translations by any of the vectors in the series $...3\mathbf{t}, 2\mathbf{t}, \mathbf{t}, -\mathbf{t}, -2\mathbf{t}...$ leave the figure invariant. The pattern is unchanged by any translation $\pm k\mathbf{t}$, where k is any integer. It is easy to verify that the set of all such translations form a group.

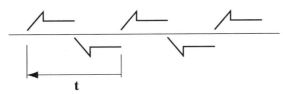

Figure 5.36 Symmetry group for a frieze pattern.

A similar argument holds for the parquet pattern we see in Figure 5.37. Again, assume that the pattern continues indefinitely over the plane. If we combine two translations, then we produce another mapping of the pattern onto itself. $\mathbf{t}_1$, $\mathbf{t}_2$, $\mathbf{t}_3$, and $\mathbf{t}_4$ are examples of translations that map the pattern onto itself, and all symmetry translations have vectors of the general form

$$\mathbf{t} = k_1 \mathbf{t}_1 + k_2 \mathbf{t}_2 \tag{5.58}$$

where k_1 and k_2 are integers.

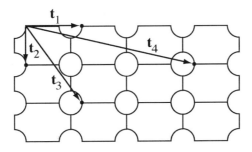

Figure 5.37 Symmetry group for a parquet pattern.

The Cyclic Groups

If a group has exactly n elements, then it is finite and has order n; otherwise it is infinite. Similarly, if there is a smallest positive integer n such that $\alpha^n = \iota$, then the transformation α has order n; otherwise α has an infinite order. Here is an example: Let ρ be a rotation of $\dfrac{360°}{n}$ about the origin where n is some positive integer, and let $\tau(x,y) = (x+1,y)$. Then ρ has order n, the set $\{\rho, \rho^2, \ldots, \rho^n\}$ forms a group of order n, τ has infinite order, and the set of all transformations τ^k, where k is an integer, forms an infinite group. We easily generalize this as follows: If transformation α has order n; then the set $\{\alpha, \alpha^2, \ldots, \alpha^n\}$ forms a group of order n. If α has infinite order, then the set of all integral powers of α forms an infinite group.

If every element of a group containing transformation α is a power of α, then the group is cyclic with generator α. We denote a cyclic group as $\langle \alpha \rangle$. If ρ is a rotation of $45°$, then $\langle \rho \rangle$ is a cyclic group of order eight. If ρ is a rotation of $36°$, then $\langle \rho \rangle$ is a cyclic group of order ten. Notice that this latter group is also generated by β where $\beta = \rho^3$, which means that $\langle \rho \rangle = \langle \rho^3 \rangle$ (as it happens for $\rho = 36°$), $\langle \rho \rangle = \langle \rho^3 \rangle = \langle \rho^7 \rangle = \langle \rho^9 \rangle$. Recall that an involution is a cyclic group of order two.

It is possible for a cyclic group to have more than one generator. Furthermore, since the powers of a transformation always commute (for example, $\alpha^n \alpha^m = \alpha^{n+m} = \alpha^{m+n} = \alpha^m \alpha^n$, where m and n are integers), a cyclic group is always abelian.

We generalize the notation for a cyclic group as follows: If $G = \langle \alpha, \beta, \gamma, \ldots \rangle$, then we can express every element of G as a product of powers of $\alpha, \beta, \gamma, \ldots$ so that $\{\alpha, \beta, \gamma, \ldots\}$ generates G (for example, $\alpha^3 \beta^2 \alpha^{-2} \beta \gamma^{-5}$). For simple cyclic groups, C_n denotes the cyclic group of order n generated by ρ, where $\rho = \rho_{360°/n}$. Although this notation does not specify a center of rotation, for regular polygons and polyhedra we assume it to be the geometric center or centroid of the object. Group C_1 contains only one element, the identity element, and it is the symmetry group of a scalene triangle. C_2 contains the identity and halfturn transformations, and it is the symmetry group of a parallelogram that is not a rhombus. Finally, if $n > 2$, then C_n is the symmetry group of a certain class of $2n$-gon (see next section), otherwise it may denote a cyclic subgroup of a dihedral group.

Reflection and Symmetry

The General Cyclic Group C_n

For $n > 2$, the cyclic group C_n describes the symmetries associated with certain specially defined $2n$-gons. These polygons have no odd isometries, that is no lines of reflection. We can easily construct polygons with this characteristic. One way to do this is to use a succession of equivalently truncated sides of a regular n-gon as alternating sides of a $2n$-gon. For example, carefully examine Figure 5.38. We have systematically removed all the symmetries of reflection without destroying the cyclic or rotational symmetries of the regular polygons. Of course, there are many other figures having cyclic symmetry.

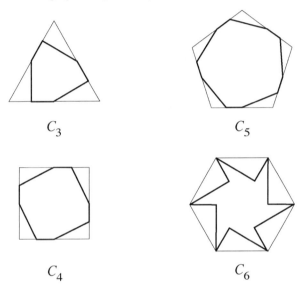

Figure 5.38 The general cyclic group C_n.

The Dihedral Groups

Every regular polygon and polyhedron has a symmetry group, and so we will begin the study of dihedral groups by considering the symmetry group of the square (Figure 5.39). The reason for this particular orientation of the square will soon become clear.

First, four distinct rotations leave the square invariant based on a cyclic subgroup of order four with generator ρ, where $\rho = 90°$. These are ρ, ρ^2, ρ^3, and ρ^4 (or $90°$, $180°$, $270°$, and $360°$). Second, four distinct reflections also leave the square invariant. These are $\sigma_x, \sigma_y, \sigma_r$ and σ_ℓ. Now we simplify the following development by letting $\sigma = \sigma_y$.

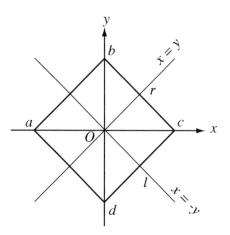

Figure 5.39 Symmetry of a square.

If a and b are adjacent vertices of the square, then under a symmetry transformation a may go to any one of the four vertices and b must go to one of the two adjacent vertices. This means that the images of the remaining vertices are then determined. We count eight permutations of symmetries for the square, and isometries ρ and σ generate the entire symmetry group. When a symmetry group consists of rotations and reflections, we denote this group as $\langle \rho, \sigma \rangle$ and call it a dihedral group. "Dihedral" means two-faced, and its use here indicates the presence of reflections in the symmetry group. (One way of interpreting reflections is to think of them as allowing us to see the backside, or second face of a plane figure that exhibits the reverse cyclic order of its vertices.) The elements of this group are the four distinct rotations ρ, ρ^2, ρ^3, and ρ^4 and the four distinct odd isometries $\rho\sigma$, $\rho^2\sigma$, $\rho^3\sigma$. and $\rho^4\sigma$. Notice that $\rho^4 = \sigma^2 = \iota$, the identity transformation. Try to verify the following relationships (Figure 5.40):

a. $\rho = \sigma_r \sigma_y = \rho^{-3}$ d. $\rho\sigma = \sigma_r$ g. $\sigma\rho = \rho^3 \sigma$

b. $\rho^2 = \sigma_x \sigma_y$ e. $\rho^2\sigma = \sigma_x$ h. $\sigma\rho^2 = \rho^2 \sigma$ (5.59)

c. $\rho^3 = \sigma_\ell \sigma_y = \rho^{-1}$ f. $\rho^3\sigma = \sigma_\ell$ i. $\sigma\rho^3 = \rho\sigma$

where ρ^{-n} is the inverse of ρ^n.

We use these relationships to construct the Cayley table for the symmetry group of a square (Table 5.1). It is a group of order eight, and we denote the particular combination of rotations and reflections as D_4. (We will soon see an explanation for this notation.)

Reflection and Symmetry

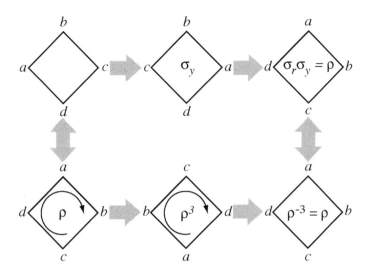

Figure 5.40 $\rho = \sigma_r \sigma_y = \rho^{-3}$.

Table 5.1 The D_4 dihedral symmetry group for a square.

D_4	ι	ρ	ρ^2	ρ^3	σ	$\rho\sigma$	$\rho^2\sigma$	$\rho^3\sigma$
ι	ι	ρ	ρ^2	ρ^3	σ	$\rho\sigma$	$\rho^2\sigma$	$\rho^3\sigma$
ρ	ρ	ρ^2	ρ^3	ι	$\rho\sigma$	$\rho^2\sigma$	$\rho^3\sigma$	σ
ρ^2	ρ^2	ρ^3	ι	ρ	$\rho^2\sigma$	$\rho^3\sigma$	σ	$\rho\sigma$
ρ^3	ρ^3	ι	ρ	ρ^2	$\rho^3\sigma$	σ	$\rho\sigma$	$\rho^2\sigma$
σ	σ	$\rho^3\sigma$	$\rho^2\sigma$	$\rho\sigma$	ι	ρ^3	ρ^2	ρ
$\rho\sigma$	$\rho\sigma$	σ	$\rho^3\sigma$	$\rho^2\sigma$	ρ	ι	ρ^3	ρ^2
$\rho^2\sigma$	$\rho^2\sigma$	$\rho\sigma$	σ	$\rho^3\sigma$	ρ^2	ρ	ι	ρ^3
$\rho^3\sigma$	$\rho^3\sigma$	$\rho^2\sigma$	$\rho\sigma$	σ	ρ^3	ρ^2	ρ	ι

Notice that since $\sigma\rho^k$ is a reflection in a line (the y axis) through the origin, it is an involution. This means that

$$\sigma\rho^k = \left(\sigma\rho^k\right)^{-1} = \rho^{-k}\sigma \tag{5.60}$$

or for our particular case

$$\rho^{-k}\sigma = \rho^{4-k}\sigma \tag{5.61}$$

Thus, we can compute all the elements of the Cayley table using the equations $\sigma\rho^k = \rho^{-k}\sigma$ and $\rho^4 = \sigma^2 = \iota$.

There is a theorem, Leonardo's Theorem, which states that a finite group of isometries is either a cyclic group C_n or a dihedral group D_n. It has the following corollary: The symmetry group for a polygon is either a cyclic group or a dihedral group. Although we will not prove this theorem here, it is not a difficult proof.

The General Dihedral Group D_n

If a regular n-gon with $n > 2$ is centered at the origin of the xy plane with one vertex on the positive y axis, then the n-gon is fixed by ρ and σ, where $\rho = 360°/n$ and σ is a reflection in the y-axis and $\rho^n = \sigma^2 = \iota$. These symmetries form a group and fix the n-gon by n distinct rotations $\rho, \rho^2, \ldots, \rho^n$ and by n distinct odd isometries $\rho\sigma, \rho^2\sigma, \ldots, \rho^n\sigma$. We see that the symmetry group of the n-gon contains at least these $2n$ symmetries. If V_a and V_b are adjacent vertices, then, under a symmetry, V_a may go to any of the n-gon's n vertices, and so V_b must go into one of the two adjacent vertices. This determines the images of the remaining vertices and tells us that there are at most $2n$ symmetries for the n-gon. Therefore, it follows that there are exactly $2n$ symmetries of the n-gon and we have identified them all, where isometries ρ and σ generate the entire group.

The symmetry group of an n-gon is a group of order $2n$ consisting of rotations and reflections $\langle \rho, \sigma \rangle$ and is denoted by D_n. We call these D_n symmetry groups the dihedral groups. If, again, we arrange the n-gon so that its center is at the origin and a vertex lies on the positive y axis, then $\sigma\rho^k$ is a reflection in the y axis and is an involution for any integer k. Therefore, from Equation 5.60, $\sigma\rho^k = \left(\sigma\rho^k\right)^{-1} = \rho^{-k}\sigma = \rho^{n-k}\sigma$. The elements of D_n are ι, ρ, ρ^2, ..., ρ^{n-1}, σ, $\rho\sigma$, $\rho^2\sigma$, ..., $\rho^{n-1}\sigma$. Using $\sigma\rho^k = \rho^{-k}\sigma$ and $\rho^n = \sigma^2 = \iota$, we easily generate the Cayley table for D_n.

The first two dihedral groups, D_1 and D_2, are the symmetry groups of an isosceles triangle that is not equilateral and of a rectangle that is not a square. $D_1 = \langle \sigma \rangle$ and $D_2 = \langle \rho, \sigma \rangle$, where ρ is a rotation of 180° about the center of an n-gon and σ is a reflection in a line through this center and a vertex.

For each positive integer n, there is a polygon with a symmetry group D_n and polygons having symmetry group C_n. Each dihedral symmetry group D_n with $n > 1$ has a cyclic subgroup C_n containing all the even isometries in D_n.

Finite Symmetry Groups on Space

Let's begin our study of finite symmetry groups on space by reviewing what we now know about symmetry in the plane. A geometric figure or set of points in space is symmetric if it is congruent to itself in more than one way. First, consider the concept of congruence from a very general point of view. Given a finite set of transformations in space that takes points into points, including a transformation that leaves every point fixed (the identity transformation), an inverse for each transformation, and products of these transformations that are themselves members of the set, then the set of transformations forms a finite group. Two figures are congruent with respect to this group if they can be taken into each other by a transformation of the group. A transformation of a given group that carries a figure (i.e., a set of points) into coincidence with itself (through a permutation of the points) is a superposition of the figure. All the superposition transformations of a given figure form a subgroup of the group. The simplest transformations in space are the isometries, or rigid motions, to which we add reflection in a plane and inversion in a point. These are the possible elements of finite symmetry groups on space.

If a set S contains all the points into which a given point P is mapped by the transformations of a group, then S is mapped onto itself by every transformation of the group and is therefore symmetric with respect to that group. If a finite group of transformations consists only of isometries, reflections, and inversions, and the set of points S is finite, then the geometric center O is invariant under all the transformations of the group. Thus, the isometries of the group are rotations about O, reflection in a plane through O followed by rotation about an axis through O and perpendicular to the plane, and inversion in O.

The finite symmetry groups on space are closely related to the regular polyhedra to which we must now necessarily refer, although Section 5.9 presents much more about polyhedral symmetry.

There are five finite rotation groups in space that, of course, produce symmetries. These are the cyclic, dihedral, tetrahedral, octahedral and icosahedral groups. We will consider each in turn. Each rotation group contains only rotations and the identity; these are the even isometries. All rotations with axes concurrent at O plus the identity form a group, since the product of two such rotations and the inverse of such a rotation are even and fix O, and since the only even isometries with a fixed point O are the identity and the rotations with axes through O.

We again let C_n denote the cyclic group produced by a rotation of order n, where $\rho = 360°/n$. C_1 is the cyclic group containing only the identity. The right prism in Figure 5.41 whose cross-section is an equilateral triangle is an example of C_3, cyclic symmetry of order 3.

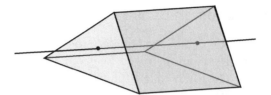

Figure 5.41 Cyclic symmetry in space.

D_n denotes the group produced by a rotation ρ of order n and a halfturn (reflection) σ; this requires that the axes of ρ and σ are perpendicular and concurrent at O. Here we interpret a plane reflection with axis l in a given plane as the restriction to this plane of the halfturn in space about line l. In other words, a rotation of 180° about l has the same effect on a plane Π through l as the plane reflection in l on Π (Figure 5.42). This is the dihedral group for space, and it has order $2n$, containing n halfturns about axes that are perpendicular to the axis of ρ at a concurrent point O and that form angles of $180°/n$ with each other. Notice that D_1 on space is superfluous since it contains the identity and halfturn transformation that make it equivalent to C_2.

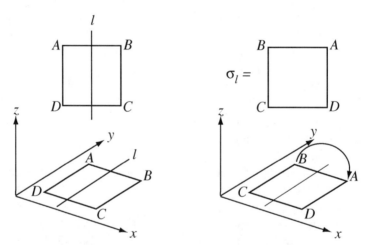

Figure 5.42 Equivalence of a reflection and halfturn.

The finite rotation groups $C_1, C_2, C_3, \ldots$ and $D_2, D_3, D_4, \ldots$ produce symmetries, but there are other rotation groups to consider, as well. It turns out, and you can easily conduct a thought experiment with the help of a pencil and paper to verify this, that the tetrahedron has rotational symmetries that cannot be generated by either the cyclic or dihedral groups. Its rotational symmetry group is denoted by T, and it is called the tetrahedral symmetry group. Further observation leads us to conclude that the octahedron, too, has rotational symmetries that cannot be generated by the cyclic and dihedral groups, nor by the tetrahedral group. This new group we call the octahe-

REFLECTION AND SYMMETRY

dral group and denote it by O. Finally, we also discover that the icosahedron has rotational symmetries not in C_n, D_n, T, or O. So we define an icosahedral symmetry group and denote it by I. Section 5.9 derives the details of these three additional symmetry groups, T, O, and I. The five symmetry groups C_n, D_n, T, O, and I, plus reflection and inversion form the basis for all the finite symmetry groups on space.

Section 5.6 Exercises

1. Verify each of the relationships of Equation 5.59 by sketching the appropriate sequence of transformations as in Figure 5.31.

2. Describe the symmetry group for each of the shapes in Figure 5.43.

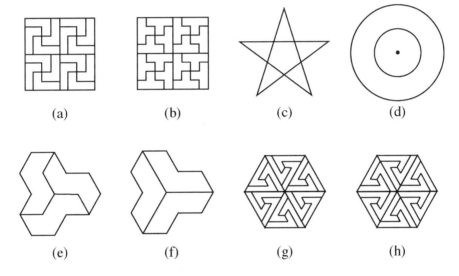

Figure 5.43 Exercise 2.

5.7 Ornamental Groups

The ornamental groups are comprised of three kinds of patterns in the plane, or three kinds of plane symmetries. The first kind is the set of seven frieze groups. These groups all exhibit symmetry along a line or strip. The second kind is the set of seventeen wallpaper groups, each member of which exhibits symmetry of a repeated pattern that fills the entire plane. In both frieze and wallpaper groups we find symmetry as the ordered repetition of a basic pattern. The third kind consists of the rosette groups, which in turn consist of the cyclic and dihedral groups C_n and D_n. All frieze groups contain a subgroup of translations generated by one translation, whereas wallpaper groups contain a subgroup of translations generated by two non-parallel translations.

The Frieze Groups

A frieze group is the symmetry group of a repeated pattern on a planar strip that is invariant under a sequence of translations along the strip. It can also be the symmetry group of a pattern over the entire plane, or half-plane, if all the translations are in one direction. Otherwise, the strip must have a finite width, although it may be infinite in length or form a closed loop. This allows the creation of a variety of patterns that may be invariant under reflection in an axis parallel to the strip's longitudinal center line or in one perpendicular to it, and also invariant under halfturns and glide reflections. (The word "frieze" is from architecture and refers to a decorative band along a wall, usually with a repetitive pattern.)

Incredible as it may seem, there are only seven possible frieze groups. Figure 5.44 shows a simple example of each one of them along with its F-notation (explained below) and a description of the pattern-generating transformations. The light dashed lines are reflection axes, and the small circle denotes the center of a halfturn.

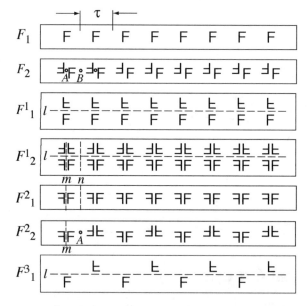

Figure 5.44 The seven frieze groups.

The F_1 group arises from a repeated translation, so that $F_1 = \langle \tau \rangle$. A frieze pattern having F_1 as its symmetry group has no point of symmetry, has no line of symmetry, and is not fixed by a glide reflection. A frieze pattern with F_2 as its symmetry group has a point of symmetry but no line of symmetry, and we denote it as $F_2 = \langle \sigma_A, \sigma_B \rangle$. F_1^1 has no point of symmetry, and the center line l is a line of symmetry, so that $F_1^1 = \langle \tau, \sigma_\ell \rangle$. F_2^1 has three distinct lines of reflection symmetry, including the center line l and $F_2^1 = \langle \sigma_\ell, \sigma_m, \sigma_n \rangle$. F_1^2 is the frieze pattern having no

Reflection and Symmetry

point of symmetry and two distinct lines of symmetry, so that $F_1^2 = \langle \sigma_m, \sigma_n \rangle$. F_2^2 has a line of symmetry and a point of symmetry, and $F_2^2 = \langle \sigma_m, \sigma_A \rangle$. Finally, the frieze pattern with F_1^3 as its symmetry group has no point of symmetry, no line of symmetry but is fixed by a glide reflection, so that $F_1^3 = \langle \gamma \rangle$. These seven groups exhaust the possibilities for generating the symmetries of frieze patterns.

Now, about the F notation: The subscript indicates whether or not there is a halfturn in the group, where 1 indicates no halfturn and 2 indicates there is a halfturn. A superscript 1 indicates that the centerline of the strip is a line of symmetry, a reflection axis. If the superscript is 2, then the center is not a line of symmetry, but there is a line of symmetry perpendicular to the center line. Finally, the superscript 3 denotes that the symmetry group is generated by a glide reflection. (This notation scheme is the work of the Hungarian mathematician Fejes-Tóth.)

To determine the kind of frieze group exhibited in a repetitive pattern, we simply ask the following questions:

1. Is there a point of symmetry (center point of a halfturn)?

2. Is the centerline (parallel to the edges of a frieze strip) a line of symmetry (an axis of reflection)?

3. Is there any line of symmetry?

4. Is the pattern fixed by a glide reflection?

Figure 5.45 shows how to arrange these questions into a flowchart or decision-tree to help identify or classify frieze patterns.

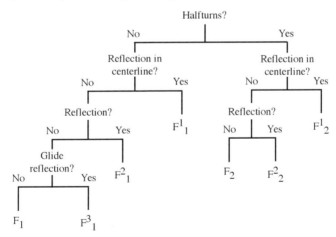

Figure 5.45 Flowchart classifying frieze patterns.

Wallpaper Groups

We now consider symmetry groups in the plane containing two nonparallel translations. These are the wallpaper groups, or sometimes the crystallographic groups of motions in the plane. There are only 17 of them. Some of these include rotations and some do not. In fact, exactly four wallpaper groups have no rotational symmetries. Five groups have 2-fold centers of rotation, three groups have 3-fold centers, three groups have 4-fold centers, and two have 6-fold centers. Thus a wallpaper group W is a group of isometries whose translations are exactly those in $\langle \tau_1, \tau_2 \rangle$, where τ_1 is not parallel to τ_2 and whose rotations, if any, are restricted to 2-, 3-, 4- and 6-fold centers of rotation. Section 5.7.3 explains why there are only four permissible categories of rotational symmetry (not counting, of course, the case of no rotation).

We identify a group by the translations and rotations it has. From the translation we can also describe a simple geometric figure that we will call the unit cell. A unit cell is any connected region in the pattern that does not contain a pair of equivalent points in its interior. If it does, then we have chosen a cell that is too large. For wallpaper groups it is usually easy to identify the unit cell, and in Figure 5.46 and those that follow, the shaded area indicates a unit cell.

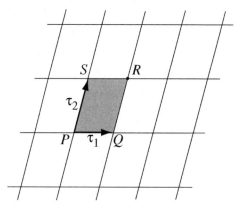

Figure 5.46 A lattice defined by two nonparallel translations.

We begin first with the four groups that we denote as W_1, W_1^1, W_1^2 and W_1^3 and that each contain two parallel translations and no rotations. For these four groups, as well as the other 13, systems of equivalent points under the symmetry transformations always form a plane lattice or a figure composed of congruent lattices in parallel positions. To demonstrate this, we start with any point P and choose a translation τ_1 that sends P to an equivalent point Q, which has the smallest distance possible from P in the pattern (Figure 5.46). Then the translations parallel to τ_1 generate a row of points equivalent to P on the straight line containing P and Q.

We assume that there are other translations in the group that are not parallel to line PQ. This means that there are other points in the pattern that are equivalent to P but not on the line containing P and Q. Next, we pick such a point R as close to P as possible but not on the line containing P and Q. We denote τ_2 as the translation in the group that sends P to R. And we know that $PR \geq PQ$. If S is the image of Q un-

REFLECTION AND SYMMETRY

der τ_2, then *PQRS* is a parallelogram. We see that all the translations of the group can be generated as combinations of τ_1 and τ_2; for example, in vector form $\mathbf{t} = m\mathbf{t}_1 + n\mathbf{t}_2$, where m and n are positive or negative integers. These generate a lattice that exhausts all the points equivalent to *P*.

A wallpaper pattern of symmetry group W_1 has no rotational symmetry (and therefore no centers of symmetry) and is not fixed by any reflection or glide reflection. Figure 5.47 and Table 5.2 define symbols used in figures of wallpaper groups.

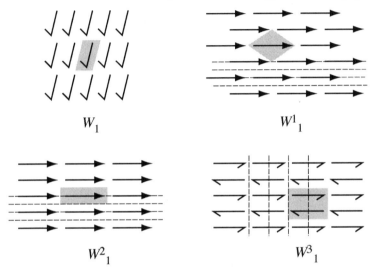

Figure 5.47 Symmetry groups W_1, W_1^1, W_1^2, and W_1^3.

Symmetry group W_1^1 has no rotational symmetry, is fixed by both reflections and glide reflections, and some axes of the glide reflections are not lines of symmetry. Symmetry group W_1^2 has no rotational symmetry, is fixed by both reflections and glide reflections, and all axes of the glide reflections are lines of symmetry. Symmetry group W_1^3 has no rotational symmetry, has no line of symmetry, but is fixed by a glide reflection. Space permits only a description of these groups and those to follow. Formal derivations of the necessary and sufficient conditions defining their distinguishing characteristics are beyond the scope of this text.

Table 5.2 Definition of symbols for wallpaper groups.

▱	A unit cell
– – – – – –	Line of reflection
- - - - - - - -	or glide reflection
○	2-fold center of rotational symmetry
△	3-fold center of rotational symmetry
□	4-fold center of rotational symmetry
⬡	6-fold center of rotational symmetry
●	○ on a line of reflection, forming group D_2
▲	△ on a line of reflection, forming group D_3
■	□ on a line of reflection, forming group D_4
⬢	⬡ on a line of reflection, forming group D_6
σ	Reflection
γ	Glide reflection

A wallpaper pattern of symmetry group W_2 has centers of 2 fold rotational symmetry, where every center is of this kind and is not fixed by any reflections or glide reflections (Figure 5.48). Symmetry group W_2^1 has centers of 2 fold rotational symmetry, where every center is of this kind and some, but not all, of these centers are on lines of symmetry. Symmetry group W_2^2 has centers of 2 fold rotational symmetry, where every center is of this kind and every center is on a line of symmetry. Symmetry group W_2^3 has centers of 2 fold rotational symmetry, where every center is of this kind, has lines of symmetry, and all lines of symmetry are parallel. Symmetry group W_2^4 has centers of 2 fold rotational symmetry, where every center is of this kind, has no lines of reflection symmetry, but is fixed by a glide reflection.

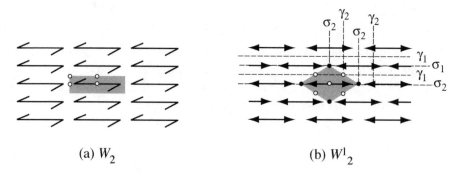

(a) W_2 (b) W^1_2

Figure 5.48ab Symmetry groups W_2 and W_2^1.

Reflection and Symmetry

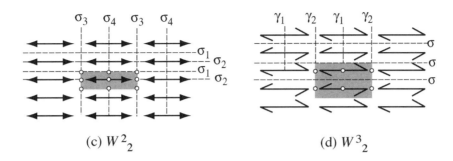

(c) W_2^2 (d) W_2^3

Figure 5.48cd Symmetry groups W_2^2 and W_2^3.

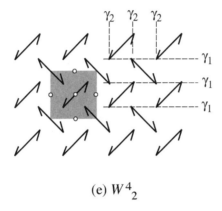

(e) W_2^4

Figure 5.48d Symmetry group W_2^4.

A wallpaper pattern of symmetry group W_3 has centers of 3-fold, but no 6-fold, rotational symmetry, and has no line of symmetry (Figure 5.49). Symmetry group W_3^1 has centers of 3 fold, but no 6 fold, rotational symmetry, and every center is on a line of symmetry. Symmetry group W_3^2 has centers of 3-fold, but no 6 fold, rotational symmetry, and has some centers that do not lie on the lines of symmetry.

A wallpaper pattern of symmetry group W_4 has centers of 2 and 4 fold rotational symmetry and no lines of symmetry (Figure 5.50). Symmetry group W_4^1 has a line of symmetry on a 4 fold center as well as 2 fold centers on axes of glide reflection symmetry. Symmetry group W_4^2 has centers of 2 and 4 fold rotational symmetry and lines of reflection symmetry that are not concurrent with any 4 fold centers.

A wallpaper pattern of symmetry group W_6 has centers of 2-, 3- and 6-fold rotational symmetry but no lines of symmetry (Figure 5.51a). Symmetry group W_6^1 has centers of 2, 3 and 6 fold rotational symmetry and lines of reflection and glide reflection symmetry (Figure 5.51b). This completes the description of all 17 wallpaper symmetry groups.

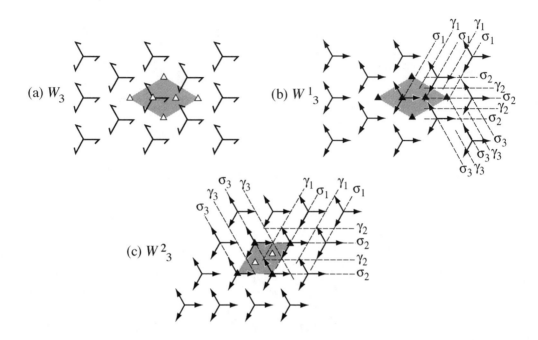

Figure 5.49 Symmetry groups W_3, W_3^1, and W_3^2.

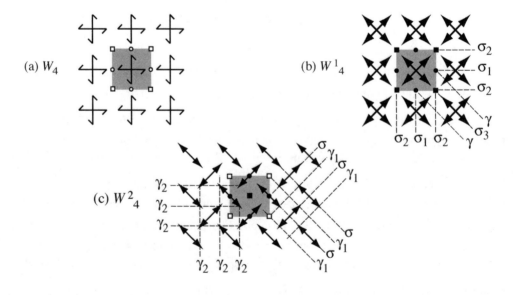

Figure 5.50 Symmetry groups W_4, W_4^1, and W_4^2.

REFLECTION AND SYMMETRY

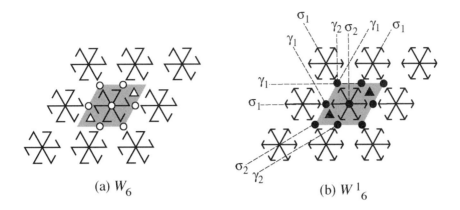

(a) W_6

(b) W_6^1

Figure 5.51 Symmetry groups W_6 **and** W_6^1.

The Crystallographic Restriction

Four of the 17 wallpaper groups do not contain rotations. The other 13 do, and these in turn belong to one of four possible classes according to the angles of rotation in the group. All possible angles are of the form $2\pi/n$, where n is an integer. The only permissible values of n, other than 1, are 2, 3, 4 and 6. A proof runs as follows:

Choose the smallest possible translation of the group, and denote it $\mathbf{t}_1$ (Figure 5.52). Assume point A is an n-fold center of rotation of the group and that $\mathbf{t}_1$ translates A to B. A rotation of $2\pi/n$ about A sends B to some point B'. It follows from group properties that there is also a translation $\mathbf{t}_1'$ that can send A to B' as well. So the transformation $-\mathbf{t}_1 - \mathbf{t}_1'$, where $-\mathbf{t}_1$ is the inverse of $\mathbf{t}_1$, must send B into B'. Since the magnitude of $\mathbf{t}_1$ is as small as possible in the group, it follows that $BB' \geq AB$ and therefore $\angle BAB' = (2\pi/n) \geq (\pi/3)$. This means that $n \leq 6$.

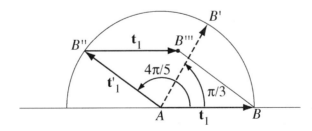

Figure 5.52 The crystallographic restriction.

If we can show that $n \neq 5$, then we have proved the assertion. To do this we assume that A is a 5-fold center of rotation and apply the rotation $2(2\pi/5)$ about A to map B into B''. Again, from the properties of such groups, we know that the group must contain the translation $\mathbf{t}_1''$ that sends A into B''. But this means that the translation $\mathbf{t}_1 + \mathbf{t}_1''$ would send point A to point C. By inspection it is obvious that distance AC is shorter than AB, which contradicts our requirement that $\mathbf{t}_1$ must be the smallest possible translation in the group. So we conclude that $n = 5$ is not allowed and that the wallpaper groups contain rotations restricted to those with 2, 3, 4, and 6 fold centers of rotation. This, then, is the crystallographic restriction. It tells us that the very nature of space dictates the number and kind of symmetry groups allowed in the plane, much as it dictates that there are only five regular polyhedra in space.

Plane Lattices

A regular system of points in the plane, or in a space of three or more dimensions, for that matter, forms what we call a lattice. Lattices offer a simple geometric model for understanding the structure of space, crystallographic structures (that is, how atoms organize themselves into regular repetitive patterns), the ornamental symmetry groups, and tilings in the plane. We can generate a lattice in the plane as follows: Starting from any fixed point O in the plane, the images of O under the set of translations $m\mathbf{a} + n\mathbf{b}$, where m and n are integers and $\mathbf{a}$ and $\mathbf{b}$ are nonparallel vectors, form a lattice (Figure 5.53).

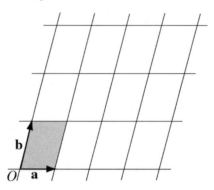

Figure 5.53 Plane lattices.

A unit cell of a lattice is the smallest unit of the lattice that can generate the whole pattern under repetitive translations. We may also define the unit cell of a lattice as the parallelogram formed by two nonparallel line segments with lattice points at the vertices and no lattice points within it (Figure 5.54). Notice that all the possible unit cells in a square lattice enclose an equal area. Depending on the magnitudes and relative directions of the vectors $\mathbf{a}$ and $\mathbf{b}$, a lattice may be square (or rectangular) or triangular (equilateral or not).

REFLECTION AND SYMMETRY

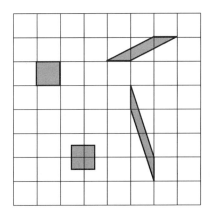

Figure 5.54 Unit cells in a lattice.

The properties of an equilateral triangle lattice are indicative of the properties of lattices in general (Figure 5.55). The first property we notice is the periodicity of the lattice. The translation vectors are responsible for this characteristic. There are a multitude of lattice planes (this nomenclature is from lattices in space). Two kinds of order are apparent in the lattice, long-range translational order and long-range orientation order. The families of parallel lines (lattice planes) demonstrate the former, while two kinds of unit cells that can tile the lattice without change in orientation demonstrate the latter.

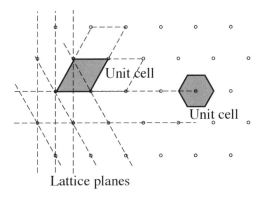

Figure 5.55 More plane lattices.

Patterns tiling the plane by translation in two nonparallel directions have the symmetry of the lattice, at least as a subgroup. Only the parallelogram and hexagon tile the plane with the symmetry group of the underlying lattice as a necessary and sufficient description.

Section 5.7 Exercises

1. Determine the frieze group for each of the following patterns.

   a. IIIIIII
   b. AAAAAAA
   c. MWMWMWM
   d. FFFFFFF
   e. DWDMDWDM
   f. SSSSSSS
   g. DDDDDDD

2. Give the symmetry group for each of the following patterns (Figure 5.56).

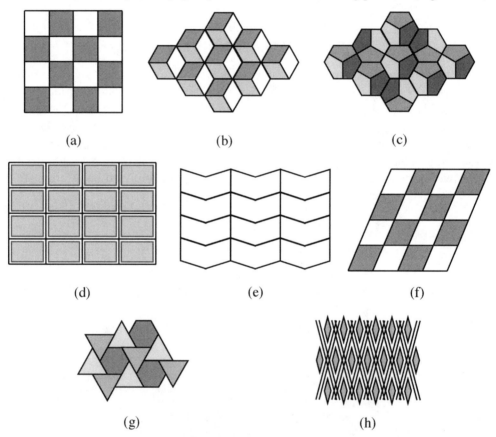

Figure 5.56 Exercise 2.

3. Give the symmetry group for each of the following wallpaper patterns and show all centers of rotational symmetry, lines of symmetry and glide reflection axes. Identify the centers as 2-, 3-, 4- or 6-fold centers, and shade the unit cell (Figure 5.57).

REFLECTION AND SYMMETRY 247

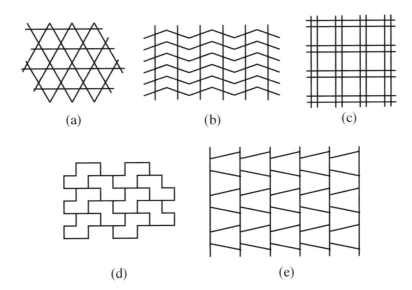

Figure 5.57 Exercise 3.

5.8 Polygonal Symmetry and Tiling

A tiling of the plane is a set $\{T_1, T_2, \ldots\}$ of polygonal regions of finite area covering the plane without gaps or overlap, where the T_i are the tiles (or, frequently, the "prototiles"). A tiling is monohedral if only a single size and shape tile is sufficient to cover the plane. A tiling is dihedral, trihedral or r hedral if two, three or r different tiles are necessary to cover the plane. See Figure 5.58 for some examples.

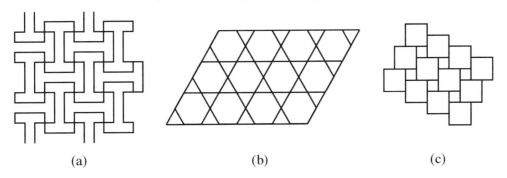

Figure 5.58 Tilings.

The variety of tiles and tilings is virtually unlimited, and there are several excellent books describing their categories and characteristics, including Grünbaum and Shephard's seminal work. Some tilings may be without any symmetry whatsoever, and, of course, we know many exhibit a wallpaper symmetry group. The polygons defining the individual tiles, or prototiles, may themselves be without symmetry. But we also know that for each positive integer n there is a polygon having symmetry group C_n as well as a polygon having symmetry group D_n, cyclic and dihedral symmetry. We will restrict our discussion to tilings based on regular polygons, where adjacent tiles share a common edge terminated at each end by a common vertex. This is an edge-to-edge tiling. Notice that although the dihedral tiling of Figure 5.58c consists of regular polygons (squares) they do not share common edges terminated by common vertices.

Regular Tilings

There are only three regular tilings of the plane (Figure 5..59). These are based on the equilateral triangle, the square and the hexagon. Each vertex of these tilings is surrounded identically with congruent faces. The regular pentagon, for example, cannot tile the plane without gaps or overlaps since three of them around a vertex leave a gap, and four must necessarily overlap.

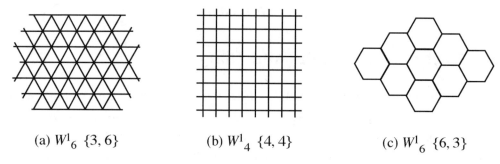

(a) W^1_6 {3, 6}   (b) W^1_4 {4, 4}   (c) W^1_6 {6, 3}

Figure 5.59 Regular tilings with symmetry group and Schläfli symbol.

Schläfli (1814-1895) invented a notation scheme to describe the geometry of networks of polygons: $\{p,q\}$ indicates there are q p-gonal faces surrounding each vertex. The Schläfli symbol for the tiling based on the equilateral triangle is {3,6}, indicating that there are six triangles surrounding each vertex. For the regular tiling of squares the Schläfli symbol is {4,4}, indicating that there are four squares surrounding each vertex. And for the regular tiling of hexagons the Schläfli symbol is {6,3}, indicating that there are three hexagons surrounding each vertex.

The {3,6}, {4,4}, and {6,3} tilings are the only edge-to-edge tilings possible with congruent polygons, not necessarily regular, of any kind. Thus, any triangle can tile the plane as {3,6} by rotating the triangle around the midpoint of any side to form a parallelogram (Figure 5.60a). Any four-sided polygon tiles the plane as {4,4} by successively inverting it in the midpoints of its sides (Figure 5.60b). Finally, any hexagon, with opposite sides parallel and equal tiles the plane as {6,3} (Figure 5.60c).

Reflection and Symmetry

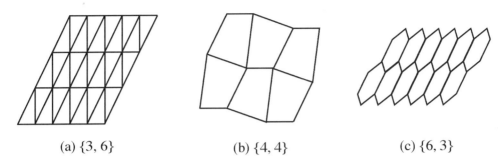

(a) {3, 6}   (b) {4, 4}   (c) {6, 3}

Figure 5.60 Congruent tilings.

Each regular tiling has another tiling associated with it, its dual. To construct the dual of a regular tiling, connect the geometric centers of polygons sharing a common edge (Figure 5.61). This produces the following three pairs of mutually dual tilings: $\{3,6\} \leftrightarrow \{6,3\}$, $\{4,4\} \leftrightarrow \{4,4\}$ and $\{6,3\} \leftrightarrow \{3,6\}$.

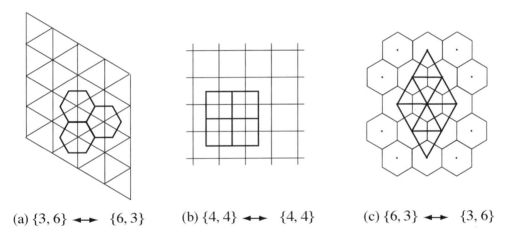

(a) {3, 6} ⟷ {6, 3}   (b) {4, 4} ⟷ {4, 4}   (c) {6, 3} ⟷ {3, 6}

Figure 5.61 Duals of regular tilings.

Three Regular Tilings Theory

There are three and only three regular tilings on the plane, and these are the equilateral triangle {3,6}, the square {4,4} and the hexagon {6,3}. We prove this as follows: Consider a p sided regular polygon $\{p\}$. Each of its interior angles is

$$\theta = \frac{180(p-2)}{p}$$

If the typical vertex of a regular tiling is surrounded by q regular p-sided polygons, then the sum of the interior angles around the vertex is

$$\frac{180q(p-2)}{p} = 360$$

which we can rewrite as

$$(p-2)(q-2) = 4$$

The positive-integer solutions to this equation are $\{p,q\} = \{3,6\}$, $\{4,4\}$, and $\{6,3\}$. These are also the only regular edge-to-edge tilings possible with congruent, but not necessarily regular, polygons of any kind. (We do not prove the more general case here.)

Semiregular Tilings

If we eliminate the restriction that only one kind of regular polygon can be used to tile the plane, but continue to require that each vertex be surrounded identically, then we extend the number of tilings to include the eight semiregular tilings. We will require that each vertex in the tiling be surrounded alike in number of each kind of regular polygon and alike in the cyclic order of these polygons around the vertex. Let's determine and describe all the tilings that meet these requirements.

The Schläfli symbol that we used above is inadequate to describe the conditions at a vertex when there is more than one kind of vertex surrounding it. So we use an expanded Schläfli scheme in this situation. If a vertex is surrounded by regular n-gons in cyclic order $n_1, n_2, \ldots, n_r$, then the vertex is type $n_1, n_2, \ldots, n_r$. (Notice that we have used both n and p to indicate the number of sides in a regular polygon. Both are commonly used in the literature, as well as e for edges.) The tiling in Figure 5.62a exhibits two kinds of vertices: 3.3.6.6 and 3.6.3.6, which we abbreviate as $3^2.6^2$ and $(3.6)^2$, respectively. All the vertices in Figure 5.62b are of the type 3.6.3.6. The tiling in Figure 5.62c contains three different kinds of vertices, 3.3.6.6, 3.6.3.6 and 6.6.6. Clearly, the tilings in Figures 5.62a and c are not admissible as semiregular tilings, since this category requires that all the vertices must be surrounded alike.

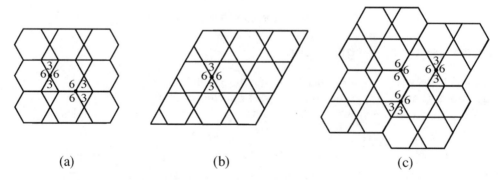

Figure 5.62 A variety of vertices in tilings.

Now, let's suppose that a given vertex is type $n_1, n_2, \ldots, n_r$. The equilateral triangle has the smallest vertex angle, so that $r \leq 6$ and $r = 6$ if each $n_i = 3$. We

REFLECTION AND SYMMETRY

know that the vertex angle of a regular n-gon is $180(n-2)/n$ and that the sum of the r terms $180(n_i - 2)/n_i$ must be 360. That is, $\sum_{i=1}^{r} 180(n_i - 2)/n_i = 360$. In other words, we require the positive integer solutions to the equation

$$\sum_{i=1}^{r} \frac{1}{n_i} = \frac{r-2}{2} \tag{5.62}$$

where $3 \leq r \leq 6$. There are 17 solutions, four of which are of two types. It turns out that not all of these solutions are realizable as edge-to-edge tilings by regular polygons where each vertex is identically surrounded. The 17 solutions and four variations are listed below. Solutions 1, 7 and 17 correspond to the three regular tilings. Now we must investigate the rest to determine their admissibility.

1. 3.3.3.3.3.3	10. 3.9.18
2. 3.3.3.3.6	11. 3.10.15
3. 3.3.3.4.4 and 3.3.4.3.4	12. 3.12.12
4. 3.3.4.12 and 3.4.3.12	13. 4.5.20
5. 3.3.6.6 and 3.6.3.6	14. 4.6.12
6. 3.4.4.6 and 3.4.6.4	15. 4.8.8
7. 4.4.4.4	16. 5.5.10
8. 3.7.42	17. 6.6.6
9. 3.8.24	

We quickly eliminate solutions 8, 9, 10 and 11, since any vertex of type $3.x.y$ produces the contradiction that $x = y$ when we cycle around an equilateral triangle and see that its sides must alternate as the common sides of an x-gon and y-gon. Since 3 is odd, after one circuit we find $x = y$. A similar argument holds for solutions of the form $x.5.y$ unless $x = y$, which is a contradiction since 5 is also odd. This eliminates solutions 13 and 16. If two vertices A and B of an equilateral triangle ABC are both of type $3.x.y.z$, then at vertex C the triangle lies either between the two x-gons or between two z-gons. This is impossible since all three vertices must be type 3.3.4.12, 3.4.3.12, 3.3.6.6 or 3.4.4.6. So now we can eliminate these solutions. The eight that remain are the semiregular tilings we seek (Figure 5.63). Of these, type 4.6.12 is unique in that it has two distinct cyclic orientations, and it turns out that both are necessary for the tiling to succeed.

Although this ends the discussion of polygonal symmetry and tilings, it by no means exhausts the subject. Those who pursue this subject further will find a wealth of mathematical beauty awaiting them.

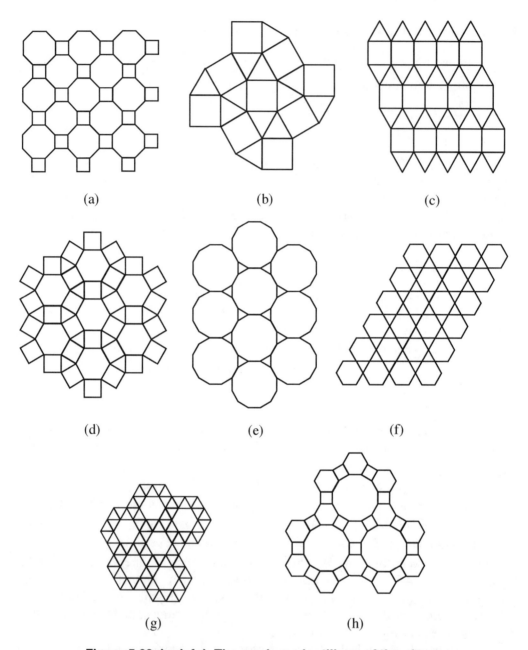

Figure 5.63abcdefgh The semi-regular tilings of the plane.

Section 5.8 Exercises

1. Investigate and discuss the symmetries of each of the eight semiregular tilings.

5.9 Polyhedral Symmetry

In Section 5.6 we studied the finite symmetry groups of space and found there a direct relationship to the Platonic solids in the tetragonal, octahedral and icosahedral symmetry groups. Now we will pursue this subject further by studying the symmetries of the 31 regular, semiregular and dual polyhedra.

Many kinds of polyhedra are possible, in various combinations of straight edges defining plane polygonal faces. We can construct an infinite number of polyhedra using regular polygons as faces. However, only five polyhedra are possible given the following four conditions:

1. All polygons are regular.
2. All polygons in a given polyhedron are congruent.
3. All vertices are identical.
4. All dihedral angles are equal.

Only the five Platonic solids, the tetrahedron, cube, octahedron, dodecahedron, and icosahedron satisfy all these conditions. If we relax conditions 2 and 4 so that a polyhedron may have two or more kinds of faces and two or more different-valued dihedral angles, then an infinite set of polyhedra is possible, including the 13 semiregular or Archimedean polyhedra whose faces are regular polygons. If we relax conditions 1 and 3 so that a polyhedron may have nonregular faces and more than one kind of vertex, then another infinite set of polyhedra is possible, including the special group of 13 dual polyhedra.

Two polyhedra are duals if the vertices of one have a one-to-one correspondence with the centers of the faces of the other. For example, the cube and the octahedron are dual polyhedra, since by appropriate scaling the six vertices of the octahedron can be made to correspond to the six faces of the cube, and the eight vertices of the cube can be made to correspond to the eight faces of the octahedron (Figure 5.64). The Archimedean polyhedra have identical vertices and their duals have congruent faces. Truncating the five regular polyhedra generates all the semiregular polyhedra except two snub forms.

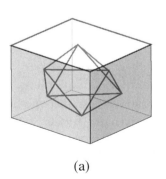 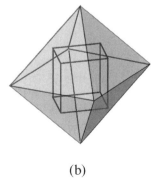

(a) (b)

Figure 5.64 The cube and octahedron duals.

Rotational Symmetry

In Section 5.6 under the discussion of finite symmetry groups on space we asserted that the tetrahedron, octahedron and icosahedron have rotational symmetries that cannot be generated by the cyclic or dihedral groups. It turns out that all the regular, semiregular and dual polyhedra fall into one of these three symmetry groups. We designate these groups as 3.2, 4.3.2, and 5.3.2, indicating that they exhibit combinations of rotational symmetry with 2, 3, 4, and 5 fold axes.

The 4.3.2 Symmetry Group

This is the symmetry group of the cube, which has the highest symmetry of all the polyhedra. It is also the symmetry group of the octahedron. The 4.3.2 notation indicates that the cube has 4, 3 and 2 fold axes of symmetry. Let's see what they are and what they mean.

We begin with the cube, not only because of our familiarity with it, but also because it holds a surprise or two. The cube is invariant under rotations of $90°$, $180°$, $270°$, and $360°$ (the identity transformation) about an axis through the center of opposite faces (Figure 5.65a). There are three of these 4 fold axes. The cube is also invariant under rotations of $120°$ and $240°$ about each of its four body diagonals, also known as the 3 fold axes of symmetry It is the existence of 3 fold axes of symmetry in a cube that is surprising (Figure 5.65b). Finally, the cube is invariant under $180°$ rotation about axes passing through the centers of pairs of opposite edges (Figure 5.65c). These are the six 2 fold axes. A cube has 24 rotations including the identity transformation that leave it invariant. There are other transformations, reflection and inversion, that increase the total number of symmetry transformations to 48.

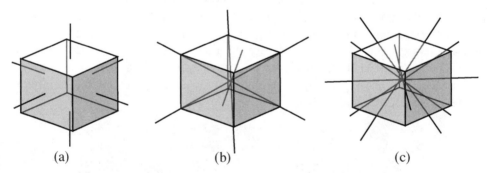

Figure 65 The rotational symmetry of a cube.

Truncating each of the six vertices of the octahedron so that one-third of each edge remains, generates the truncated octahedron. Truncating further so that the midpoint of each of the original 12 edges of the octahedron becomes a vertex generates the cuboctahedron. Further truncations generate the truncated cube and then the cube itself (Figure 5.66). Notice that we can reverse this process so that it is possible to begin with the cube and end with the octahedron.

Reflection and Symmetry

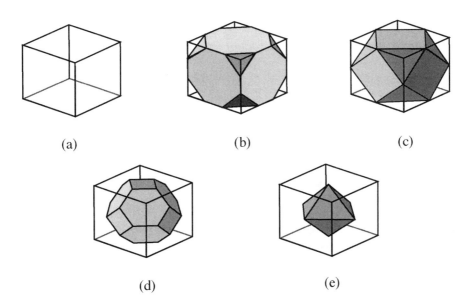

Figure 5.66 Some members of the 4.3.2 symmetry group.

Once we generate the cuboctahedron, it is possible to truncate its vertices to generate distorted forms of the great rhombicuboctahedron and then the small rhombicuboctahedron. From the latter it is possible to continue to the truncated cube and finally the cube. All of these truncation transformations produce polyhedra in the 4.3.2 symmetry group.

The 5.3.2 Symmetry Group

The icosahedron and its dual, the dodecahedron, are the basic or root polyhedra with 5.3.2 rotational symmetry (Figure 5.67). This means that they have 5-fold, 3-fold and 2-fold axes of symmetry. They are fairly easy to identify.

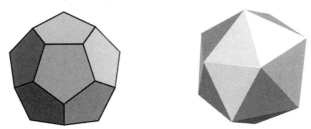

Figure 5.67 The icosahedron and dodecahedron duals.

The 5-fold axes of symmetry pass through the center of each face and the center of the icosahedron. The 3 fold axes pass through each vertex and the center, and the 2 fold axes pass through the center of each edge and the center of the icosahedron. Truncating the 12 vertices of the icosahedron so that one-third of each edge remains, generates the truncated icosahedron. Truncating further so that each of the

30 edges of the icosahedron becomes a vertex generates the icosidodecahedron. Further truncations generate the truncated dodecahedron and finally the dodecahedron, itself. As in the 4.3.2 symmetry group, we can reverse this process.

Once we have defined the icosidodecahedron, we can truncate its vertices to generate the great rhombicosidodecahedron and the small rhombicosidodecahedron. From the small one we continue the truncation process to produce the truncated dodecahedron and finally the dodecahedron itself. All of these derivative polyhedra belong to the 5.3.2 symmetry group.

The 3.2 Symmetry Group

The tetrahedron is unique in that it is its own dual. It has four 3-fold axes of symmetry, where each of these axes passes through a vertex and the center of the opposite face. It also has three 2 fold axes, each passing through the center points of opposite edges (Figure 5.68).

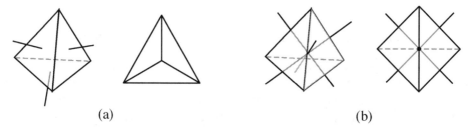

(a)   (b)

Figure 5.68 The tetrahedron and 3.2 symmetry.

Truncating each of the four vertices of the tetrahedron so that one-third of each edge remains, generates the truncated tetrahedron (Figure 5.69). Truncating further, so that each of the six edges of the original tetrahedron becomes a vertex, generates the octahedron, which of course is in a different symmetry group.

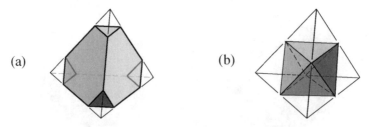

Figure 5.69 Truncations of the tetrahedron.

Section 5.9 Exercises

1. Describe the 4.3.2 axes of rotational symmetry for the octahedron.

2. Describe the 5.3.2 axes of symmetry for the dodecahedron.

6 MORE LINEAR TRANSFORMATIONS

Having considered the rigid body motions, reflection, and symmetry, we now explore the rest of the linear transformations. These are the more general affine transformations and the projection transformations. We'll consider the affine transformations of dilation and shear, where parallel lines remain parallel, and a variety of projections, where parallelism is not invariant but at least straight lines remain straight. We revisit appropriate Cartesian equations for these transformations presented in Chapter 2 and then express them as equivalent vector-matrix equations.

6.1 Isotropic Dilation

An isotropic dilation is a uniform expansion or contraction of the plane or space about some fixed point or center (Figure 6.1). If this point is the origin, then the transformation is a homogeneous isotropic dilation. (We will also study anisotropic dilation and shear.) Although an isotropic dilation is an affine transformation and preserves angle-size, it is not an isometry. However, the product of such a dilation and an isometry is a similarity. Recall that a similarity transformation is a one-to-one mapping in which all distances or lengths are multiplied by the same number, k, where k is the ratio of the similarity transformation. Physicists and engineers often refer to this ratio as strain. Also, recall that two figures are similar independent of their location and orientation. Mathematicians sometimes use the term homothety in place of dilation. A pure dilation (or homothety) produces an image parallel to the original figure. Thus, if corresponding sides of similar triangles are parallel, then the triangles are homothetic; but if we rotate one of them, then they are not homothetic.

If $k > 1$, then the dilation is an expansion. If $0 < k < 1$, then the dilation is a contraction. However, if $k < 0$, then the transformation is a dilation followed by an inversion or halfturn about the center. Figure 6.1a shows an k_a of a vector space, and Figure 6.1b shows a contraction k_b of this space. The inverse of dilation k is $1/k$.

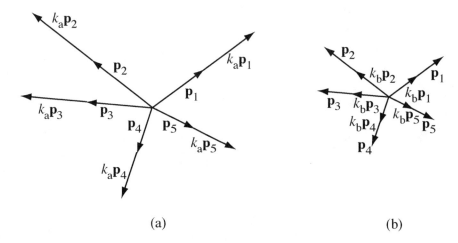

Figure 6.1 Isotropic dilation.

Isotropic Dilation Fixing the Origin

The Cartesian equations for a dilation fixing the origin are

$$x' = kx, \quad y' = ky, \quad z' = kz \tag{6.1}$$

In terms of vectors, this is simply a scalar multiplication: $\mathbf{p}' = k\mathbf{p}$. In matrix form we

write $\quad \mathbf{p}' = \mathbf{D}\mathbf{p} \tag{6.2}$

where $\quad \mathbf{D} = \begin{bmatrix} k & 0 & 0 & 0 \\ 0 & k & 0 & 0 \\ 0 & 0 & k & 0 \\ 0 & 0 & 0 & 1 \end{bmatrix} \tag{6.3}$

We see that the diagonal elements control the dilation transformation as they also did for reflections. The inverse of any dilation, $\mathbf{D}$, where $\mathbf{D}$ is given by Equation 6.3, is simply

$$\mathbf{D}^{-1} = \begin{bmatrix} 1/k & 0 & 0 & 0 \\ 0 & 1/k & 0 & 0 \\ 0 & 0 & 1/k & 0 \\ 0 & 0 & 0 & 1 \end{bmatrix} \tag{6.4}$$

Just what do we mean when we use the terms "uniform expansion"? Here is a definition we can test: If every pair of points separated by a distance d before a dilation k is separated by kd after the dilation, then the plane or space has undergone a uniform expansion, and similarly for a uniform contraction. The proof is simple. We'll use vector algebra to do it (Figure 6.2).

More Linear Transformations

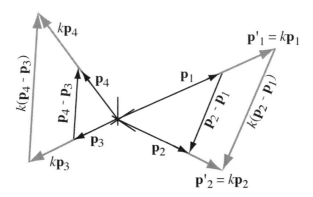

Figure 6.2 Uniform expansion.

The distance between two points $\mathbf{p}_1$ and $\mathbf{p}_2$ is $|\mathbf{p}_2 - \mathbf{p}_1|$, where $d = |\mathbf{p}_2 - \mathbf{p}_1|$. After an expansion k, the distance between $\mathbf{p}_1$ and $\mathbf{p}_2$ is kd, since $|\mathbf{p}'_2 - \mathbf{p}'_1| = k|\mathbf{p}_2 - \mathbf{p}_1|$. We apply a similar argument to any pair of points where $|\mathbf{p}_i - \mathbf{p}_j| = d$.

Centerless Uniform Expansion

We speak about a fixed point or center of expansion under a dilation transformation, but this point is merely an artifact of the Cartesian formulation of this transformation. Here is a slightly different way of looking at uniform expansion.

Using coordinate-free vector geometry, we can demonstrate that the expansion under an isotropic dilation k looks the same from any point $\mathbf{p}$ independent of the fixed point or center of expansion used in the Cartesian formulation of the transformation. An observer at $\mathbf{p}$ stakes out, or in some way marks, a point at a distance and direction $\mathbf{s}$ from $\mathbf{p}$, and another in the opposite direction, $-\mathbf{s}$ (Figure 6.3). The observer is thus at the midpoint of a line joining points $\mathbf{p} + \mathbf{s}$ and $\mathbf{p} - \mathbf{s}$ whose length is $l = 2|\mathbf{s}|$.

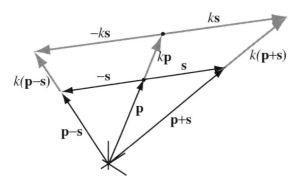

Figure 6.3 Centerless uniform expansion.

After a uniform expansion, k, the observer is at $\mathbf{p}' = k\mathbf{p}$, and the two staked points are now at $\mathbf{p}' + \mathbf{s}' = k(\mathbf{p}+\mathbf{s})$ and $\mathbf{p}' - \mathbf{s}' = k(\mathbf{p}-\mathbf{s})$. The length of the line joining them is $l' = 2k|\mathbf{s}|$. The observer compares l to l' and finds $l'/l = k$. Clearly, this is true no matter what the observer's position (including at the so-called center of expansion) and no matter where the observer places the stakes. Measurements made at or relative to any point $\mathbf{p}$ before and after expansion will yield the same (or uniform) expansion factor k. A good metaphor to represent a centerless expansion is the points on the surface of a balloon at successive stages of inflation. All move uniformly farther away from each other, and there is no privileged point.

Isotropic Dilation Fixing an Arbitrary Point

Uniform dilation fixing an arbitrary point in space, $\mathbf{p}_c$, is a three-step process. First, we must execute a translation $\mathbf{T}_c^{-1}$ that maps $\mathbf{p}_c$ to the origin. Next, we execute the actual expansion $\mathbf{D}$. Finally, we execute a translation $\mathbf{T}_c$ that reverses the first translation and returns $\mathbf{p}_c$ to its original position. In matrix form we write

$$\mathbf{H} = \mathbf{T}_c \mathbf{D} \mathbf{T}_c^{-1} \tag{6.5}$$

where

$$\mathbf{T}_c^{-1} = \begin{bmatrix} 1 & 0 & 0 & -x_c \\ 0 & 1 & 0 & -y_c \\ 0 & 0 & 1 & -z_c \\ 0 & 0 & 0 & 1 \end{bmatrix} \tag{6.6}$$

$$\mathbf{D} = \begin{bmatrix} k & 0 & 0 & 0 \\ 0 & k & 0 & 0 \\ 0 & 0 & k & 0 \\ 0 & 0 & 0 & 1 \end{bmatrix} \tag{6.7}$$

and

$$\mathbf{T}_c = \begin{bmatrix} 1 & 0 & 0 & x_c \\ 0 & 1 & 0 & y_c \\ 0 & 0 & 1 & z_c \\ 0 & 0 & 0 & 1 \end{bmatrix} \tag{6.8}$$

The resulting transformation matrix $\mathbf{H}$ is

$$\mathbf{H} = \begin{bmatrix} k & 0 & 0 & (1-k)x_c \\ 0 & k & 0 & (1-k)y_c \\ 0 & 0 & k & (1-k)z_c \\ 0 & 0 & 0 & 1 \end{bmatrix} \tag{6.9}$$

More Linear Transformations

From the final transformation equation in matrix form

$$\mathbf{p}' = \mathbf{T}_C \mathbf{D} \mathbf{T}_C^{-1} \mathbf{p} \qquad (6.10)$$
$$= \mathbf{H}\mathbf{p}$$

we obtain the Cartesian transformation equations

$$\begin{aligned} x' &= kx + (1-k)x_c \\ y' &= ky + (1-k)y_c \\ z' &= kz + (1-k)z_c \end{aligned} \qquad (6.11)$$

The Product of Two Dilations with the Same Center

The product of two dilations k_1 and k_2 with the same center (e.g., origin) is

$$\mathbf{H} = \mathbf{D}_2 \mathbf{D}_1 \qquad (6.12)$$

in the order k_1 followed by k_2. From a repeated application of Equation 6.3, with proper notation added, we find

$$\mathbf{H} = \begin{bmatrix} k_1 k_2 & 0 & 0 & 0 \\ 0 & k_1 k_2 & 0 & 0 \\ 0 & 0 & k_1 k_2 & 0 \\ 0 & 0 & 0 & 1 \end{bmatrix} \qquad (6.13)$$

and the corresponding Cartesian equations are

$$x' = k_1 k_2 x, \ y' = k_1 k_2 y, \text{ and } z' = k_1 k_2 z \qquad (6.14)$$

The Product of Two Dilations with Different Centers

Now things become a little more complicated. The matrix representing the product of two dilations $k_1 k_2$ with different centers $\mathbf{p}_{c_1}$ and $\mathbf{p}_{c_2}$ is (using Equation 6.9)

$$\begin{aligned} \mathbf{H} &= \begin{bmatrix} k_2 & 0 & 0 & (1-k_2)x_{c_2} \\ 0 & k_2 & 0 & (1-k_2)y_{c_2} \\ 0 & 0 & k_2 & (1-k_2)z_{c_2} \\ 0 & 0 & 0 & 1 \end{bmatrix} \begin{bmatrix} k_1 & 0 & 0 & (1-k_1)x_{c_1} \\ 0 & k_1 & 0 & (1-k_1)y_{c_1} \\ 0 & 0 & k_1 & (1-k_1)z_{c_1} \\ 0 & 0 & 0 & 1 \end{bmatrix} \\ &= \begin{bmatrix} k_1 k_2 & 0 & 0 & k_2(1-k_1)x_{c_1} + (1-k_2)x_{c_2} \\ 0 & k_1 k_2 & 0 & k_2(1-k_1)y_{c_1} + (1-k_2)y_{c_2} \\ 0 & 0 & k_1 k_2 & k_2(1-k_1)z_{c_1} + (1-k_2)z_{c_2} \\ 0 & 0 & 0 & 1 \end{bmatrix} \end{aligned} \qquad (6.15)$$

Notice the diagonal elements of this matrix. They are identical to those of Equation 6.13. The actual expansion or contraction ratio k_1k_2 is the same for the product of two dilations with the same or different centers. This affects the metric properties in the same way, too.

Section 6.1 Exercises

1. Show that parallelism is preserved under isotropic dilation in the plane.
2. Show that any polygon goes into a similar polygon under a dilation transformation.
3. Show that angle size is preserved under a dilation k.
4. Show that an isotropic dilation k multiplies the area of every triangle and circle by k^2.
5. Find the center of dilation for the transformation given by the matrix
$$\begin{bmatrix} k & 0 & c_1 \\ 0 & k & c_2 \\ 0 & 0 & 1 \end{bmatrix}$$
6. Show that $\mathbf{DD}^{-1} = \mathbf{I}$.

6.2 Anisotropic Dilation

An anisotropic dilation is an expansion or contraction of the plane or space whose ratio is dependent on orientation. This means that although two line segments are of equal length, if their directions are different, then, in general, under an anisotropic dilation the lengths of their images are not equal. We are free to impose different dilation ratios in different directions.

Anisotropic Dilation Fixing the Origin

The simplest example of an anisotropic dilation is that of three mutually perpendicular dilations centered at the origin and aligned with the coordinate axes. In other words, the dilation or strain axes correspond to the coordinate axes. The orthogonality condition creates a state of pure strain, as a physicist or engineer would say. The Cartesian transformation equations for this are

$$\begin{aligned} x' &= k_x x \\ y' &= k_y y \\ z' &= k_z z \end{aligned} \qquad (6.16)$$

Let $\mathbf{D}$ denote an anisotropic dilation; then its matrix is

More Linear Transformations

$$\mathbf{D} = \begin{bmatrix} k_x & 0 & 0 & 0 \\ 0 & k_y & 0 & 0 \\ 0 & 0 & k_z & 0 \\ 0 & 0 & 0 & 1 \end{bmatrix} \quad (6.17)$$

The analog in the plane is obvious.

Let's now study anisotropic dilations in the plane to understand how the expansion ratio varies with direction. Again, we assume that the dilation axes correspond to the x and y axes. Let k_ϕ denote the ratio in the direction ϕ with respect to the x axis (Figure 6.4).

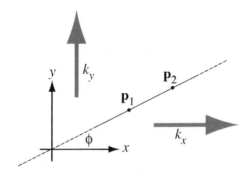

Figure 6.4 Anisotropic dilation in the plane.

Our approach here will be to compare lengths (distances between a pair of points $\mathbf{p}_1, \mathbf{p}_2$ on a line through the origin), before and after an anisotropic dilation, using $k_\phi = l'/l$. The distance between two points $\mathbf{p}_1$ and $\mathbf{p}_2$ is

$$l = \sqrt{(x_2 - x_1)^2 + (y_2 - y_1)^2} \quad (6.18)$$

After an anisotropic dilation with ratios k_x and k_y the distance between these same points is

$$l' = \sqrt{(x_2' - x_1')^2 + (y_2' - y_1')^2} \quad (6.19)$$

Substituting from Equation 6.16 for x' and y' we find

$$l' = \sqrt{k_x^2(x_2 - x_1)^2 + k_y^2(y_2 - y_1)^2} \quad (6.20)$$

so that

$$k_\phi = \frac{l'}{l} = \sqrt{\frac{k_x^2(x_2 - x_1)^2 + k_y^2(y_2 - y_1)^2}{(x_2 - x_1)^2 + (y_2 - y_1)^2}} \quad (6.21)$$

This form of expressing the ratio obscures the effect of orientation. We can lift the fog by using the relationship $y = x \tan \phi$. Appropriately substituting this into Equation 6.21, we obtain

$$k_\phi = \sqrt{k_x^2 \cos^2 \phi + k_y^2 \sin^2 \phi} \tag{6.22}$$

This is much better. Since k_x and k_y are orthogonal, we immediately see by inspection the maximum and minimum dilation and the sought after relationship to direction, all in a single expression. We also recognize that a graph of k_ϕ and ϕ produces an ellipse (Figure 6.5), where k_x and k_y are the major and minor axes depending on their relative magnitudes.

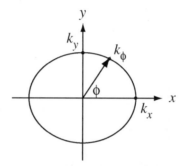

Figure 6.5 Variation of k_ϕ in the plane.

Now let's look at some slightly more complex dilations. Although we will continue to focus our attention on these transformations as they apply to geometry in the plane, the extension to three dimensions is fairly obvious and direct.

There is no reason to constrain the dilation axes to a collinear correspondence with the coordinate axes. We can consider just as easily two mutually orthogonal dilations k_1 and k_2, where k_1 makes an angle α with the x axis (Figure 6.6). In this case, the final transformation matrix $\mathbf{H}$ is a product of three other matrices: $\mathbf{R}_\alpha^{-1}$ rotates k_1 into the x axis, $\mathbf{D}$ is the dilation, and $\mathbf{R}_\alpha$ rotates k_1 back to its original orientation. Thus

$$\mathbf{H} = \mathbf{R}_\alpha \mathbf{D} \mathbf{R}_\alpha^{-1} \tag{6.23}$$

or
$$\mathbf{H} = \begin{bmatrix} k_1 \cos^2 \alpha + k_2 \sin^2 \alpha & (k_1 - k_2) \sin \alpha \cos \alpha & 0 \\ (k_1 - k_2) \sin \alpha \cos \alpha & k_1 \sin^2 \alpha + k_2 \cos^2 \alpha & 0 \\ 0 & 0 & 1 \end{bmatrix} \tag{6.24}$$

MORE LINEAR TRANSFORMATIONS

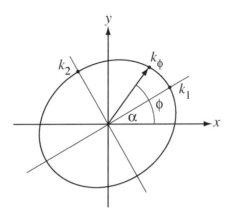

Figure 6.6 General orthogonal dilations in the plane.

The net dilation k_ϕ in a direction ϕ lies on an ellipse rotated by α with respect to the coordinate axes, where k_1 and k_2 are the major and minor axes of the ellipse, again depending on their relative magnitudes.

We can also remove another constraint and generalize this transformation even further. If we no longer require an orthogonal relationship between k_1 and k_2, then we must specify the orientation of each one. Let α_1 and α_2 denote the orientation of k_1 and k_2, respectively, relative to the x axis (Figure 6.7). Then we write

$$\mathbf{H} = \mathbf{R}_{\alpha_2} \mathbf{D}_2 \mathbf{R}_{\alpha_2}^{-1} \mathbf{R}_{\alpha_1} \mathbf{D}_1 \mathbf{R}_{\alpha_1}^{-1} \tag{6.25}$$

where $\mathbf{D}_1 = \begin{bmatrix} k_1 & 0 & 0 \\ 0 & 1 & 0 \\ 0 & 0 & 1 \end{bmatrix}$, and $\mathbf{D}_2 = \begin{bmatrix} 1 & 0 & 0 \\ 0 & k_2 & 0 \\ 0 & 0 & 1 \end{bmatrix}$.

Now determining the effect of orientation becomes far more complex, as is the case for finding maximum and minimum dilations. We will not attempt this here. It is, however, a challenge worth the effort.

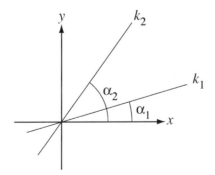

Figure 6.7 General nonorthogonal dilation in the plane.

Anisotropic Dilation Fixing an Arbitrary Point

So far, in this section, we have discussed anisotropic dilations fixing the origin. Now we move on to anisotropic dilations fixing an arbitrary point, say $\mathbf{p}_c$. To do this we simply introduce a translation $\mathbf{T}_c$ and its inverse $\mathbf{T}_c^{-1}$. The general transformation matrix then becomes

$$\mathbf{H} = \mathbf{T}_c \mathbf{D} \mathbf{T}_c^{-1} \tag{6.26}$$

or

$$\mathbf{H} = \begin{bmatrix} 1 & 0 & 0 & x_c \\ 0 & 1 & 0 & y_c \\ 0 & 0 & 1 & z_c \\ 0 & 0 & 0 & 1 \end{bmatrix} \begin{bmatrix} k_x & 0 & 0 & 0 \\ 0 & k_y & 0 & 0 \\ 0 & 0 & k_z & 0 \\ 0 & 0 & 0 & 1 \end{bmatrix} \begin{bmatrix} 1 & 0 & 0 & -x_c \\ 0 & 1 & 0 & -y_c \\ 0 & 0 & 1 & -z_c \\ 0 & 0 & 0 & 1 \end{bmatrix} \tag{6.27}$$

so that

$$\mathbf{H} = \begin{bmatrix} k_x & 0 & 0 & (1-k_x)x_c \\ 0 & k_y & 0 & (1-k_y)y_c \\ 0 & 0 & k_z & (1-k_z)z_c \\ 0 & 0 & 0 & 1 \end{bmatrix} \tag{6.28}$$

Anisotropic Dilation Fixing a Line in the Plane

A special kind of anisotropic dilation in the plane emerges from setting $k_y = 1$. In Cartesian form we have

$$\begin{aligned} x' &= k_x x \\ y' &= y \end{aligned} \tag{6.29}$$

This is a dilation fixing a line in the plane; here the transformation fixes the y axis (Figure 6.8).

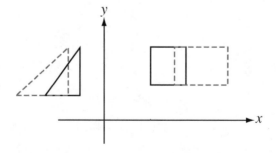

Figure 6.8 Dilation fixing a line in the plane: the y axis.

MORE LINEAR TRANSFORMATIONS

A more general form of this transformation fixes an arbitrary line in the plane. If we assume that the line passes through the origin at an angle ϕ with the x axis, then the transformation matrix is

$$\mathbf{H} = \mathbf{R}_\phi \mathbf{D} \mathbf{R}_\phi^{-1} \tag{6.30}$$

where $\mathbf{D} = \begin{bmatrix} 1 & 0 & 0 \\ 0 & k_y & 0 \\ 0 & 0 & 1 \end{bmatrix}$.

Anisotropic Dilation Fixing a Line in Space

We find other interesting anisotropic dilations simply by manipulating the ratios in Equation 6.17 and pre- and post-multiplying by appropriate rotation and translation matrices. For example, this transformation matrix fixes the z axis:

$$\mathbf{D} = \begin{bmatrix} k_x & 0 & 0 & 0 \\ 0 & k_y & 0 & 0 \\ 0 & 0 & 1 & 0 \\ 0 & 0 & 0 & 1 \end{bmatrix} \tag{6.31}$$

To fix an arbitrary line in space we use the now-familiar technique: apply a translation that puts the line through the origin, rotate the line into coincidence with a principal axis, dilate the points or the space, and reverse the earlier rotations and translation.

The Product of Anisotropic Dilations

The product of two sets of anisotropic dilations k_{x_1}, k_{y_1} and k_{x_2}, k_{y_2} that fixes the origin is

$$\mathbf{H} = \begin{bmatrix} k_{x_2} & 0 & 0 & 0 \\ 0 & k_{y_2} & 0 & 0 \\ 0 & 0 & k_{z_2} & 0 \\ 0 & 0 & 0 & 1 \end{bmatrix} \begin{bmatrix} k_{x_1} & 0 & 0 & 0 \\ 0 & k_{y_1} & 0 & 0 \\ 0 & 0 & k_{z_1} & 0 \\ 0 & 0 & 0 & 1 \end{bmatrix} \tag{6.32}$$

or $\mathbf{H} = \begin{bmatrix} k_{x_1} k_{x_2} & 0 & 0 & 0 \\ 0 & k_{y_1} k_{y_2} & 0 & 0 \\ 0 & 0 & k_{z_1} k_{z_2} & 0 \\ 0 & 0 & 0 & 1 \end{bmatrix} \tag{6.33}$

which is just what we should expect.

If the two sets of dilations have different centers, then by making use of Equation 6.28 we have

$$\mathbf{H} = \begin{bmatrix} k_{x_1} k_{x_2} & 0 & 0 & A_{1,4} \\ 0 & k_{y_1} k_{y_2} & 0 & A_{2,4} \\ 0 & 0 & k_{z_1} k_{z_2} & A_{3,4} \\ 0 & 0 & 0 & 1 \end{bmatrix} \qquad (6.34)$$

where

$$A_{1,4} = k_{x_2}\left(1 - k_{x_1}\right)x_{c_1} + \left(1 - k_{x_2}\right)x_{c_2}$$
$$A_{2,4} = k_{y_2}\left(1 - k_{y_1}\right)y_{c_1} + \left(1 - k_{y_2}\right)y_{c_2}$$
$$A_{3,4} = k_{z_2}\left(1 - k_{z_1}\right)z_{c_1} + \left(1 - k_{z_2}\right)z_{c_2}$$
$$A_{4,4} = 1$$

Again, the elements on the principal diagonal are identical to those in Equation 6.33.

Section 6.2 Exercises

1. Find **H** for an anisotropic dilation that fixes an arbitrary line $y = mx + b$ in the plane.
2. Find **D** for an anisotropic dilation that fixes the xy plane.

6.3 Shear

A slippery deck of cards provides an appropriate physical model of the shear transformation. Place a deck of cards on a table. Pushing down on the deck with sufficient pressure to keep it in tact, while pushing horizontally induces the kind of deformation shown in Figure 6.9. If we assume that the bottom card is fixed, then each card moves horizontally a distance directly proportional to its vertical position in the deck. The top card moves the farthest. It should be easy to abstract this idea to sets of points in the plane.

Figure 6.9 A slippery deck of cards.

Shear in the Plane Fixing the x Axis

The Cartesian equations for a shear transformation in the xy plane that fixes the x axis are (Figure 6.10a)

$$x' = x + k_y y \qquad (6.35)$$
$$y' = y$$

This transformation in matrix form is

$$\mathbf{p}' = \mathbf{S}_x \mathbf{p} \qquad (6.36)$$

where the subscript x on $\mathbf{S}$ indicates that the x axis is fixed and where

$$\mathbf{S}_x = \begin{bmatrix} 1 & k_y & 0 \\ 0 & 1 & 0 \\ 0 & 0 & 1 \end{bmatrix} \qquad (6.37)$$

We can express the shear coefficient k_y in terms of the angle ϕ (Figure 6.10b). Thus $k_y = -\tan\phi$.

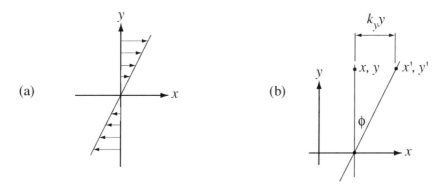

Figure 6.10 Shear in the plane fixing the x axis.

Line segments parallel to the x axis do not change length under this shear, while those parallel to the y axis change length according to the ratio $l'/l = \sqrt{1 + k_y^2}$ (see Exercise 2).

Rotation as the Product of Three Shears

We can represent every rotation in the plane as the product of three shears parallel to the x and y axes. (Gauss was the first to describe and analyze this transformation, and we must credit A. W. Paeth for the recent incarnation of it.) This transformation is particularly useful for its computation speed and economy in rotating raster images of a computer graphic display. Here we direct our attention to rotation about the origin. The more general case is easily reduced to this. From Equation 4.6 in Chapter 4 we have

$$\mathbf{R} = \begin{bmatrix} \cos\phi & -\sin\phi & 0 \\ \sin\phi & \cos\phi & 0 \\ 0 & 0 & 1 \end{bmatrix}$$

and from Equation 6.37 and its y axis counterpart, we have

$$\mathbf{S}_x = \begin{bmatrix} 1 & k_y & 0 \\ 0 & 1 & 0 \\ 0 & 0 & 1 \end{bmatrix} \quad \text{and} \quad \mathbf{S}_y = \begin{bmatrix} 1 & 0 & 0 \\ k_x & 1 & 0 \\ 0 & 0 & 1 \end{bmatrix}$$

Next, we alternate three arbitrary and distinct x and y shears and equate their product matrix to the rotation matrix. These three shears plus the orthogonality condition allow us to solve the resulting matrix equation. Thus, we have

$$\begin{bmatrix} 1 & a & 0 \\ 0 & 1 & 0 \\ 0 & 0 & 1 \end{bmatrix} \begin{bmatrix} 1 & 0 & 0 \\ b & 1 & 0 \\ 0 & 0 & 1 \end{bmatrix} \begin{bmatrix} 1 & c & 0 \\ 0 & 1 & 0 \\ 0 & 0 & 1 \end{bmatrix} = \begin{bmatrix} \cos\phi & -\sin\phi & 0 \\ \sin\phi & \cos\phi & 0 \\ 0 & 0 & 1 \end{bmatrix}$$

or

$$\begin{bmatrix} ab+1 & a+abc+c & 0 \\ b & bc+1 & 0 \\ 0 & 0 & 1 \end{bmatrix} = \begin{bmatrix} \cos\phi & -\sin\phi & 0 \\ \sin\phi & \cos\phi & 0 \\ 0 & 0 & 1 \end{bmatrix}$$

Solving this equation we obtain

$$a = c = (\cos\phi - 1)/\sin\phi$$
$$b = \sin\phi$$

When $\phi \to 0$ the expression $(\cos\phi - 1)/\sin\phi$ becomes computationally unreliable. However, from elementary trigonometry we find that $-\tan(\phi/2) = (\cos\phi - 1)/\sin\phi$. We conclude that a rotation by ϕ about the origin is equivalent to the product of shears

$$\begin{bmatrix} 1 & -\tan(\phi/2) & 0 \\ 0 & 1 & 0 \\ 0 & 0 & 1 \end{bmatrix} \begin{bmatrix} 1 & 0 & 0 \\ \sin\phi & 1 & 0 \\ 0 & 0 & 1 \end{bmatrix} \begin{bmatrix} 1 & -\tan(\phi/2) & 0 \\ 0 & 1 & 0 \\ 0 & 0 & 1 \end{bmatrix}$$

Shear in Space

We will consider two kinds of shear in space that fix a plane. The formulation is simple if we choose to fix the xy plane. Study Figure 6.11, where the z axis is

MORE LINEAR TRANSFORMATIONS

normal to the page. Here is that slippery deck of cards again, and we are looking down at the top of it. The shear deformations we induce in Figures 6.11a and b are quite similar, but in 6.11c something different is happening.

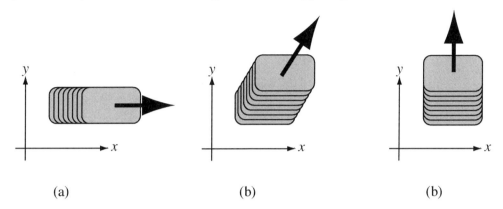

(a)   (b)   (b)

Figure 6.11 Shear in space.

The two shears of a and b are now working together to produce the deformation. The Cartesian transformation equations describing each of them are simply:

In Figure 6.11a

$$x' = x + k_1 z$$
$$y' = y \qquad (6.38)$$
$$z' = z$$

In Figure 6.11b

$$x' = x$$
$$y' = y + k_2 z$$
$$z' = z$$

In figure 6.11c

$$x' = x + k_1 z$$
$$y' = y + k_2 z$$
$$z' = z$$

Equiareal Transformations

Finally, a brief comment about an invariance property of the shear transformation: An equiareal transformation is an affine transformation under which any triangle corresponds to another triangle of the same area. The test for this is $|\mathbf{H}| = 1$. For shear in the plane we find

$$|\mathbf{H}| = \begin{vmatrix} 1 & k_y & 0 \\ 0 & 1 & 0 \\ 0 & 0 & 1 \end{vmatrix} = 1 \tag{6.39}$$

For the product of two shears in the plane we find

$$|\mathbf{H}| = \begin{vmatrix} 1 & k_y & 0 \\ k_x & k_x k_y + 1 & 0 \\ 0 & 0 & 1 \end{vmatrix} = k_x k_y + 1 - k_x k_y = 1 \tag{6.40}$$

Therefore, we conclude that shear transformations in the plane are equiareal.

Section 6.3 Exercises

1. Write the Cartesian equations for a shear transformation in the plane that fixes the y axis.
2. Find the ratio l'/l for line segments parallel to the y axis under the shear transformation of Equation 6.35.
3. Derive the general equation for l'/l as a function of k_y and ϕ, where ϕ is the angle the line segment makes with the x axis.
4. Derive the transformation matrix $\mathbf{H}$ for a shear in the plane fixing an arbitrary line $y = mx + b$.
5. Derive the transformation matrix $\mathbf{H}$ for the product of two shears in the plane, one of which fixes the x axis and the other the y axis. Does the order of the product matter? Defend your answer.
6. Write the transformation matrices for the three shears described by Equations 6.38.

6.4 Projective Geometry

During the four hundred years that passed between Brunelleschi's discussions of architectural perspective drawing and Klein's work establishing its firm algebraic foundations, what we now call projective geometry slowly matured. In the sixteenth century, during the time Descartes was formulating his own revolutionary ideas, Gerard Desargues (1593-1662), an architect from Lyons, extended the Euclidean interpretation of points, lines and planes and their relationship to each other. This work, as well as that of Pascal and Poncelet, created the basis for modern projective geometry.

Projection transformations preserve points, straight lines, intersections, the degree of a curve, and some other geometric properties. The properties not preserved

MORE LINEAR TRANSFORMATIONS

by these transformations are equally important and interesting; they include parallels, angles and the distance between points. These transformations are important for many reasons, not the least of which is that they allow us to produce and understand two-dimensional images of three-dimensional objects. Study the shadows cast by variously shaped objects to gain an intuitive understanding of these transformations, and verify that the metric properties of their shadows are not invariant as you change the position of the light source. For example, we see in the shadows of objects the lengthening and foreshortening of edges, although straight edges do remain straight; we see angles distorted, where the shadows of angles are greater or less than corresponding angles on the object itself; and we see parallel edges converge.

Every theorem of projective geometry also holds for affine geometry, since every property preserved by all the transformations of projective geometry are also preserved by any more restrictive subset of these transformations. For example, projection transformations of the plane onto itself include all of the affine transformations as well as certain transformations that are not affine. The distinguishing characteristic of the projection of one plane onto another is simple to state: the lines joining points on one plane to their projected images on another plane are either parallel or concurrent.

A projective transformation of a plane onto itself does this in a one-to-one way, except that the points of one line may have no images and the points of another line may have no originals. Here are two examples of projection of the plane that are not one-to-one. The first is represented by the transformation matrix equation

$$\begin{bmatrix} x' \\ y' \end{bmatrix} = \begin{bmatrix} 1 & 0 \\ 0 & 0 \end{bmatrix} \begin{bmatrix} x \\ y \end{bmatrix} \tag{6.41}$$

which is an example of a mapping of the set of points on the plane into, but not onto, itself (Figure 6.12). It maps points on the plane onto the x axis. It has no inverse, since we cannot map points from the line back to the plane.

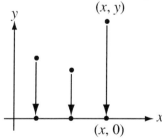

Figure 6.12 Projecting points on the plane onto the *x* axis.

The second example is the transformation

$$\begin{bmatrix} x' \\ y' \end{bmatrix} = \begin{bmatrix} 1 & 0 \\ 1 & 0 \end{bmatrix} \begin{bmatrix} x \\ y \end{bmatrix} \tag{6.42}$$

which again is an example of a mapping of the plane into, but not onto, itself (Figure 6.13). It maps points on the plane onto the line $y = x$. Moreover, it, too, has no inverse.

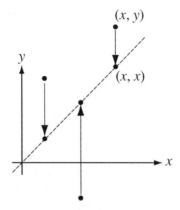

Figure 6.13 Projecting points on the plane onto the line *y = x*.

Although we are usually more concerned with the projection of solid objects in space onto an image plane, rather than with the more abstract considerations associated with the projection of the plane onto itself, we will examine each of these projections and more in this and the following sections. In Section 2.7, we briefly examined the equations of projective transformations and the central projection of figures in the plane and in space. Here we will take a closer look at the nature of projective geometry itself, parallel projection, central projection, map projections and some display projections. Before moving into these areas, study the next two sections; each presents a projection you are very likely to be familiar with already.

Projection Theorems of Trigonometry

The purpose of trigonometry is to deduce relationships among the sides and angles of a triangle. In the course of interpreting some of these relationships, we find a subtle connection with projections. Given triangle ABC (Figure 6.14), we define the vectors $\mathbf{a} = BC$, $\mathbf{b} = CA$, and $\mathbf{c} = AB$, and the angles $\alpha = (\mathbf{c}, -\mathbf{b})$, $\beta = (\mathbf{a}, -\mathbf{c})$, and $\gamma = (\mathbf{b}, -\mathbf{a})$. By definition, we know that $\mathbf{a} + \mathbf{b} + \mathbf{c} = 0$. If $\mathbf{u}_a$, $\mathbf{u}_b$, and $\mathbf{u}_c$ are the unit vectors of $\mathbf{a}$, $\mathbf{b}$, and $\mathbf{c}$, respectively, then the scalar products of the equation $\mathbf{a} + \mathbf{b} + \mathbf{c} = 0$ with each of the unit vectors yields the three projection equations of trigonometry:

$$a = b \cos \gamma + c \cos \beta$$
$$b = c \cos \alpha + a \cos \gamma$$
$$c = a \cos \beta + b \cos \alpha$$

More Linear Transformations

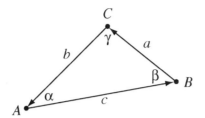

Figure 6.14 Projection theorems of trigonometry.

The length of side AB is composed of the algebraic sum of the orthogonal projections of sides BC and CA onto it, and similarly for the other two sides.

Vector Projection

The component of vector **a** along the line of vector **b** (or the projection of **a** onto **b**) is computable in scalar or vector form (Figure 6.15). The scalar projection of **a** onto **b** is $|\mathbf{a}'| = \mathbf{a} \bullet \mathbf{b}/|\mathbf{b}|$; the vector projection of **a** onto **b** is $\mathbf{a}' = (\mathbf{a} \bullet \mathbf{b})\mathbf{b}/|\mathbf{b}|^2$.

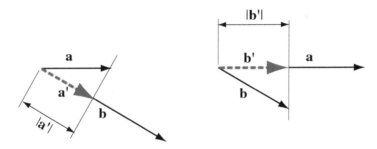

Figure 6.15 Vector projection.

The scalar projection of **b** onto **a** is $|\mathbf{b}'| = \mathbf{a} \bullet \mathbf{b}/|\mathbf{a}|$, and the vector projection is $\mathbf{b}' = (\mathbf{a} \bullet \mathbf{b})\mathbf{a}/|\mathbf{a}|^2$. In general,

$$\mathbf{a} \bullet \mathbf{b}/|\mathbf{b}| \neq \mathbf{a} \bullet \mathbf{b}/|\mathbf{a}| \text{ and } (\mathbf{a} \bullet \mathbf{b})\mathbf{b}/|\mathbf{b}|^2 \neq (\mathbf{a} \bullet \mathbf{b})\mathbf{a}/|\mathbf{a}|^2$$

The component of the vector **a** along a unit vector **u** is $a_x u_x + a_y u_y + a_z u_z$.

Projective Collineations of the Plane

In projective geometry, we study geometric properties that we derive without recourse to measurement or comparison of distances or angles. When we project a figure in one plane through a point and onto another plane, the distances and angles change and parallel lines do not necessarily remain parallel (Figure 6.16). Of course, some properties do remain invariant, or else we would have no geometry at all.

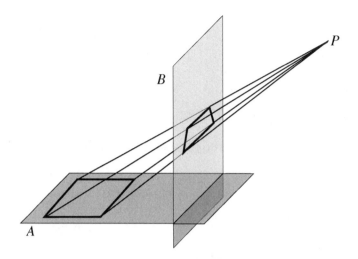

Figure 6.16 Projection of *A* onto *B* through *P*.

Points, lines and planes are the basic elements of projective geometry, and incidence relationships between these elements form the fundamental theorems. Homogeneous coordinates allow us to express projective properties as algebraic equations; otherwise, we would be restricted to making synthetic statements only. Before passing to the abstractions of projective geometry, we next consider projective collineations of the plane.

Let Π, Π' be two distinct planes and P, P' two distinct points not on Π or Π'. If we project the points of Π through P onto Π' and then project these image points on Π' back onto Π through P', then we have produced a collineation of Π. In other words, to each point A in Π we assign the point A'', where $A'' = (AP \cap \Pi')P' \cap \Pi$ (Figure 6.17). It is, in fact, a central collineation, since every point on the line $\Pi \cap \Pi'$ and every line through $PP' \cap \Pi$ is left unchanged (Figure 6.18).

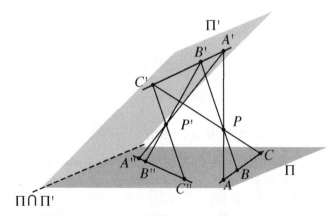

Figure 6.17 Projective collineation of the plane.

Figure 6.17 shows a collinear set of points $A, B, C, \ldots$ in Π projected onto Π' through P, creating collinear images $A', B', C', \ldots$. These are subsequently projected back onto Π through P', creating the collinear images $A'', B'', C''\ldots$.

Figure 6.18 shows that the image x'' of any point x on any line l in Π, containing the point $PP' \cap \Pi$, also lies on l. The proof of this is simple and develops from the observation that all the projective constructions lie in the plane containing l and the line through PP'.

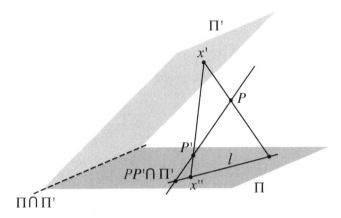

Figure 6.18 Special properties of a projective collineation of the plane.

Equations of Projective Geometry

The equations of projective geometry in the plane are (from Chapter 2, Equation 2.34)

$$x' = \frac{ax + by + h}{ex + fy + g}$$
$$y' = \frac{cx + dy + k}{ex + fy + g} \qquad (6.43)$$

with the following condition on the coefficients:

$$\begin{vmatrix} a & b & h \\ c & d & k \\ e & f & g \end{vmatrix} \neq 0.$$

This leads us to examine the common denominator of Equations 6.43, where we see that $ex + fy + g = 0$ represents a line. Any point not on this line has an image, whereas any point on the line does not have an image. Conversely, solving these equations for x and y we obtain equations of the form

$$x = \frac{Ax' + By' + H}{Ex' + Fy' + G}$$
$$y = \frac{Cx' + Dy' + K}{Ex' + Fy' + G} \tag{6.44}$$

We quickly realize that $Ex' + Fy' + G = 0$ is another special line. Points on this line have no original, and all points not on it have corresponding original points. These equations and the interpretation of their special conditions lead us to explore the projective plane with its ideal points and lines, and the usefulness of homogeneous coordinates.

The Projective Plane

In the sixteenth and seventeenth century, geometers began to suspect that the Euclidean plane was somehow incomplete, that there was a certain lack of symmetry in the conventionally stated relationships between points and lines. One of the driving forces behind this suspicion was the idea that a set of parallel lines can be thought of as passing through a point at infinity. This allowed them to extend the notion of the Euclidean plane by adding to it points at infinity. Moreover, they went further and considered these points to lie on a line (the ideal line) also at infinity.

If we now assume that all points and lines, including ideal points and lines, are accorded equal status, then we dramatically change the nature of the plane. We create the projective plane. In doing this we recognize that points and lines now satisfy the following two axioms:

1. Two distinct points in the plane determine a unique line on which they both lie.

2. Two distinct lines in the plane determine a unique point through which they both pass.

Notice that we no longer admit the possibility of nonintersecting parallel lines. We have thus created a geometry on an extended plane where points, lines, and intersections (or incidences) have a crucial role. This is projective geometry.

How do we assign coordinates to points in the projective plane, since there are no pairs of numbers left over for points at infinity when we use the real number pairs (x, y)? The set of real number pairs $\{x, y\}$ merely generates the familiar Euclidean plane. The new coordinates we must use are the homogeneous coordinates. If $\{x, y, z\}$ is the set of all real number triples, where at least one of the numbers x, y or z is nonzero, then we consider two triples to be identical if they differ only by a constant k such that $\{x, y, z\} = \{kx, ky, kz\}$. If $z \neq 0$, then $\{x, y, z\} = \{x/z, y/z, 1\}$ corresponds to the Euclidean point with Euclidean coordinates $(x/z, y/z)$. If $z = 0$, then $\{x, y, 0\}$ corresponds to the point at infinity through which all the lines with direction numbers (x, y) pass.

MORE LINEAR TRANSFORMATIONS

Here is an example of a natural mapping of Euclidean three-space onto the projective plane (Figure 6.19). It sends a point with coordinates x, y, z to a point on the projective plane with coordinates $x' = x/z$, $y' = y/z$, and it defines every point except the origin.

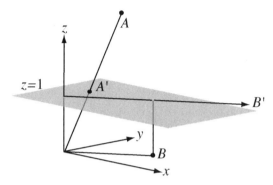

Figure 6.19 Euclidean three-space mapped onto the projective plane.

We consider two categories of Euclidean points. First, if $z \neq 0$, then point A, or (x, y, z), transforms to the point of intersection A' of the plane $z = 1$ and the line through A and the origin. Second, if $z = 0$, then point B, or $(x, y, 0)$ transforms to the point B' at infinity on the projective plane in the direction x, y. The inverse of this transformation does not produce a unique point. This is true for both points A' and B'. Therefore, the transformation is not one-to-one. Felix Klein offers us some insight on both homogeneous coordinates and the projective plane, in the next section.

Felix Klein on Projective Transformations

Felix Klein observed that "…operation with homogeneous coordinates produces, at least with beginners, something like physical discomfort." He offered some comments and geometric interpretations aimed at relieving this condition. Here is a somewhat abbreviated representation of them (Figure 6.20).

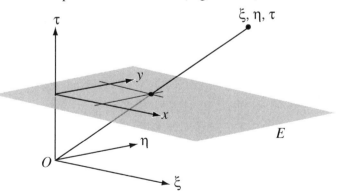

Figure 6.20 Homogeneous coordinates and projective transformations.

We begin with an assertion: The rectangular coordinates of points in plane E are $x = \xi/\tau$, $y = \eta/\tau$. We may interpret ξ, η, τ as rectangular coordinates in space. Now choose the plane $\tau = 1$, parallel to the ξ, η-plane, as the plane E. In E let $x = \xi$, $y = \eta$. Join the point x, y of E to O by a straight line. Then for points on this line ξ/τ and η/τ are constant and $\xi/\tau = x$, $\eta/\tau = y$, since for $\tau = 1$ we find that $\xi = x$, $\eta = y$. The use of homogeneous coordinates signifies the intersection of E with the set of lines radiating from O. Thus, the homogeneous coordinates of a point in E are simply the space coordinates of the points of the projecting ray of that point.

We see that the indefiniteness of the homogeneous coordinates of the points of E arises because to each point of E there are correspondingly infinitely many points on the ray. We excluded the point $\xi, \eta, \tau = 0$, since the point O does not determine a ray and thus no point in E. We do not need infinite values of ξ, η, τ, since we obtain all rays by joining O with finite points. Finally, we avoid infinitely large values of the coordinates by replacing the infinite region of E by parallel rays through O given by $\tau = 0$.

The concept of a line at infinity arises from the condition that all infinitely distant points satisfy the linear equation $\tau = 0$, just as every infinite line has a linear equation. Therefore, we designate the infinitely distant line as the set of points in E that satisfies $\tau = 0$.

Homogeneous Coordinates of Points

We have just reviewed the equations giving the projective transformation of the Euclidean plane onto itself and confronted the natural extension of the Euclidean plane to the more inclusive projective plane. Now it is appropriate that we further explore the nature of homogeneous coordinates. We already know that the equations applying to Euclidean points do not apply to the ideal points, and we know that the projective plane consists of both Euclidean points and ideal points. It is not unreasonable to seek equations and a system of coordinates that applies to all points in the projective plane. Therefore, we introduce a new set of coordinates that we call the homogeneous coordinates. Sometimes we refer to the ordinary rectangular coordinates of the plane as nonhomogeneous coordinates. The most striking characteristic of the new system is that each point on the projective plane has three numbers associated with it instead of the usual two.

First, let's define the new coordinates for Euclidean points. Later we will extend this to include ideal points. Three numbers h_1, h_2, and h_3, in this order, are the homogeneous coordinates of the Euclidean point (x, y) if and only if

$$\frac{h_1}{h_3} = x, \quad \frac{h_2}{h_3} = y, \text{ where } h_3 \neq 0 \tag{6.45}$$

We write these homogeneous coordinates as the real-number triple (h_1, h_2, h_3).

More Linear Transformations

The homogeneous coordinates (4,3,1) represent the point (4,3) in the plane, as is also true for (8,6,2), (12,9,3) and $(4k, 3k, k)$, where $k \neq 0$. This example tells us that every Euclidean point in the projective plane has infinitely many sets of homogeneous coordinates, where any two sets are proportional. See the subsection below on projective geometry in space for a view of this in terms of the projective geometry of space.

If (a,b) is a Euclidean point, then one set of homogeneous coordinates is $(a,b,1)$, and other sets are (ka, kb, k), where $k \neq 0$. Conversely, any three given numbers h_1, h_2, h_3 are homogeneous coordinates of a Euclidean point if $h_3 \neq 0$. Two triples (a_1, a_2, a_3) and (b_1, b_2, b_3) represent the same Euclidean point if and only if the determinants

$$\begin{vmatrix} a_1 & a_2 \\ b_1 & b_2 \end{vmatrix} = \begin{vmatrix} a_2 & a_3 \\ b_2 & b_3 \end{vmatrix} = \begin{vmatrix} a_3 & a_1 \\ b_3 & b_1 \end{vmatrix} = 0 \tag{6.46}$$

The proof of this is straightforward. Hint: Assume that the triples represent the same Euclidean point so that the homogeneous coordinates are proportional (see Exercise 1).

If we relax the condition $h_3 \neq 0$, then a set of homogeneous coordinates of the form $(h_1, h_2, 0)$ represents an ideal point. Thus, (4,1,0) and (–2,5,0) represent ideal points on the projective plane. To sharpen our conception and intuition of ideal points, we must investigate the lines on which these points fall.

Parallel Lines, Ideal Points and Homogeneous Coordinates

Consider a family of parallel straight lines in a plane with a slope equal to 3 (Figure 6.21). The general Cartesian equation for any line of this family is $y = 3x + b$.

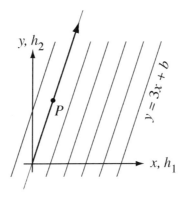

Figure 6.21 A family of parallel straight lines.

The lines of this family are all associated with the same ideal point I. Let's see how this arises in terms of homogeneous coordinates. To determine the homogeneous coordinates of I, consider the Euclidean point P on a line of the family. (Since b is an arbitrary constant, we will choose the line through the origin, for which $b = 0$.) Let P recede indefinitely far along the line. We see that the coordinates of P approach definite limits, and in particular h_3 approaches zero. It is reasonable to take these limits as the homogeneous coordinates of I. The proof proceeds as follows:

If (ξ, η) are the Euclidean coordinates of P, then $(\xi, \eta, 1)$ are the corresponding homogeneous coordinates. This is also true for $(1, \eta/\xi, 1/\xi)$ for $\xi \neq 0$ (that is, if P is not on the y axis). Now, since $\eta = 3\xi + b$, substituting we can rewrite the last set of coordinates as $\{1, (3\xi+b)/\xi, 1/\xi\}$ or $\{1, 3+(b/\xi), 1/\xi\}$. As P recedes indefinitely far in either direction along the line, we find that $|\xi| \to \infty$ while b is constant. Therefore b/ξ and $1/\xi$ approach zero, and the homogeneous coordinates of P approach $(1,3,0)$; these are the coordinates of the ideal point I. It is easy to see how we generalize this so that $(1, m, 0)$ are the homogeneous coordinates of the ideal point associated with the family of parallel straight lines whose slope is m.

If every ideal point in the projective plane, like every Euclidean point, is to have an infinite number of proportional coordinates, then it is clear that coordinates $(k, km, 0)$ represent the same ideal point. We can summarize all this as follows: The ideal point associated with the family of straight lines whose slope is m has homogeneous coordinates $(1, m, 0)$ or, more generally, $(k, km, 0)$, where $k \neq 0$.

Special Points on the Projective Plane

We have now assigned homogeneous coordinates to all but three special points on the projective plane (Figure 6.22). The first of these, the origin, we assign the coordinates $(0,0,1)$ or, more generally, $(0,0,k)$. For lines parallel to the x axis we have $m = 0$, so that the homogeneous coordinates of the ideal point associated with these lines or, more particularly, the x axis is $(1,0,0)$ or $(k,0,0)$. As an exercise show that the coordinates of the ideal point associated with the y-axis and lines parallel to it are $(0,1,0)$ or $(0,k,0)$. We call the axes with the addition of the associated ideal points the extended x and y axis. Any ordered number triple corresponds to a point in the projective plane, with the exception of $(0,0,0)$ which has no meaning in this context and is unassigned.

More Linear Transformations

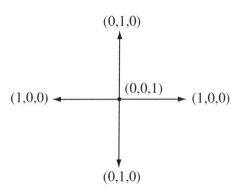

Figure 6.22 Special points on the projective plane.

Equations of Projective Lines

A polynomial equation is a homogeneous equation if all of its nonzero terms are of the same degree. Thus $x^2 + y^2 + z^2 = 0$ is a homogeneous equation of the second degree, whereas $x^2 + y^2 + z^2 = 8$ is a nonhomogeneous equation of the same degree. In general, most polynomial equations of curves expressed in rectangular Cartesian coordinates x, y are nonhomogeneous. We can make these equations homogeneous in, say, h_1, h_2, h_3 by substituting $x = h_1/h_3$ and $y = h_2/h_3$ and eliminating fractions. For example, $2x - y = 3$ becomes $2h_1 - h_2 = 3h_3$. This provides us with further justification for the distinction between homogeneous and nonhomogeneous coordinates.

Equations in h_1, h_2, h_3 apply to both Euclidean and ideal points, whereas equations in x and y only apply to Euclidean points. We shall see next that a homogeneous equation of the first degree represents a projective line. Conversely, we can represent every projective line by a homogeneous equation of the first degree. Let l be a straight line with the equation

$$4x - 2y + 3 = 0 \tag{6.47}$$

Substitute $x = \dfrac{h_1}{h_3}$ and $y = \dfrac{h_2}{h_3}$ to obtain

$$4\frac{h_1}{h_3} - 2\frac{h_2}{h_3} + 3 = 0 \tag{6.48}$$

or $\quad 4h_1 - 2h_2 + 3h_3 = 0 \tag{6.49}$

This last is an equation for l in homogeneous coordinates. What are the coordinates of the ideal point(s) satisfying Equation 6.49? If $h_3 = 0$, then $4h_1 - 2h_2 = 0$, or $h_1/h_2 = 1/2$, and any ideal point for which the ratio of h_1 to h_2 is 1/2 has coordinates that satisfy Equation 6.49. However, there is just one such ideal point and that is $(2k,k,0)$, where $k \neq 0$. This is the ideal point associated with any line l whose slope is 2. Equation 6.49 is the equation of the projective line l' associated with l. In fact, it is the equation for both l and l'.

Let's generalize the above example. Given the equation of any straight line g in nonhomogeneous coordinates

$$ax + by + c = 0 \tag{6.50}$$

where a and b are not both zero, the equation of g in homogeneous coordinates is

$$a\frac{h_1}{h_3} + b\frac{h_2}{h_3} + c = 0 \tag{6.51}$$

or $\quad ah_1 + bh_2 + ch_3 = 0 \tag{6.52}$

Every Euclidean point with coordinates (x, y) satisfying Equation 6.50 has homogeneous coordinates (h_1, h_2, h_3) in the projective plane satisfying Equation 6.52, and every Euclidean point with coordinates (h_1, h_2, h_3) in the projective plane satisfying Equation 6.52 has nonhomogeneous coordinates satisfying Equation 6.50. Thus, at least as far as Euclidean points are concerned, the locus of Equation 6.52 is identical to that of Equation 6.50.

Now let $h_3 = 0$ to obtain $ah_1 + bh_2 = 0$, or $h_1/h_2 = a/b$ if $a \neq 0$, and $h_2/h_1 = -a/b$ if $b \neq 0$. There is only one point which will satisfy these conditions, and that is $(-b, a, 0)$ or, more generally, $(-kb, ka, 0)$, where $k \neq 0$. This is the ideal point associated with g, and Equation 6.52 is the equation of the projective line also associated with g.

The ideal line is the locus of all the ideal points in the projective plane, and it has the equation $ch_3 = 0$. Writing this in the form $0 \cdot h_1 + 0 \cdot h_2 + ch_3 = 0$ we see that it is a special instance of Equation 6.52 where $a = b = 0$.

In summary then, every projective line has an equation of the form $ah_1 + bh_2 + ch_3 = 0$, where a, b, and c are not all zero. Conversely, the locus of each instance of this equation, with a, b, and c not all zero, is a projective line.

Equations of Projectivity in the Projective Plane

Now we will derive the equations for the projective transformations of the projective plane onto itself, writing Equation 6.41 in a more convenient form:

$$x' = \frac{a_1 x + a_2 y + a_3}{c_1 x + c_2 y + c_3}, \quad y' = \frac{b_1 x + b_2 y + b_3}{c_1 x + c_2 y + c_3} \tag{6.53}$$

where $\begin{vmatrix} a_1 & a_2 & a_3 \\ b_1 & b_2 & b_3 \\ c_1 & c_2 & c_3 \end{vmatrix} \neq 0$.

We use the following substitutions to express Equation 6.53 in terms of homogeneous coordinates: $x = h_1/h_3$, $y = h_2/h_3$ and $x' = h'_1/h'_3$, $y' = h'_2/h'_3$. After we make these substitutions and simplify the results, we obtain

$$\frac{h'_1}{h'_3} = \frac{a_1 h_1 + a_2 h_2 + a_3 h_3}{c_1 h_1 + c_2 h_2 + c_3 h_3}, \quad \frac{h'_2}{h'_3} = \frac{b_1 h_1 + b_2 h_2 + b_3 h_3}{c_1 h_1 + c_2 h_2 + c_3 h_3} \tag{6.54}$$

These equations imply that if P is any point in the projective plane with homogeneous coordinates (h_1, h_2, h_3), then under a projective transformation the coordinates of its image P' are (h'_1, h'_2, h'_3), where

$$\begin{aligned} h'_1 &= (a_1 h_1 + a_2 h_2 + a_3 h_3)k \\ h'_2 &= (b_1 h_1 + b_2 h_2 + b_3 h_3)k \\ h'_3 &= (c_1 h_1 + c_2 h_2 + c_3 h_3)k \end{aligned} \tag{6.55}$$

where $k \neq 0$. Since we may assume $k = 1$, Equations 6.55 simplify to

$$\begin{aligned} h'_1 &= a_1 h_1 + a_2 h_2 + a_3 h_3 \\ h'_2 &= b_1 h_1 + b_2 h_2 + b_3 h_3 \\ h'_3 &= c_1 h_1 + c_2 h_2 + c_3 h_3 \end{aligned} \tag{6.56}$$

where the condition on the determinant of the coefficients still holds. Furthermore, the set of all such transformations forms a group.

Projective Conics

Recall that the general equation for a conic in the plane is $Ax^2 + Bxy + Cy^2 + Dx + Ey + F = 0$. By rewriting this in terms of homogeneous coordinates we obtain

$$Ah_1^2 + Bh_1 h_2 + Ch_2^2 + Dh_1 h_3 + Eh_2 h_3 + Fh_3^2 = 0 \tag{6.57}$$

The difference between a locus of points defined by these two equations is that the locus of Equation 6.57 may contain ideal points. Thus, Equation 6.57 is the general equation for a projective conic.

Let's look at some examples. First, consider the parabola $y = x^2$. Expressing this in homogeneous coordinates, we obtain

$$h_2 h_3 = h_1^2 \tag{6.58}$$

To determine what, if any, ideal points this projective parabola may contain, we let $h_3 = 0$ to yield $h_1^2 = 0$, or $h_1 = 0$. Since in this case h_2 can take on any real value, we find that $(0, k, 0)$ is an ideal point on the projective parabola. If we let

$k = 1$, then the ideal point becomes $(0,1,0)$. This is the ideal point we associate with the extended y-axis in the projective plane, the axis of the given parabola.

Next consider the hyperbola $xy = 1$ or, in homogeneous coordinates in the projective plane,

$$h_1 h_2 = h_3^2 \tag{6.59}$$

Again, letting $h_3 = 0$, we obtain $h_1 h_2 = 0$. This is satisfied by either $h_1 = 0$ or $h_2 = 0$. These solutions produce two ideal points $(1,0,0)$ and $(0,1,0)$, the ideal points we associate with the extended x and y axes, respectively.

We have considered two nondegenerate conics where the coefficients A, B, and C are not all zero. These clearly have equations of the form expressed in Equation 6.57. Certain sets of values of the coefficients produce degenerate conics consisting of straight lines or single points.

We have found that projective conics have homogeneous equations of the second degree in h_1, h_2, and h_3, that is, equations of the same form as Equation 6.57. If the conic is nondegenerate, then A, B, and C are not all zero. Conversely, the loci of such equations in the projective plane are projective conics. These are nondegenerate if and only if the loci of the corresponding nonhomogeneous equations are nondegenerate. The property of being a degenerate or a nondegenerate projective conic, respectively, is preserved by all projectivities in the plane. Furthermore, any two nondegenerate projective conics in the plane are projectively equivalent. See the next section for comments on projective geometry in space.

Projective Geometry in Space

We may divide the set of all quadruples of real numbers into subsets or classes of quadruples such that two quadruples (a_1, a_2, a_3, a_4) and (b_1, b_2, b_3, b_4) are in the same class if and only if there is a real number $k \neq 0$ such that $a_1 = kb_1$, $a_2 = kb_2$, $a_3 = kb_3$ and $a_4 = kb_4$ or, in a more abbreviated notation, such that $a_i = kb_i$, where $i = 1, 2, 3, 4$. We exclude the class $(0,0,0,0)$ and define the remaining classes as the points of projective space. Let (h_1, h_2, h_3, h_4) represent the homogeneous coordinates of a point in projective space. The corresponding nonhomogeneous coordinates are

$$x = \frac{h_1}{h_4}, \ y = \frac{h_2}{h_4}, \ z = \frac{h_3}{h_4}$$

A system of linear homogeneous equations defines transformations in projective space. Thus,

$$h_1' = a_{11}h_1 + a_{12}h_2 + a_{13}h_3 + a_{14}h_4$$
$$h_2' = a_{21}h_1 + a_{22}h_2 + a_{23}h_3 + a_{24}h_4$$
$$h_3' = a_{31}h_1 + a_{32}h_2 + a_{33}h_3 + a_{34}h_4 \quad , \text{ where } |a_{ij}| \neq 0$$
$$h_4' = a_{41}h_1 + a_{42}h_2 + a_{43}h_3 + a_{44}h_4$$

If we let $x' = h_1'/h_4'$, $y' = h_2'/h_4'$, $z' = h_3'/h_4'$, and substitute appropriately into the set of four equations above, then we obtain the projective transformation in ordinary Euclidean space in terms of the nonhomogeneous coordinates x, y, and z.

Section 6.4 Exercises

1. Show that two triples (a_1, a_2, a_3) and (b_1, b_2, b_3) represent the same point if the determinants $\begin{vmatrix} a_1 & a_2 \\ b_1 & b_2 \end{vmatrix} = \begin{vmatrix} a_2 & a_3 \\ b_2 & b_3 \end{vmatrix} = \begin{vmatrix} a_3 & a_1 \\ b_3 & b_1 \end{vmatrix} = 0$.

2. Derive the homogeneous coordinates of the ideal point associated with lines parallel to the y-axis.

3. Write the homogeneous equations for the following straight lines.
   a. $x = 3$ b. $y = 5$ c. $x + y = 0$ d. $2x + 6y = 3$ e. $x - y = 1$

6.5 Parallel Projection

Here we review briefly and then expand upon the ideas of parallel projection introduced in Chapter 2, Section 2.6. First, we make the obvious observation that parallel projection differs from central projection in the geometric relationship between the lines of projection. An example of the former will help to clarify this. Let Π and Π' be two planes in space (Figure 6.23). A family of parallel lines (not parallel to either Π or Π') produces a correspondence between the points of intersection with Π and Π'. This unique mapping is a parallel projection from Π to Π', and it is obviously invertible.

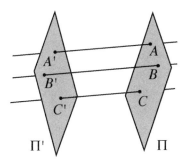

Figure 6.23 Parallel projection.

A family of parallel lines can also define a correspondence between points in space and points in a plane. In this case, we associate with every point P in space the point of intersection of Π with the line passing through P. If the parallel lines of projection are perpendicular to Π, then the mapping is the orthogonal projection of space onto the plane.

There are two categories of parallel projection, based on the relationship between the direction of projection and the normal to the projection plane. In orthographic parallel projections, the direction of projection is normal to the projection plane, while in oblique parallel projection it is not.

Parallel projections of points in space occupy a gray area categorically, since we can think of them as singular or noninvertible affinities. Consider Figure 6.24 where **d** is the direction of projection, Π is the plane of projection, and A, B, C, and $\hat{C}$ are arbitrary points in space. The direction d determines the location of the image points A', B', and C' on Π. Thus, the extended lines AA', BB', CC' are parallel to **d**. But look; C' is also the image of $\hat{C}$, since $\hat{C}$, C and C' are collinear. We cannot uniquely recover C from the image point C'. The most we can say is that C lies on a line through C' that is parallel to **d**.

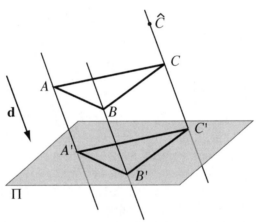

Figure 6.24 Parallel projection onto a plane.

An analogous situation prevails when we construct the parallel projection of points onto a line (Figure 6.25). If l is a line in space and the plane Π represents a two-dimensional direction in space such that Π_1, Π_2, and Π_3 are equivalent directions if they are parallel to one another, then we find that the image A' of A is the intersection point of l and Π_1. Furthermore, if B and C are coplanar in Π_2, then their images B' and C' are coincident. Lines AA', BB', and CC' are all parallel to Π. We notice that B and C are not uniquely recoverable (invertible) given their common image point. This, of course, is likewise true for all points in a particular plane, say Π_2.

MORE LINEAR TRANSFORMATIONS

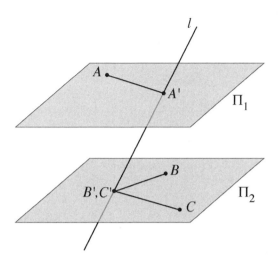

Figure 6.25 Parallel projection onto a line.

Parallel Projection of a Line

Let's review the notion of projecting a line l onto the x axis. If A is any point on l and the direction of projection is perpendicular to the x axis, then A' is the orthogonal projection of A onto this axis (Figure 6.26).

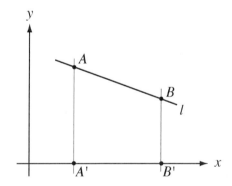

Figure 6.26 Orthogonal projection of a line.

A similar relationship holds for the other points of l. Thus, B' is the intersection of a line through B and perpendicular to the x axis. All lines of projection, such as AA' and BB', are parallel. The equations describing this mapping are simply $x' = x$ and $y' = 0$, and the matrix equation in homogeneous coordinates for this transformation is

$$\begin{bmatrix} x' \\ y' \\ 1 \end{bmatrix} = \begin{bmatrix} 1 & 0 & 0 \\ 0 & 0 & 0 \\ 0 & 0 & 1 \end{bmatrix} \begin{bmatrix} x \\ y \\ 1 \end{bmatrix} \quad (6.60)$$

To continue this example, if the lines of projection are not perpendicular to the x axis but intersect it at some angle α (Figure 6.27), then the transformation equations are $x' = x - y \cot \alpha$ and $y' = 0$, or

$$\begin{bmatrix} x' \\ y' \\ 1 \end{bmatrix} = \begin{bmatrix} 1 & -\cot \alpha & 0 \\ 0 & 0 & 0 \\ 0 & 0 & 1 \end{bmatrix} \begin{bmatrix} x \\ y \\ 1 \end{bmatrix} \tag{6.61}$$

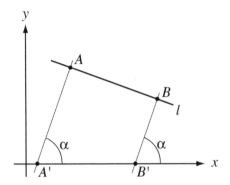

Figure 6.27 Nonorthogonal parallel projection.

Do you see why we cannot use a suggestive sequence of three transformations (a rotation of $90 - \alpha$, an orthogonal projection, and a rotation of $\alpha - 90$) to find this projection?

A more general transformation asks for the parallel projection of the points on line l onto a coplanar line l' (Figure 6.28). The Cartesian equations for this mapping are somewhat tedious to derive, so we will omit the algebraic details and merely describe how to proceed: First, rotate and translate the system so that one of the lines is coincident with the x axis. Next, perform the mapping defined by Equation 6.61. Finally, reverse the initial rotation and translation. Notice that this transformation of l onto l' is one-to-one and multiplies all distances by the same factor, not necessarily 1. Thus, it preserves ratios and betweenness.

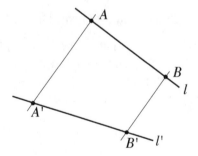

Figure 6.28 Parallel projection of points on ℓ onto ℓ'.

Parallel Projection of a Plane

Projective properties become even subtler when we study the parallel projection of one plane onto another. Consider two planes Π and Π' and a direction of projection generating a family of parallel projection lines that is not parallel to Π or Π'. If A is any point on Π, then there is a unique projection line through A that intersects Π' in A'. A' is the image of A under the given parallel projection (Figure 6.29).

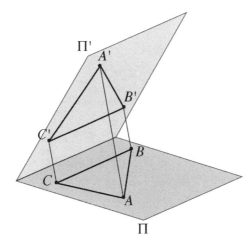

Figure 6.29 Parallel projection of a plane.

In general, distance is not invariant under the parallel projection of one plane onto another. In fact, this transformation usually does not even multiply all distances by the same constant factor. A study of the figure quickly convinces us that this is true. Let $\triangle ABC$ be an equilateral triangle on Π, where line BC is parallel to $\Pi \cap \Pi'$. Clearly, its image on Π is not equilateral. We see that $A'B' = A'C' > B'C'$. Furthermore, if we rotate slightly $\triangle ABC$ in Π, then we see that $A'B' \neq B'C' \neq C'A'$. If and only if Π and Π' themselves are parallel will distances be multiplied by a common factor, and then it will be all distances, and the factor is 1.

We summarize without proof the properties of a parallel projection from one plane onto another as follows: Π maps onto Π' in a one-to-one way, sending lines into lines and preserving parallelism, concurrence, betweenness and ratios of line segments. Distance is not multiplied by a constant factor except when it is preserved. A mapping of a plane onto itself, which is the result of a parallel projection, is an affine transformation. We can show that every affine transformation of a plane onto itself is the result of one or more parallel projections. (It turns out that not more than six parallel projections are required to generate any affine transformation.)

Parallel Projection of Points in Space

The problem of representing all the points in space upon a single plane by a parallel projection is second only in importance to that of a similar central projection. Consider first the orthogonal projection of any point **p** onto an arbitrary plane Π, given by $Ax + By + Cz + D = 0$ (Figure 6.30).

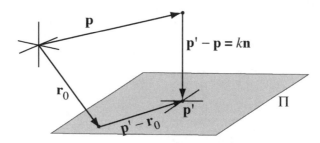

Figure 6.30 Orthogonal projection of a point onto a plane.

We find the projected image **p'** by constructing a line through **p** normal to the plane. We may use vector geometry to solve this problem. Find any point $\mathbf{r}_0$ on Π and the unit normal **n** using the plane's coefficients A, B, C, and D. Then

$$\mathbf{p}' = \mathbf{p} + k\mathbf{n} \tag{6.62}$$

and $\quad (\mathbf{p}' - \mathbf{r}_0) \bullet (\mathbf{p}' - \mathbf{p}) = 0 \tag{6.63}$

producing four equations in four unknowns, namely k, p'_x, p'_y, p'_z. Solving these equations and substituting appropriately yields **p'**. There are several variations of this, all equally efficient.

Next, consider the oblique parallel projection of a point onto an arbitrary plane (Figure 6.31). Again, we use vector geometry to expedite this transformation.

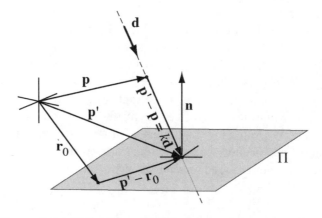

Figure 6.31 Oblique projection of a point onto a plane.

Let the vector **d** denote the oblique direction of projection, **n** the unit normal to the plane, and $\mathbf{r}_0$ any arbitrary point on the plane. Note: **n** and $\mathbf{r}_0$ are functions of the plane's coefficients (or, of course, may be directly defined by vectors). We have

$$\mathbf{p}' = \mathbf{p} + k\mathbf{d} \tag{6.64}$$

and $\quad \mathbf{n} \bullet (\mathbf{p}' - \mathbf{r}_0) = 0 \tag{6.65}$

producing four equations in four unknowns, namely k, p'_x, p'_y, p'_z, from which we obtain $\mathbf{p}'$.

Of course, the simplest procedure is to transform the original point set and coordinate system so that the *xy* plane becomes coincident with the plane of projection and then merely execute an orthogonal projection onto it using the following transformation equation (Figure 6.32):

$$\mathbf{p}' = \begin{bmatrix} 1 & 0 & 0 & 0 \\ 0 & 1 & 0 & 0 \\ 0 & 0 & 0 & 0 \\ 0 & 0 & 0 & 1 \end{bmatrix} \mathbf{p} \tag{6.66}$$

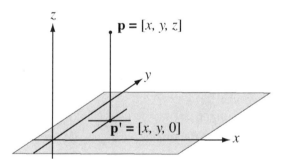

Figure 6.32 Plane of projection coincident with the *xy* plane.

Section 6.8 discusses how to use parallel projections to create a picture or computer graphics display.

Section 6.5 Exercises

1. If lines *l* and *l'* intersect, show that there are just two parallel projections of *l* onto *l'* that preserve distance.

2. If *l* and *l'* are parallel what can you say about the parallel projection of one onto the other?

6.6 Central Projection

We saw earlier that parallel projections are transformations in which lines joining points to their images are all parallel. In what follows, we will consider a transformation where all the lines joining points to their images are concurrent. We call this kind of transformation a central projection or perspective projection. The subject of central projection reaches back to the fifteenth century artists Leon Battista Alberti (1404-1472), Leonardo da Vinci (1452-1519) and Albrecht Dürer (1471-1528). They laid the aesthetic and empirical foundations for its subsequent analytical development. The following section presents an intuitive introduction to perspective projection.

Central Projection and Perspective Drawing

Parallel projections cannot represent an object in a completely natural way. In nature, we think of the projection lines as visual rays extending from points on an object to the eye of the observer. For these rays to be parallel, the observer must be infinitely far from the object. This is not the situation in normal viewing. At best, parallel projections only roughly simulate the true picture of an object. Perspective projection recognizes and accounts for the fact that an observer is at a finite, measurable distance from any object he or she sees.

The Figure 6.33 shows a simple arrangement in space of a perspective projection. Here a house-like object is viewed, with the eye of the observer at O. The visual rays are the straight lines drawn from O through various key defining points of the house. We insert projection planes Π_1 and Π_2. The points of intersection of these visual rays with each plane define the images of the object on Π_1 and Π_2. Obviously, the projective image of the object on Π_1, which is between the observer and the object, is smaller that the object. On Π_2 the perspective projection is larger than the object. For a picture Π_3 (not shown) located behind the observer, the perspective projection is inverted as in an image formed by a lens.

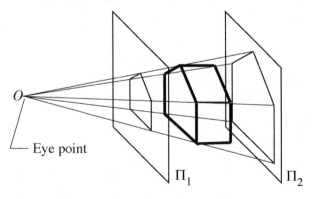

Figure 6.33 Perspective drawings.

MORE LINEAR TRANSFORMATIONS

To begin in the simplest way possible, imagine a figure in a plane Π, a point O the center of projection, and a second plane Π' (Figure 6.34). If we draw straight lines from O through each point of the figure in Π, then each line intersects Π' in a corresponding image point. The image of a straight line is another straight line, since the set of lines through O and each point on the original straight line all lie in a plane. The intersection of this plane with this projection plane is, of course, a straight line. We conclude that points and straight lines are invariant under a central projection. Of course, there is more to it than this.

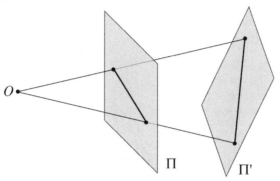

Figure 6.34 Central projection.

Let's start again, this time with the central projection of the points of one line onto another line, where the lines are coplanar with the center of projection. Here is a simple example of a central projection in the plane: Let l and l' be parallel lines and O a point in the plane but not on l or l' (Figure 6.35). Given any point P on l, if we extend line OP to intersect l' at P', then P' is the image of P under the central projection through O, where we call point O the center of projection. Notice that P is the inverse of P' under this transformation. Continuing this process for a series of points A, B, and C we discover, from similar triangles, that $A'B'/AB = A'C'/AC = B'C'/BC$. If we let this ratio equal k, then we have $A'B' = kAB$, $A'C' = kAC$, and $B'C' = kBC$. All distances are multiplied by the same constant, k. The transformation is one-to-one and preserves betweenness.

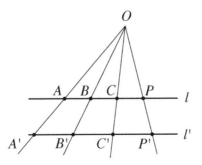

Figure 6.35 Central projection of parallel lines.

Central Projection of Intersecting Lines

Now we will again consider the central projection of two coplanar lines, this time intersecting instead of parallel. Figure 6.36 shows two intersecting lines l and l' and a center point of projection O. Let M be the point of intersection of l and a line through O parallel to l'. Similarly, let N' be the intersection point of l' and a line through O parallel to l. We notice several interesting features about this transformation, beginning with points M and N'. All of the points on l have corresponding image points on l', except M. Similarly, N' does not have an image point on l. Unlike the projection transformation between parallel lines, above, this one is not capable of mapping a complete line onto a complete line. By neglecting M and N', however, it is one-to-one.

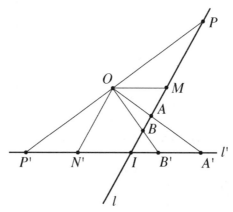

Figure 6.36 Central projection of intersecting lines.

As we choose points closer and closer to M, we find that the corresponding image points become increasingly more distant. This is why we call M and N' the vanishing points on l and l', respectively. Conversely, points far from M, beyond P for example, go into points near N'.

Betweenness is not always preserved, for we see that point A is between B and P on l; but with regard to their images, A' is not between B' and P' on l'. Furthermore, we see that distances cannot all be multiplied by the same constant. These observations tell us that the ratio of division is not invariant.

Central Projection of Parallel Planes

Here is another example, this time in space (Figure 6.37). Given parallel planes Π and Π' and a center of projection O, points A', B', C' on Π' are the images of points $A, B,$ and C on Π under a projective transformation through O.

More Linear Transformations

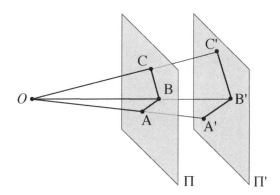

Figure 6.37 Central projection of parallel planes.

If AB is any line in Π, then the projection lines joining the points of AB to O form a plane which intersects Π' in a line $A'B'$, and similarly for lines BC and $B'C'$. Thus, under the special conditions we stated above, this transformation is one-to-one, maps lines into lines, and preserves parallelism and concurrence.

Triangles OAB and $OA'B'$ are similar, so that $OA'/OA = OB'/OB = A'B'/AB$. Triangles OBC and $OB'C'$ are similar, so that $OB'/OB = OC'/OC = B'C'/BC$. Thus, $A'B'/AB = B'C'/BC = OA'/OA = k$. All distances on Π are multiplied by a constant k, which is the same for all lines. Notice that k equals the ratio of the perpendicular distances from O to Π' and Π.

We can summarize these results as follows: The central projection of a plane Π onto a parallel plane Π' is one-to-one, sends lines into lines, multiplies all distances by the same constant k, and preserves parallelism, concurrence, ratios of division (not discussed), and betweenness (obvious by inspection).

Central Projection of Intersecting Planes

Given intersecting planes Π and Π' and a center of projection O, we find their properties of projection similar to those mentioned above for intersecting lines (Figure 6.38). There is a set of collinear points on Π that have no image points on Π'. Points B and C are examples. Lines of projection through these points are parallel to Π'. Their common plane intersects Π in a line m parallel to $\Pi \cap \Pi'$. There is an analogous set of collinear points on Π' with no corresponding originals on Π. D' and E' are examples. We call m and the points on it (such as B and C) the vanishing line and points, respectively, on Π. We call n' and the points on it (such as D' and E') the vanishing line and points, respectively, on Π'.

What about the projection of a line other than the vanishing line? Consider this example (Figure 6.39). If l is any nonvanishing line on Π, then the plane determined by l and O intersects Π' in line l', the projected image of l on Π'.

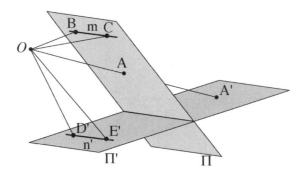

Figure 6.38 Central projection of intersecting planes.

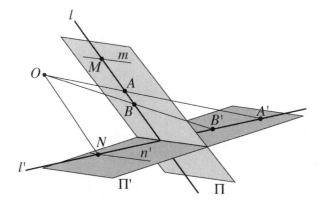

Figure 6.39 Central projection of a line.

There are many interesting properties which devolve from this transformation, the details of which we will not discuss here beyond summarizing the results: The central projection of a plane onto an intersecting plane does so with a one-to-one correspondence between the points of the planes, except for the points on the vanishing lines in each plane. It establishes a one-to-one correspondence between the lines of the planes, again with the exception of the vanishing lines. It preserves collinearity and cross-ratio, but not betweenness or ratio of division, and complete lines do not necessarily go into complete lines. Also, we find that parallel lines may go into intersecting lines, and conversely intersecting lines may go into parallel lines. Here is a summary of all the projective properties we have described thus far.

A projection from one plane onto another:

a. Preserves collinearity

b. Maps rectilinear figures onto rectilinear figures

c. Maps conics onto other conics, not necessarily of the same type

d. Preserves cross-ratios

e. Establishes a one-to-one correspondence between points of one plane and those of another, except for the points of one line in each plane

f. Does not preserve parallels

g. Does not preserve angles

h. Does not preserve distance

Central Projection of Points in Space

The central, or perspective, projection of points on an object in space approximates the way we form a visual image of that object. The basic geometry of a perspective transformation includes the position of the observer, usually denoted as point O, the plane of projection (or picture plane), and λ the normal distance from the observer to the plane. The simplest arrangement of these elements (Figure 6.35) places the observer on the z axis at $z = \lambda$, the distance from the xy plane, which we define as the picture plane. Using the properties of similar triangles, we find

$$\frac{x'}{\lambda} = \frac{x}{\lambda - z}, \quad \frac{y'}{\lambda} = \frac{y}{\lambda - z}, \quad z' = 0 \tag{6.67}$$

or $\quad x' = \dfrac{\lambda x}{\lambda - z}, \quad y' = \dfrac{\lambda y}{\lambda - z}, \quad z' = 0 \tag{6.68}$

In matrix form these become

$$\begin{bmatrix} x' \\ y' \\ z' \\ 1 \end{bmatrix} = \begin{bmatrix} 1 & 0 & 0 & 0 \\ 0 & 1 & 0 & 0 \\ 0 & 0 & 0 & 0 \\ 0 & 0 & -1/\lambda & 1 \end{bmatrix} \begin{bmatrix} x \\ y \\ z \\ 1 \end{bmatrix} \tag{6.69}$$

Here for the first time we use a position in the last row of the matrix (other than a_{44}, of course). It is important to note that in a left-hand coordinate system $a_{34} = 1/\lambda$. Try to verify this. If we cannot arrange the basic geometry as it is in Figure 6.40, then we may use two- or three-point perspective to achieve a realistic representation. Here the other elements in the last row of the transformation matrix, a_{41} and a_{42}, come into play. Although we will not address these variations here. We conclude with the vector solution to a general perspective projection of any point $\mathbf{p}$ onto an arbitrary plane relative to the center of projection $\mathbf{p}_0$ (Figure 6.41).

If the plane equation is $Ax + By + Cz + D = 0$, then

$$\mathbf{p}' = k(\mathbf{p} - \mathbf{p}_0) \tag{6.70}$$

and $\quad (\mathbf{p}' - \mathbf{d}) \bullet \mathbf{d} = 0 \tag{6.71}$

where $d_x = \dfrac{AD}{A^2 + B^2 + C^2}$, $d_y = \dfrac{BD}{A^2 + B^2 + C^2}$, and $d_z = \dfrac{CD}{A^2 + B^2 + C^2}$

from elementary analytic geometry.

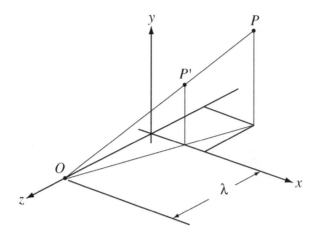

Figure 6.40 Central projection of points in space.

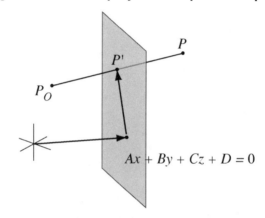

Figure 6.41 Vector solution for a general perspective projection.

Vector Equations 6.70 and 6.71 produce four algebraic equations in four unknowns, k, p'_x, p'_y, and p'_z, which we easily solve to find $\mathbf{p}'$. The vector $\mathbf{d}$ is of course the normal vector to the plane from the origin. Equation 6.70 ensures that $\mathbf{p}'$ lies along the line joining $\mathbf{p}$ and $\mathbf{p}_0$, and Equation 6.71 ensures that $\mathbf{p}'$ lies on the projection plane.

Telescopic Projection

A telescope makes distant objects appear closer, and a microscope makes small objects appear larger. We can mathematically model both of these phenomena using the thin-lens equations from elementary optics (Figure 6.42).

$$\frac{1}{f} = \frac{1}{z} + \frac{1}{z'} \tag{6.72}$$

More Linear Transformations

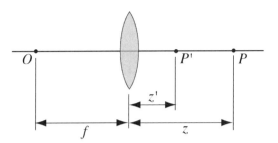

Figure 6.42 Thin-lens geometry.

where f is the focal length of the lens, z is the location of the original point, and z' is the location of the image point. We rearrange this equation to obtain

$$z' = \frac{fz}{z-f} \qquad (6.73)$$

In matrix form this becomes

$$\begin{bmatrix} x' \\ y' \\ z' \\ 1 \end{bmatrix} = \begin{bmatrix} 1 & 0 & 0 & 0 \\ 0 & 1 & 0 & 0 \\ 0 & 0 & 1 & 0 \\ 0 & 0 & 1/f & -1 \end{bmatrix} \begin{bmatrix} x \\ y \\ z \\ 1 \end{bmatrix} \qquad (6.74)$$

This equation is obviously similar to the results we obtained for a central projection. However, notice that a telescopic projection preserves the third coordinate (in this case z), which is usually lost in the perspective projection. Furthermore, it transforms from three dimensions into three dimensions, whereas the perspective projection transforms from three into two.

Section 6.6 Exercises

1. Under a central projection of the points on a line onto a parallel line, are distances between these points preserved?

6.7 Map Projections

The most familiar kind of projection is a cartographic projection, ... a map. Maps are ubiquitous; we see them in the newspaper, on television, and we use them in our travels. Yet, with all this exposure, unless one is a cartographer, how much do we really understand about creating a map? In this section, we survey some of the common types of mapping. Since we are most familiar with maps representing portions of the earth's surface we will focus on the projection of a sphere onto a plane or developable geometric shape such as a cylinder or cone. We will investigate both parallel and central projections.

Over a limited area, we can assume that the earth's surface is essentially flat, and we have little difficulty in creating satisfactory maps of such areas. When we create maps of relatively large areas, we must take care to minimize distortions.

The most true-to-life image of a surface is the result of an isometric mapping. The distance between any two points is equal or proportional to the distance between their image points, all angles are unchanged, and geodesic lines are mapped into geodesic lines. Two arbitrary surfaces cannot, in general, be mapped onto each other isometrically. Differential geometry tells us that the Gaussian curvature at corresponding points of two surfaces related by an isometric mapping must be equal. This means that only surfaces that transform isometrically into a part of the plane are surfaces whose Gaussian curvature is everywhere zero, and this excludes any portion of a sphere. As a consequence, no geographic map is free of distortions.

Area-Preserving Mappings

Equiareal transformations are less restrictive but also less accurate. Their distinguishing characteristic is that the area within the image of a closed curve is equal to the area within the original closed curve. Area-preserving mappings are most simply characterized in terms of differential geometry.

We often use area-preserving maps in geography. Here is a simple method for creating an area-preserving map of a part of a sphere by projecting it onto a cylinder (Figure 6.43):

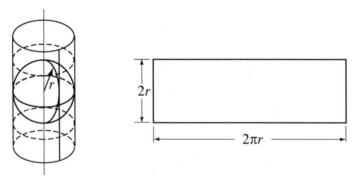

Figure 6.43 An area-preserving mapping.

Construct a cylindrical surface of radius r around a sphere of the same radius. Next, project the points of the sphere onto the cylindrical surface along the normals to the cylinder. We may now slip the cylindrical surface along a generator and unroll it into a plane. Note that the image is more distorted the farther the object and consequently its image points are from the common tangent circle of the sphere and cylinder.

The entire spherical area is mapped onto a cylindrical area that is $2r$ wide by $2\pi r$ long. The cylindrical area is identical to the area of the sphere: $4\pi r^2$.

Geodesic Mappings

Geodesic mappings are another type of cartographic transformation. In geographical application, these are most frequently used to produce charts for navigation. Here we map the geodesics of one surface into the geodesics of another. Isometric mappings are a special case of geodesic mapping. If we project a sphere from its center onto the plane, then great circles on the sphere transform into straight lines in the plane, and the map is geodesic. We can show that only surfaces of constant Gaussian curvature transform geodesically on to the plane. In general, however, determining whether or not two surfaces can be mapped geodesically onto each other is a difficult problem involving differential geometry, and it is one we do not address here.

Conformal Mappings

Conformal mappings preserve the angle at which two curves intersect. These, of course, have much in common with isometric mappings. A stereographic projection is an example of a conformal mapping.

A stereographic projection describes a one-to-one correspondence between the points on a sphere and the plane, where the north pole on the sphere correspond to points at infinity on the plane (Figure 6.44). Consider a sphere and a plane tangent to it at its south pole. Let P be any point on the sphere except N, the point at the north pole. Under a stereographic projection, the points N and P and its image P' on the plane are collinear. That is, P' is the point of intersection of the plane with the line containing N and P. Although we won't prove it here, a stereographic projection preserves angles on the sphere and is therefore conformal. And finally, the set of all circles on the sphere corresponds to the set of all circles and straight lines in the plane, which means that stereographic projection is circle-preserving.

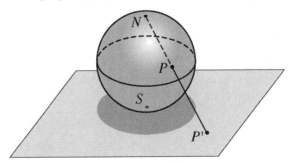

Figure 6.44 Stereographic projection.

Continuous Mapping

The most general form of map projection is the continuous mapping. The only restriction is that the mapping be one-to-one and that neighboring points go over into neighboring points. The transformation must be topologically invariant. That is, it may distort a figure in anyway except by tearing or joining separate regions together. As general as this transformation is, it does not make it possible to map any two arbitrary surfaces into each other. For example, it is impossible to map a circular disk and a plane circular annulus onto each other, since they are topologically different classes. We can achieve a good approximation of a spherical surface by projecting limited areas of it onto a developable surface, such as a cylinder or cone (Figure 6.45).

Figure 6.45 Projection of a sphere onto a developable surface.

Projection onto a Tangent Plane

If it is necessary to transform a small region of a sphere, then construct a tangent plane to the sphere centered on the region. Figure 6.46a shows an orthographic projection of a portion of the sphere representing the surface of the earth. It produces an acceptable visual image over a limited area surrounding the point of tangency with the projection plane.

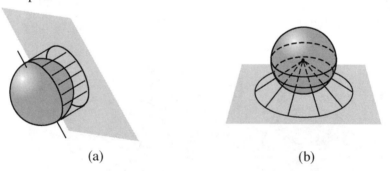

(a)   (b)

Figure 6.46 Projection onto a tangent plane.

In gnomonic projection, the projection lines radiate from the center of the sphere. Figure 6.46b shows a polar gnomonic projection with projection lines radiating from the center to a plane tangent at a pole. The primary advantage of a gnomonic projection is that all great circles are projected as straight lines. This is important in navigation since the shortest distance between two points on the earth's surface lies along a great circle through them.

Projection onto a Cone

Another way to project the surface of the earth is onto a cone tangent to it. The cone is then unrolled or developed onto a plane surface. Figure 6.47 shows the parallels of latitude projected as concentric circles and the meridians of longitude converging as straight lines. The parallel of latitude on the sphere in contact with the cone is projected true length and is called the standard parallel. Other features of the earth do not project as true, and distortions increase in proportion to the distance from the standard parallel. We limit distortion of the area projected onto the cone to that between the 0° and 60° parallels.

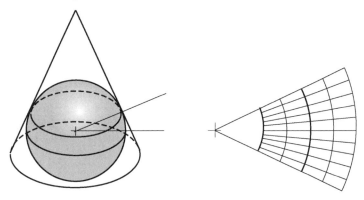

Figure 6.47 Conic projection with one standard parallel.

A variation on this theme is to locate a cone so that it intersects the sphere along two parallels. Each then becomes a standard parallel along which distances are preserved (Figure 6.48). Using this technique, we are able to project a somewhat greater area without inducing excessive scale distortion.

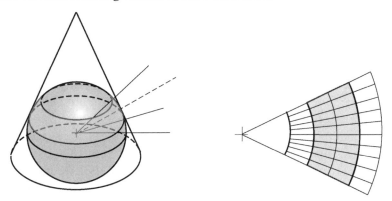

Figure 6.48 Conic projection with two standard parallels.

Cylindrical Projection

We can wrap the sphere in a cylindrical surface, project points on the sphere onto the cylinder along lines radiating from the sphere's center, and then unroll the cylinder into a planar map (Figure 6.49). Notice that both parallels of latitude and meridians of longitude project as straight lines. The equator projects at its true scale, but other parallels exhibit a scale distortion that increases with distance from the equator. If we increase the scale of the meridians in the same proportion as the scale of the parallels increases, then we produce a Mercator map. At any point on the map, the scale is the same in every direction, but areas are much exaggerated in the extreme north and south latitudes. The advantage of a Mercator map to navigation is that a constant compass bearing appears as a straight line, or rhumb line.

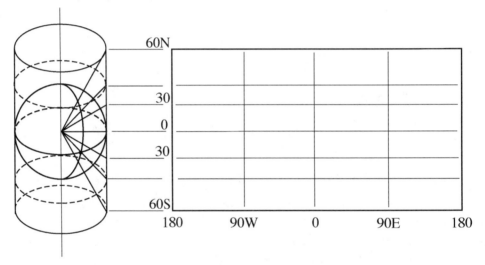

Figure 6.49 Cylindrical projection.

6.8 Display Projection

Computer graphics displays use a variety of two- and three-dimensional coordinate systems. Five of these are shown in Figure 6.50. Creating a display requires that we execute a series of transformations, including projective transformations, to pass points from a world coordinate system to a screen coordinate system.

The three-dimensional global coordinate system forms the basis for defining and locating in space all objects in a computer graphics scene, including the observer's position, picture or projection plane, and the line of sight. This system serves as the master reference system in which we define all other coordinate systems.

A local coordinate system defines the geometry of an object independently of the global system. We use it to define an object without giving it a specific location in the global system. Once we define an object locally, we can position it in the global system simply by specifying the location and orientation of the origin of the

MORE LINEAR TRANSFORMATIONS 307

local system within the global system, and then transforming the point coordinates defining the object from the local to the global system.

A view coordinate system locates objects in three-dimensions relative to the location of the observer. We use it, when necessary, to simplify the projection transformation creating the image of an object on the picture plane (the plane of projection). Of course, the location and orientation of the view coordinate system change each time the view changes.

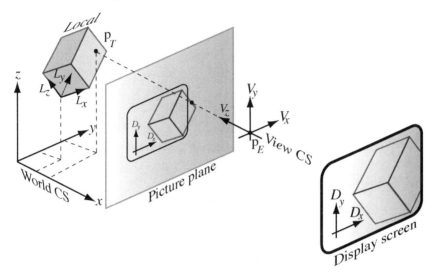

Figure 6.50 Computer graphics coordinate systems.

The Picture-Plane Coordinate System

A two-dimensional picture plane coordinate system locates object points on the picture plane or projection plane. A two-dimensional screen coordinate system locates points on the screen of a computer graphics display monitor (or on the bed of a plotting device, for that matter). In this section, we are concerned with obtaining the picture-plane coordinates only.

Three vectors define a picture plane and a coordinate system embedded in it (Figure 6.51). The vector $\mathbf{p}_0$ defines the origin, and $\mathbf{u}_1, \mathbf{u}_2$ define an orthogonal (usually) set of coordinate axes in the plane. This means that any point $\mathbf{q}$ in the plane must satisfy the vector equation

$$\mathbf{q} = \mathbf{p}_0 + a\mathbf{u}_1 + b\mathbf{u}_2 \tag{6.75}$$

where a and b are independent variables. When necessary, we define a third axis $\mathbf{u}_3$ such that $\mathbf{u}_3 = \mathbf{u}_1 \times \mathbf{u}_2$, where $\mathbf{u}_1$, $\mathbf{u}_2$, and $\mathbf{u}_3$ form a right-hand system. If we define $\mathbf{u}_3 = -\mathbf{u}_1 \times \mathbf{u}_2$, then we obtain a left-hand system.

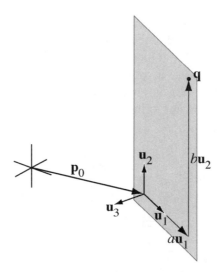

Figure 6.51 The picture plane coordinate system.

Orthographic Projection

We compute the picture plane coordinates $\mathbf{p}'_i$ of any point $\mathbf{p}_i$ under a general orthographic projection (Figure 6.52) as follows: Compute a_i, b_i, and c_i from

$$\mathbf{p}_i = \mathbf{p}_0 + a_i\mathbf{u}_1 + b_i\mathbf{u}_2 + c_i\mathbf{u}_3 \qquad (6.76)$$

This equation yields three simultaneous linear equations in three unknowns a_i, b_i, and c_i. Then compute $\mathbf{p}'_i$ from

$$\mathbf{p}'_i = \mathbf{p}_0 + a_i\mathbf{u}_1 + b_i\mathbf{u}_2 \qquad (6.77)$$

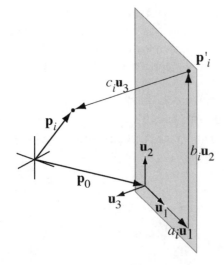 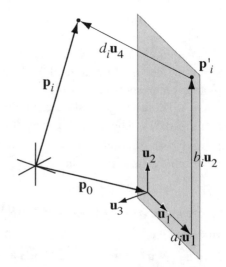

Figure 6.52 Orthographic projection.   **Figure 6.53** Oblique projection.

Oblique Projection

We compute the picture plane coordinates $\mathbf{p}'_i$ of any point $\mathbf{p}_i$ under a general oblique projection (Figure 6.53) as follows: First, compute a_i, b_i, and d_i from

$$\mathbf{p}_i + d_i \mathbf{u}_4 = \mathbf{p}_0 + a_i \mathbf{u}_1 + b_i \mathbf{u}_2 \tag{6.78}$$

where $\mathbf{u}_4$ is a unit vector in the direction of the oblique projectors. This equation yields three simultaneous linear equations in three unknowns: a_i, b_i, and d_i.

Next, compute $\mathbf{p}'_i$ from

$$\mathbf{p}'_i = \mathbf{p}_0 + a_i \mathbf{u}_1 + b_i \mathbf{u}_2 \tag{6.79}$$

or $\quad \mathbf{p}'_i = \mathbf{p}_i + d_i \mathbf{u}_4 \tag{6.80}$

Perspective Projection

We compute the picture plane coordinates $\mathbf{p}'_i$ of any point $\mathbf{p}_i$ under a perspective projection (Figure 6.54) as follows: Observe that

$$\mathbf{p}'_i = \mathbf{p}_0 + a_i \mathbf{u}_1 + b_i \mathbf{u}_2 \tag{6.81}$$

also $\quad \mathbf{p}'_i = \mathbf{p}_E + c_i (\mathbf{p}_i - \mathbf{p}_E) \tag{6.82}$

where $\mathbf{p}_E$ is the center of projection, and we arrange $\mathbf{p}_E$ and $\mathbf{p}_0$ so that $\mathbf{p}_E - \mathbf{p}_0 = \lambda \mathbf{u}_3$. Rearrange Eq.s 6.81 and 6.82 and substitute appropriately for $\mathbf{p}_E - \mathbf{p}_0$, thus

$$a_i \mathbf{u}_1 + b_i \mathbf{u}_2 - \lambda \mathbf{u}_3 - c_i (\mathbf{p}_i \bullet \mathbf{p}_E) = 0 \tag{6.83}$$

This equation produces three simultaneous linear equations in three unknowns, a_i, b_i, and c_i, from which we compute $\mathbf{p}'_i$.

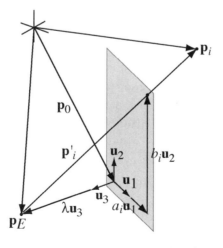

Figure 6.54 Perspective projection.

7 NONLINEAR TRANSFORMATIONS

This chapter investigates transformations that are given by nonlinear functions. We have already discussed some examples of this, including certain map projections and general topological transformations in earlier chapters. Following a brief introductory section, we will study inversion in a circle, parametric mappings, and deformation transformations.

7.1 Linear and Nonlinear Equations

Linear and nonlinear transformations are very different conceptually. For linear transformations, the input conditions are independent of the output state and it is possible to superimpose various input conditions to produce an output equivalent to the sum of the separate outputs. For example, rotation is a linear transformation and successive rotations θ_1 and θ_2 produce the same final image as the single rotation $\theta_1 + \theta_2$. Furthermore, the result of a linear transformation can be broken down and ascribed to a combination of any number of independent input states. Continuing with the example of a rotation transformation, the rotation through some angle θ_T can be achieved by a single rotation θ_T, two successive rotations θ_1 and θ_2 where $\theta_1 + \theta_2 = \theta_T$, or three successive rotations where $\theta_1 + \theta_2 + \theta_3 = \theta_T$, or…and so forth. The same is not true for nonlinear transformations. Let's look at some linear and nonlinear functions that clearly demonstrate their differences.

We begin with the linear transformation functions. Here is a specific example: $x' = 2x + y$ and $y' = x + 3y$. If $x_1 = 1$ and $y_1 = 1$, then $x'_1 = 3$ and $y'_1 = 4$. If $x_2 = 3$ and $y_2 = 2$, then $x'_2 = 8$ and $y'_2 = 9$. Now we will add the independent variables and do the transformation again: If $x = x_1 + x_2 = 4$ and $y = y_1 + y_2 = 3$, then $x' = x'_1 + x'_2 = 9$ and $y' = y'_1 + y'_2 = 13$. We easily may verify this.

Next, consider the nonlinear transformations $x' = x^2 + 2y$ and $y' = 2x + xy$. If $x_1 = 1$ and $y_1 = 2$, then $x'_1 = 5$ and $y'_1 = 4$. If $x_2 = 3$ and $y_2 = 1$, then $x'_2 = 11$ and

NONLINEAR TRANSFORMATIONS

$y'_2 = 9$. Again, we will add the independent variables and repeat the transformation: If $x = x_1 + x_2 = 4$ and $y = y_1 + y_2 = 3$, then $x' = 22$ and $y' = 20$. Clearly, $x'_1 + x'_2 \neq 22$ and $y'_1 + y'_2 \neq 20$. Here is an example of a nonlinear transformation that is more geometric in nature. This transformation maps any point P (except the origin) onto P' on the line OP (Figure 7.1) such that the product $OP \cdot OP'$ is equal to a given constant, or $OP' = k/OP$.

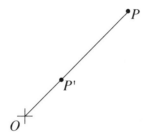

Figure 7.1 The reciprocal radii transformation.

In terms of Cartesian coordinates, we have $x':y':z' = x:y:z$. Using the analytic relationship between the distances OP and OP' and setting the constant k equal to one, we obtain

$$(x^2 + y^2 + z^2)(x'^2 + y'^2 + z'^2) = 1$$

Thus, the transformation equations are

$$x' = \frac{x}{x^2 + y^2 + z^2}$$

$$y' = \frac{y}{x^2 + y^2 + z^2}$$

$$z' = \frac{z}{x^2 + y^2 + z^2}$$

and the inverses of these equations are

$$x = \frac{x'}{x'^2 + y'^2 + z'^2}$$

$$y = \frac{y'}{x'^2 + y'^2 + z'^2}$$

$$z = \frac{z'}{x'^2 + y'^2 + z'^2}$$

It is easy to show that this transformation maps spheres onto spheres. Note also that the origin is a special point. It is the vanishing point of the transformation, since x', y', and z' become infinite as x, y, and z simultaneously approach zero. Inversion in a circle is a special instance of this transformation (Section 7.2).

7.2 Inversion In a Circle

This somewhat unusual transformation does not preserve congruence or even similarity. It does preserve the set of all circles and lines. That is, a circle always maps into either another circle or else a line, and a line always maps into another line or else a circle. It preserves angles between curves and is thus conformal. To execute this transformation we must define a center of inversion and a radius. Here is how it is done.

Given a fixed circle C with center O and radius r, we call C the circle of inversion and O the origin or center of inversion. P is any point in the plane of the circle, either inside or outside the circle (Figure 7.2). The inverse of P with respect to the circle is a point P' on the line OP whose distance from O is equal to r^2/OP. Algebraically we express this as

$$OP' = \frac{r^2}{OP} \tag{7.1}$$

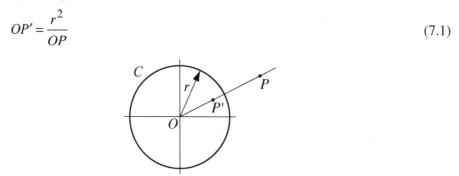

Figure 7.2 Inversion in a circle.

Clearly, P is the inverse of P' since from Equation 7.1 we may write $OP = r^2/OP'$. Inversion gets its name from the fact that as P moves OP' is inversely proportional to OP. Notice that the inverse of a point outside the circle is a point inside it, and vice versa. Furthermore, every point in the plane has an inverse. The origin is a vanishing point, and we call all infinitely distant points from the origin the point at infinity (or ideal point…see projections), which is the inverse of the origin. If P lies on the circle of inversion C, then it is its own inverse since then $OP = r$ and, from Equations 7.1, $OP' = r^2/OP = r^2/r = r$.

Inversion in a Circle by Geometric Construction

If P is a point outside the circle of inversion, draw the line OP and the two tangents from P (Figure 7.3). Denote the tangent points as A and B. From elementary geometry, $AB \perp OP$. Let $P' = AB \cap OP$. P and P' are mutually inverse with respect to the circle, since triangles OAP and $AP'O$ are similar, and thus $OP'/r = r/OP$ or $OP \times OP' = r^2$. If P is inside the circle, draw the minimum chord through P and also draw the tangents at its extremities. The tangents intersect at the inverse of P.

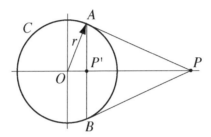

Figure 7.3 Inversion by geometric construction: Method 1.

Figure 7.4 illustrates another way to construct an inversion. Only a compass is required. Using a radius OP with P as the center, draw an arc so that it intersects C in two points D and E. Next, with D and E as centers and a radius OD, draw two intersecting arcs. One point of intersection is O and the other turns out to be P'. So that P' is the inverse of P, since ODP' is an isosceles triangle, which is similar to triangle OPD. Thus, $OP'/r = r/OP$, or $OP \times OP' = r^2$. Notice that for this construction we assume that the distance OP is such that an arc drawn with P as center and OP as radius will indeed intersect the circle C. If this is not the case, we must modify the construction (see Exercise 1).

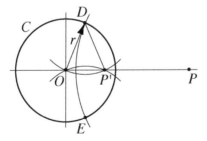

Figure 7.4 Inversion by geometric construction: Method 2.

Notice that lines through O invert into themselves, points outside the circle of inversion exchange places with points inside, and a circle concentric with O inverts into another circle also concentric to O, the radii being in inverse ratio.

Inversion in a Circle of Infinite Radius

The radius of inversion determines the scale of an inverted figure, so it is of interest to determine the effect of inverting a point with respect to a circle of infinite radius, namely a straight line. In Figure 7.5 we observe that $OP = r + DP$ and $OP' = r - P'D$. Since $OP \times OP' = r^2$, we have $(r + DP)(r - P'D) = r^2$. By expanding this last equation we find

$$DP - P'D = \frac{DP \times P'D}{r} \qquad (7.2)$$

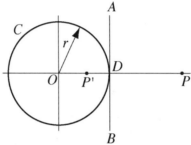

Figure 7.5 Inversion in a circle of infinite radius.

Now, holding D fixed, we move O indefinitely far to the left so that as r becomes infinite the circle C approaches the line AB and the right-hand side of Equation 7.2 approaches zero (since $P'D$ remains finite). At the limit, $DP = P'D$, and we see that the inverse of a point with respect to a straight line is its reflected image.

Circular Inversion of a Straight Line not Containing the Origin

To determine the inverse of a line l which does not pass through the center of inversion O we proceed as follows (Figure 7.6): Let P be the point on l such that OP is perpendicular to l, and let P' be the circular inverse of P. For any other point Q on l draw OQ and find its inverse Q'. Consider the triangles $OP'Q'$ and OPQ. Since $OP \times OP' = OQ \times OQ' = r^2$, we know that $OP/OQ' = OQ/OP'$. Therefore, the triangles are similar, since they have an angle in common and corresponding sides in proportion.

Since OPQ is a right angle, it follows that angle $OQ'D'$ is also a right angle. Furthermore, since the position of Q on l is arbitrary, $OQ'P'$ is always a right angle. Therefore, as Q moves along l, its inverse Q' describes a circle passing through O and having OP' as a diameter. Finally, we see that the converse must also be true: the inverse of a circle through the origin is a straight line perpendicular to the diameter through the origin.

Nonlinear Transformations

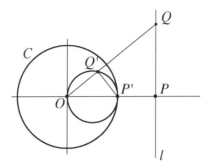

Figure 7.6 Circular inversion of a straight line not containing the origin.

The Inverse of a Circle

To determine the inverse of a circle S that does not pass through the center of inversion O we proceed as follows (Figure 7.7, where the circle of inversion is not shown): Let S denote the circle to be inverted. Construct the line l from O through the center of S and intersecting S in Q and P.

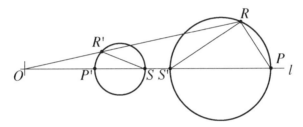

Figure 7.7 Inverse of a circle not passing through the center of inversion.

Clearly QP is a diameter of S. Let P',Q' be the inverses of P and Q, respectively. P' and Q' lie on l. It is obvious that the inverse of S must be symmetrical about l. Let R be an arbitrary point on S, and let R' be its inverse. Draw the lines $OR'R$, $Q'R'$, $R'P'$, QR, and RP. Since $OP \times OP' = OQ \times OQ' = OR \times OR'$, it must be true that $OR'/OP = OP'/OR$. Therefore, triangles ORP and $OP'R'$ are similar, $\angle OP'R' = \angle ORP$, and $\angle OQ'R' = \angle ORQ$. Finally, $\angle OP'R' - \angle OQ'R' = \angle ORP - \angle ORQ = 90°$. and hence $\angle P'R'Q' = 90°$. We see that for all positions of R on S, point R' describes a path such that $\angle P'R'Q'$ is always a right angle. Consequently, the path of R' is a circle with $P'Q'$ as a diameter, and the inverse of a circle S is another circle.

The Characteristics of Circular Inversion Summarized

1. Points invert into points.
2. A straight line through the center of inversion inverts into itself.

3. The circle of inversion inverts into itself.
4. A straight line not through the center of inversion inverts into a circle through the center of inversion.
5. A circle not through the center of inversion inverts into a circle also not through the center of inversion.
6. If one figure is the inverse of another, then the other is the inverse of the first.
7. The inverse of a figure with respect to a straight line is the reflected image of the figure in the line.
8. The center of a circle and the point at infinity are mutually inverse with respect to the circle.
9. The angles between elements of a figure are invariant in size under inversion but opposite in sense.
10. Circles concentric with the center of inversion invert into other circles similarly concentric.

Section 7.2 Exercises

1. Demonstrate a geometric construction to find the inverse of a point inside the circle of inversion when the point is too close to the origin to allow the direct application of the method of Figure 7.4.
2. Find the inverse of each of the following points, where the center of inversion is $(0,0)$ and $r = 10$.

   a. $(6,8)$ b. $(10,0)$ c. $(1,0)$

   d. $(10,10)$ e. $(2,-7)$ f. The ideal point

3. Find the inverse of each of the points given in Exercise 2, where the center of inversion is $(5,0)$ and $r = 5$.
4. Draw the inverse of a square and its diagonals.

7.3 Curvilinear Coordinate Systems

Now we will look at parametric representations of curves, surfaces and solids, which are, strictly speaking, mappings. We will try to do this without recourse to the more abstract depths of differential geometry. Having laid some groundwork in the discussion of parametric curves, surfaces and solids, we will move on to investigate deformations.

Parametric Representation of Curves

Consider a mapping that sends points on the x axis into points in space (Figure 7.8a). In its most general form, this transformation looks like

$$x' = \phi_x(x)$$
$$y' = \phi_y(x) \qquad (7.3)$$
$$z' = \phi_z(x)$$

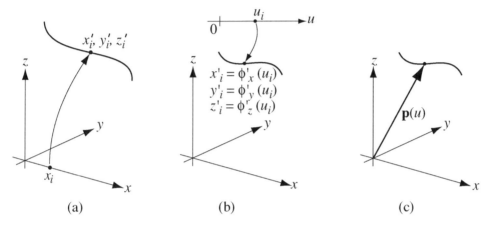

Figure 7.8 Parametric representation of curves.

Since we do not insist that $\phi_x(x)$, $\phi_y(x)$, and $\phi_z(x)$ must be linear functions of x, the transformation may produce an arbitrary curve in space. Obviously, the mapping is not, in general, one-to-one, nor does it preserve distances. As we will soon see, we control the shape of the curve by carefully choosing the $\phi_x(x)$, $\phi_y(x)$, and $\phi_z(x)$.

Points on the x axis may be replaced by points on the real line without loss of meaning. To do this we replace the independent variable x with u, which we call the parametric variable (it is still the independent variable), and drop the prime notation so that Equations 7.3 now become

$$x = \phi_x(u)$$
$$y = \phi_y(u) \qquad (7.4)$$
$$z = \phi_z(u)$$

where each value of the parametric variable u_i corresponds to a specific point (x_i, y_i, z_i) in space (Figure 7.8b).

This way of expressing the mapping is somewhat more abstract and less geometric than the first way. The emphasis is now on the roles of x, y, and z and u as dependent and independent variables and less as object and image points. An even more powerful metaphor is to treat $\phi_x(u)$, $\phi_y(u)$, and $\phi_z(u)$ as components of a position vector $\mathbf{p}(u)$ (Figure 7.8c). This restores some geometry to the concept and allows us to use vector algebra and geometry to express and control the curve's properties.

Many texts use the metaphor of the motion of a point to describe a curve in space, and in this case t (for time) is ordinarily used as the parametric or independent variable in place of u. We describe the path of a point moving through space by assigning values to the position vector $\mathbf{p}$ at successive instants of time. We denote the relationship between $\mathbf{p}$ and time t as $\mathbf{p} = \mathbf{p}(t)$, which means that $\mathbf{p}$ is a function of time. In component form, this is equivalent to

$$x = x(t)$$
$$y = y(t)$$
$$z = z(t)$$

Of course, these functions may generate almost any space curve (Figure 7.9).

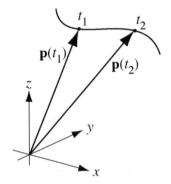

Figure 7.9 A curve as the locus of a moving point.

The positive sense on a curve defined by parametric equations is the sense in which the parameter increases. This, in turn, depends on how the parameter is used in the analytical expressions defining the curve. We can change this parameter under certain conditions without affecting the shape of the curve. For example, let's suppose that the parameter u is an arbitrary function of another parameter v so that

$$u = u(v) \tag{7.5}$$

Substitute this function into Equation 7.4 to obtain

$$x = \phi_x(u(v))$$
$$y = \phi_y(u(v)) \tag{7.6}$$
$$z = \phi_z(u(v))$$

Nonlinear Transformations

These are the parametric equations of the same curve but in terms of the new parameter v. Notice that the transformation $v = -u$ reverses the positive sense of the curve.

Now let's look at some specific examples of parametric representations of curves. The equations (Figure 7.10 and Figure 7.11)

$$x = a\cos u, \ y = a\sin u \tag{7.7}$$

and

$$x = \frac{a(1-u^2)}{1+u^2}, \ y = \frac{2au}{1+u^2} \tag{7.8}$$

both represent a circle in the xy plane with the center at $(0,0)$ and radius a.

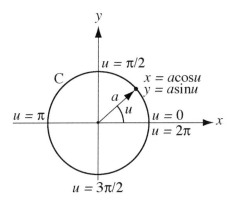

Figure 7.10 Parametric representation of a circle.

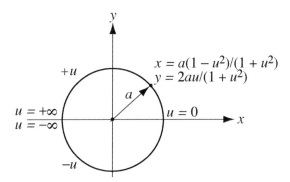

Figure 7.11 Another parametric representation of a circle.

Of course, Equations 7.4 may be linear, that is of the form

$$\begin{aligned} x &= a + lu \\ y &= b + mu \\ z &= c + nu \end{aligned} \tag{7.9}$$

where a, b, c, l, m, and n are constants. These parametric equations define a straight line through the point (a, b, c) and with direction cosines proportional to l, m, and n, respectively.

If the functions in Equations 7.4 have the form

$$x = a \cos u$$
$$y = a \sin u \qquad (7.10)$$
$$z = bu$$

where $a > 0$ and $b < 0$, then the curve is a right-handed circular helix. These parametric equations describe the locus of a point that revolves around the z axis at a constant distance a and at the same time moves parallel to it at a rate proportional to the angle of revolution, u.

A central problem here is how to change the parameterization without changing the form of the curve. In other words, what transformations of the parameter are allowed that leave the form of the curve invariant? This problem leads us firmly into differential geometry, so we will very briefly consider some guidelines and examples drawing on concepts from this discipline. A real-valued function describing a transformation of parametric variables such as $t = f(\theta)$ over an interval I_θ is an allowable change of parameter if for $f(\theta)$ the first order derivative exists and is continuous over the interval I_θ, and if $dt/d\theta \neq 0$ for all θ in I_θ. Furthermore, if $t = t(\theta)$ is an allowable change of parameter on I_θ, then $t = t(\theta)$ is a one-to-one mapping of I_θ onto an interval $I_t = t(I_\theta)$, and the inverse function $\theta = \theta(t)$ is an allowable change of parameter on I_t. Thus, for example, the function $t = (b-a)\theta + a$ in the interval $0 \leq \theta \leq 1$, where $a < b$, is an allowable transformation of the parameter θ, taking the interval $0 \leq \theta \leq 1$ onto $a \leq t \leq b$. The inverse $\theta = (t-a)/(b-a)$ is an allowable transformation of the parameter, taking $a \leq t \leq b$ onto $0 \leq \theta \leq 1$. As a final example, the function $t = \tan(\pi\theta/2)$ in the interval $0 \leq \theta \leq 1$ is an allowable transformation of parameter θ, taking $0 \leq \theta \leq 1$ onto $0 \leq t \leq \infty$. The inverse is $\theta = (2/\pi)\tan^{-1} t$, taking $0 \leq t \leq \infty$ onto $0 \leq \theta \leq 1$. Finally, we note that any property of a curve must be common to all its parametric representations, meaning that the property must be independent of any particular form of the curve's parametric representation.

Parametric Representation of Surfaces

Consider a mapping that sends points on the xy plane into points in space (Figure 7.12a). In its most general form, this transformation looks like

$$x' = \phi_x(x, y), \quad y' = \phi_y(x, y), \quad z' = \phi_z(x, y) \qquad (7.11)$$

Nonlinear Transformations

These equations are similar to Equations 7.3 for curves, except we require two independent variables for surfaces. Again, as with curves, we do not insist that $\phi_x(x,y)$, $\phi_y(x,y)$, and $\phi_z(x,y)$ must be linear functions, so that the transformation may produce an arbitrary surface in space. Any points on the xy plane may be replaced by a pair of real numbers, which we can think of as a point on a uv plane (Figure 7.12b). To do this we replace the independent variables x and y with u and v, respectively, and drop the prime notation. We call u and v the parametric variables (and they are still independent variables). Equations 7.11 now become

$$x = \phi_x(u,v), \quad y = \phi_y(u,v), \quad z = \phi_z(u,v) \tag{7.12}$$

where each pair of parametric values u_i, v_i corresponds to a specific point (x_i, y_i, z_i) in space. We treat $\phi_x(u,v), \phi_y(u,v), \phi_z(u,v)$ as components of a position vector $\mathbf{p}(u,v)$ (Figure 7.12c).

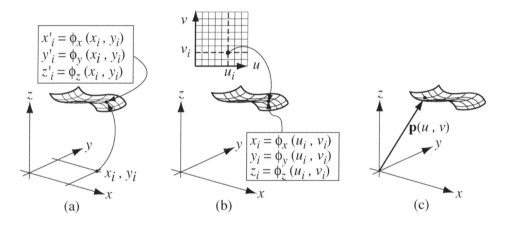

Figure 7.12 Parametric representation of surfaces.

Fixing the value of one of the parametric variables, u or v, produces a curve on the surface in terms of the other variable, which remains free (Figure 7.13). By continuing this process, first for one variable and then the other for any number of arbitrary values, we create a parametric net of two one-parameter families of curves on the surface so that just one curve of each family passes through each point $\mathbf{p}(u,v)$. Again, the positive sense on any of these curves is the sense in which the free parameter increases.

A Surface as the Locus of a Moving Deforming Curve

Just as the metaphor of a point moving in space is useful for defining and understanding curves, so too is the idea of a curve moving and deforming in space so that a surface is produced (Figure 7.14). Consider a curve $\mathbf{p} = \mathbf{p}(u)$ that moves

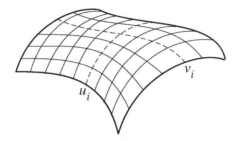

Figure 7.13 Curvilinear coordinate system on a surface.

through space. As it moves its shape changes. Successive positions and shapes of the curve generate a surface where each point is characterized by the time t at which the moving curve passes through it and the parameter u of the point on the moving curve. Thus $\mathbf{p} = \mathbf{p}(u,t)$ describes a surface in space.

The geometric properties of a surface are independent of the particular parametric net defining it. In fact, we may elect to change or transform the parametric net to suit some analytical objective. The constraints on such transformations, however, require an understanding of differential geometry that is beyond the scope of our studies. Usually we must work with restricted regions or patches of a surface to meet the requirements for a transformation of parameters.

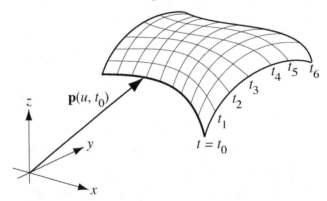

Figure 7.14 A surface as the locus of a moving deforming curve.

Parametric Representation of a Solid

Consider a mapping that sends points in space into other points in space (Figure 7.15a). In its most general form, this transformation looks like

$$x' = \phi_x(x,y,z)$$
$$y' = \phi_y(x,y,z) \quad\quad\quad (7.13)$$
$$z' = \phi_z(x,y,z)$$

Nonlinear Transformations

These equations, too, are similar to Equations 7.3 for curves, as well Equations 7.11 for surfaces. Since we do not insist that $\phi_x(x,y,z)$, $\phi_y(x,y,z)$, and $\phi_z(x,y,z)$ must be linear, they may produce a highly nonlinear, three-dimensional curvilinear coordinate system or solid.

Points in space may be replaced with a triplet of real numbers that we can think of as points in a u,v,w parametric space (Figure 7.15b). We proceed as we did with curves and surfaces, by replacing the independent variables x,y,z with the parametric variables u,v,w, respectively, and drop the prime notation. Equations 7.13 now become

$$x = \phi_x(u,v,w), \quad y = \phi_y(u,v,w), \quad z = \phi_z(u,v,w) \tag{7.14}$$

where each set of values u_i, v_i, w_i corresponds to a specific point in space (x_i, y_i, z_i).

Parametric Variable Transformation

If we transform the parameterization of a surface $\mathbf{p}(u,v)$ by the equations

$$u' = u'(u,v), \quad v' = v'(u,v) \tag{7.15}$$

then the partial derivatives of the surface with respect to the new parameters u', v' are

$$\frac{\partial \mathbf{p}}{\partial u'} = \frac{\partial \mathbf{p}}{\partial u}\frac{\partial u}{\partial u'} + \frac{\partial \mathbf{p}}{\partial v}\frac{\partial v}{\partial u'}$$

$$\frac{\partial \mathbf{p}}{\partial v'} = \frac{\partial \mathbf{p}}{\partial u}\frac{\partial u}{\partial v'} + \frac{\partial \mathbf{p}}{\partial v}\frac{\partial v}{\partial v'} \tag{7.16}$$

so that $\mathbf{A}' = \begin{bmatrix} \dfrac{\partial \mathbf{p}}{\partial u'} & \dfrac{\partial \mathbf{p}}{\partial v'} \end{bmatrix} = \mathbf{AJ}$ (7.17)

where $\mathbf{J} = \begin{bmatrix} \dfrac{\partial u}{\partial u'} & \dfrac{\partial u}{\partial v'} \\ \dfrac{\partial v}{\partial u'} & \dfrac{\partial v}{\partial v'} \end{bmatrix}$ (7.18)

$\mathbf{J}$ is the Jacobian matrix of the transformation.

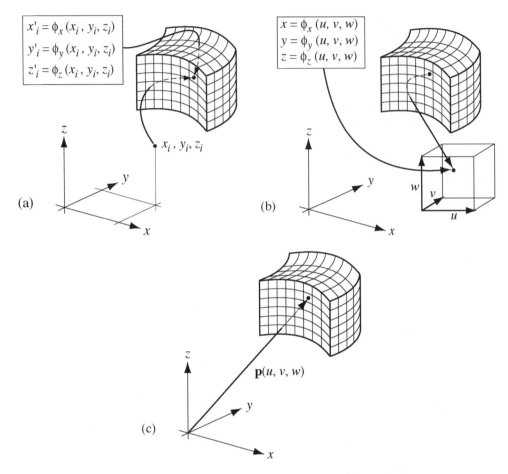

Figure 7.15 Parametric representation of a solid.

7.4 Deformations

Here is one way to deform a curve $\mathbf{p}(u)$: Define a deformation of the x axis by $\mathbf{r}(t)$ (Figure 7.16a). Next, define a relationship between the parametric variables u and t, $t = t(u)$. Then the deformed curve is given by

$$\mathbf{p}'(u) = \mathbf{r}(t(u)) + p_y(u)\mathbf{m} \qquad (7.19)$$

where $\mathbf{r}^t = d\mathbf{r}(t)/dt$, $\mathbf{l} = \mathbf{r}^t/|\mathbf{r}^t|$, and $\mathbf{m}$ is a unit vector perpendicular to $\mathbf{l}$, so that $\mathbf{l} \bullet \mathbf{m} = 0$. In effect, what we have done is to transform the rectilinear Cartesian coordinate system into a curvilinear system of a specific kind. The lines of constant y are curved and "parallel" to $\mathbf{r}(t)$, and the lines of constant x are straight and orthogonal to the lines of constant y (Figure 7.16b). Notice that the y = constant curves are not congruent. Do you see why?

Nonlinear Transformations

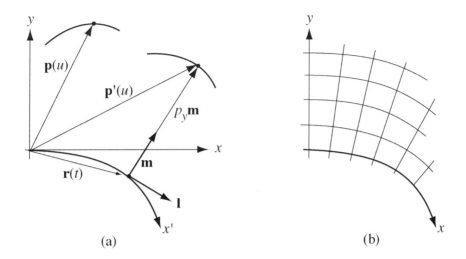

Figure 7.16 Curvilinear deformation.

A Bivariate Deformation

Define a curve $\mathbf{c}(t)$ in the parametric plane u, v (Figure 7.17a). Then define a bivariate surface $\mathbf{p}(u, v)$ in space. Map the curve onto the x, y, z space as it lies in the bivariate surface (Figure 7.17b). Thus

$$\mathbf{c}(t) = \mathbf{p}\big(u(t), v(t)\big) \tag{7.20}$$

In effect, we define the deformation by defining the embedding space. This is the way the geometric object is located, oriented and deformed in x, y, z space.

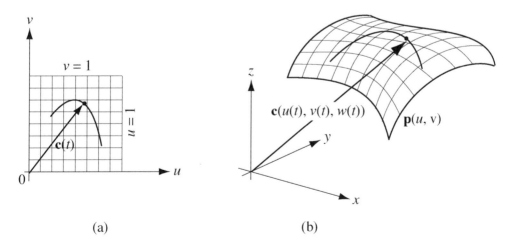

Figure 7.17 Bivariate deformation.

A Trivariate Deformation

It is easy to extend the univariate and bivariate deformation schemes. First, define a curve $\mathbf{c}(t)$ in a u,v,w parametric space (Figure 7.18a). Next, define a trivariate solid $\mathbf{p}(u,v,w)$, and then map the curve onto x,y,z space (Figure 7.18b).

Thus, $\quad \mathbf{c}(t) = \mathbf{p}(u(t), v(t), w(t))$ \hfill (7.21)

Of course, we are not limited to deforming curves. This procedure also applies to deforming surfaces or solids.

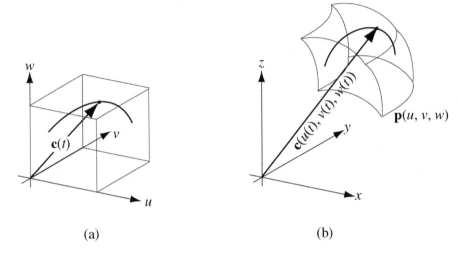

(a) \hspace{4cm} (b)

Figure 7.18 Trivariate deformation.

Sweep Transformation and Deformation

Given an open or closed curve in the l, m, n Cartesian coordinate system, we can sweep the curve through space while scaling or otherwise deforming it to produce a surface or solid shape (Figure 7.19a). First, we define a parametric curve $\mathbf{p}(u)$ that describes the successive positions of the cross section, and define another parametric function $\mathbf{d}(u)$ that describes the directed orientation of the cross section. Three mutually orthogonal unit vectors $\mathbf{l}$, $\mathbf{m}$, and $\mathbf{n}$ give the orientation of the l, m, n coordinate system at any point $\mathbf{p}(u)$, and we define these as

$$\mathbf{l} = \frac{\mathbf{p}^u}{|\mathbf{p}^u|}, \quad \mathbf{m} = \frac{\mathbf{d} \times \mathbf{p}^u}{|\mathbf{d} \times \mathbf{p}^u|}, \quad \mathbf{n} = \mathbf{l} \times \mathbf{m} \hfill (7.22)$$

where $\mathbf{p}^u = d\mathbf{p}/du$. Notice that $\mathbf{p}(u)$ and $\mathbf{d}(u)$ are coordinated through their common parametric variable, u. The vector function $\mathbf{d}(u)$ determines the rotational orientation of $\mathbf{m}$ and $\mathbf{n}$ about an axis coincident with $\mathbf{l}$. Equations 7.22 ensure that $\mathbf{l}$, $\mathbf{m}$, and $\mathbf{n}$ do indeed form a mutually orthogonal triad (Figure 7.19b).

NONLINEAR TRANSFORMATIONS

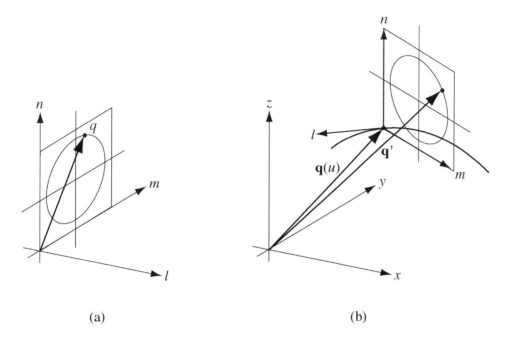

(a) (b)

Figure 7.19 Sweep transformation and deformation.

Thus, the sweep transformation of any point **q** in the initial l, m, n coordinate system onto the x, y, z system at $\mathbf{p}(u)$ is $\mathbf{q'}$, where

$$\mathbf{q'} = \mathbf{p}(u) + q_l \mathbf{l} + q_m \mathbf{m} + q_n \mathbf{n} \tag{7.23}$$

This relationship reconstructs or maps the initial curve in l, m, n space at any point $\mathbf{p}(u)$ in x, y, z space. If the initial curve is one we define by a set of control points, then these points are swept into new positions according to Equation 7.23. The complete curve is then regenerated based on the transformed control points.

We may also continuously change the scale of the curve as it sweeps along $\mathbf{p}(u)$ by simply adding another parametric function, say $\mathbf{s}(u)$, constructed to yield the appropriate scale factors (rescaled in the l, m, n system). Now all these functions $\mathbf{p}(u)$, $\mathbf{d}(u)$ and $\mathbf{s}(u)$ operate in a coordinated fashion to position, direct and scale the initial curve. Since $\mathbf{s}(u)$ is also a vector function it can be used to produce three separate scale factors s_l, s_m, s_n.

Tapering

Consider a transformation that accomplishes a global tapering along the z axis (Figure 7.20). We can express this tapering effect as follows:

$$x' = f_1(z)x, \quad y' = f_2(z)y, \quad z' = z \tag{7.24}$$

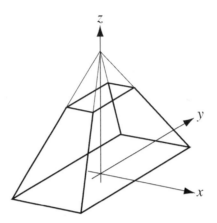

Figure 7.20 Global tapering along the z-axis.

If we assume a linear tapering function, say $f_1(z)$, $f_2(z) = -(z-z_T)/z_T$, then Equations 7.24 become

$$x' = -x(z-z_T)/z_T$$
$$y' = -y(z-z_T)/z_T \qquad (7.25)$$
$$z' = z$$

In matrix form we have

$$\begin{bmatrix} x' \\ y' \\ z' \\ 1 \end{bmatrix} = \begin{bmatrix} -(z-z_T)/z_T & 0 & 0 & 0 \\ 0 & -(z-z_T)/z_T & 0 & 0 \\ 0 & 0 & 1 & 0 \\ 0 & 0 & 0 & 1 \end{bmatrix} \begin{bmatrix} x \\ y \\ z \\ 1 \end{bmatrix} \qquad (7.26)$$

Notice that the value of some elements of the transformation matrix depend on the value of the input point coordinates. This means the matrix must be recalculated each time the z coordinate of the input point changes.

Twisting

Here is a transformation producing a global twist about the z axis (Figure 7.21):

$$x' = x\cos f(z) - y\sin f(z)$$
$$y' = x\sin f(z) + y\cos f(z) \qquad (7.27)$$
$$z' = z$$

In matrix form this becomes

$$\begin{bmatrix} x' \\ y' \\ z' \\ 1 \end{bmatrix} = \begin{bmatrix} \cos f(z) & -\sin f(z) & 0 & 0 \\ \sin f(z) & \cos f(z) & 0 & 0 \\ 0 & 0 & 1 & 0 \\ 0 & 0 & 0 & 1 \end{bmatrix} \begin{bmatrix} x \\ y \\ z \\ 1 \end{bmatrix} \qquad (7.28)$$

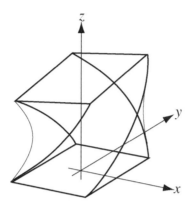

Figure 7.21 Global twisting about the z axis.

ANSWERS TO SELECTED EXERCISES

Chapter 2

Section 2.1

1. a. $\Re$, $\Re$
   b. $\Re$, $\Re$
   c. $\Re$, $f(x) \geq 0$
   d. $\Re$, $\Re$
   e. $\Re$, $f(x) \leq 2$
   f. $\Re$, $-1 \leq f(x) \leq +1$
   g. $x \geq -1$, $f(x) \geq 0$
   h. $x \neq \pm 1$, $f(x) \geq 1$

3. It is a geometric transformation because it is "onto" and "one-to one."

5. a. $f'(x') = -x'$
   b. $f'(x') = \dfrac{x'-1}{2}$
   c. $f'(x') = \dfrac{x'+5}{3}$
   d. $f'(x') = -x' + 1$
   e. $f'(x') = (x')^{\frac{1}{3}}$

7. They are identical mappings if the center of the circle corresponds to the origin of the Cartesian coordinate system on the plane, and if lines a and b correspond to the x and y axes.

9. a. Yes
   b. Yes
   c. No, an opposite orientation is produced by $\rho_\Pi \circ \sigma_d$ for which no single transformation element exists.
   d. No, a reflection in the circle's center, σ_0, is required to complete the group.
   e. Yes

Section 2.2

1. a. An expansion or contraction parallel to the y axis followed by a reflection in the x axis if u is negative.
   b. Expansions and/or contractions parallel to the coordinate axes followed by reflections in these axes if a and/or b are negative.
   c. Uniform expansion or contraction about the origin followed by an inversion in the origin if a is negative.
   d. A shearing displacement parallel to the y axis.
   e. Rotation of the plane through $\pi/4$.

3. There are an infinite number of solutions of the form $x = -\frac{15}{14}z - \frac{17}{7}$ and $y = \frac{5}{7}z - \frac{26}{7}$, where the value of z may be assigned arbitrarily, thus fixing the values of x and y.

5. There are an infinite number of solutions of the form $x = 2z + 3$ and $y = -z + 1$, where the value of z may be assigned arbitrarily, thus fixing the values of x and y. Again, since $\det \mathbf{M} = 0$, no unique solution exists.

7. $\mathbf{M}$ is a singular linear homogeneous transformation, and the points of the plane are mapped onto the line $x = 3y$.

Section 2.4

1. Given $B' = \sigma_m(B)$, then $BB' \perp m$, $OB = OB'$, and $\triangle ABO \cong \triangle AB'O$ from SAS in elementary geometry (Figure 2.81). Therefore, $AB = AB'$. A similar proof applies to the case where A is not on m.

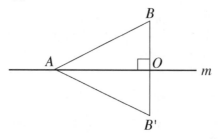

Figure 2.81

3. a. $(3, 3)$ d. (b, a)
   b. $(-3, 3)$ e. (y, x)
   c. $(1, 5)$

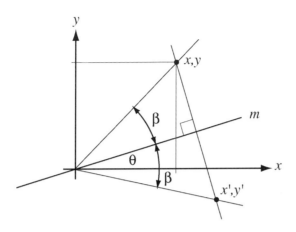

Figure 2.83

5. The axis of reflection m is given by $ax+by+c=0$ (Figure 2.84). The line through x,y and x',y' is perpendicular to m and is given by $a(y'-y)=b(x'-x)$. Also, $(x+x')/2$, $(y+y')/2$ is the midpoint and is on m, so that $a(x+x')/2+b(y+y')/2+c=0$. Rewrite these equations as $bx'-ay'=bx-ay$ and $ax'+by'=-ax-by-2c$, and solve them for x' and y'.

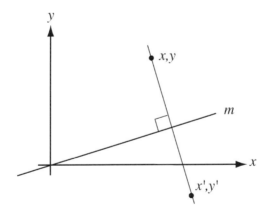

Figure 2.84

7. Since the product of three halfturns is a halfturn, and the four points form a parallelogram (Figure 2.86), we find that $\sigma_R\sigma_Q\sigma_P = \sigma_S$ and $\sigma_P\sigma_Q\sigma_R = \sigma_S$; therefore, $\sigma_R\sigma_Q\sigma_P = \sigma_P\sigma_Q\sigma_R$.

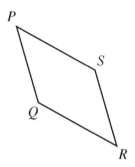

Figure 2.86

9. Since $PA \parallel NB$, $AB = BP'$, and $AP \perp NB$, then $\angle PAP' = \angle NBP' = 90°$, $\angle PP'A = \angle NP'B$, and $\angle APP' = \angle BNP'$, $\triangle PAP'$ is similar to $\triangle NBP$. Thus, since $BP'/AP' = 1/2$, therefore, $NP'/PP' = 1/2$ (Figure 2.87).

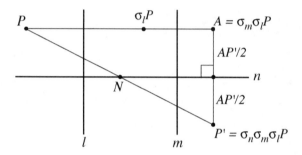

Figure 2.87

11. Show that $A'B' = AB$ using congruent triangles (Figure 2.89). Remember, the plane Π is the perpendicular bisector of AA' and BB'.

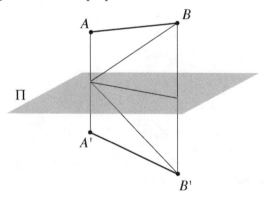

Figure 2.89

13. These are $\sigma_h, \sigma_v, \sigma_d$, and σ_{d_2}.

15. The inversion transformation and all reflections in planes.

ANSWERS TO SELECTED EXERCISES

Section 2.5

1. A dilation in the origin with scale factor of -1 is equivalent to inversion (or reflection) in the origin.

3. Here is an example for similarities in the plane (Figure 2.92). Given line segments AB and CD, find an isometry that makes $AB \parallel CD$, say $\rho_{A\alpha}AB = AE$. Find the intersection, O, of lines through AC and ED, and the ratio OC/OA. Then the similarity is given by the product $\delta_{O,OC/OA}\rho_{A\alpha}$

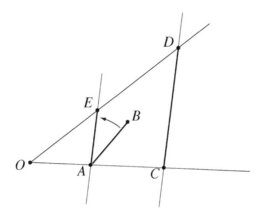

Figure 2.92

5. Given a center of dilation O, and two scale factors k_1 and k_2, then show that $\delta_{Ok_2}\delta_{Ok_1} = \delta_{Ok_1}\delta_{Ok_2}$ (Figure 2.93). We see that $OC = k_2 k_1 OA$ and $OE = k_1 k_2 OA$, but $k_1 k_2 = k_2 k_1$ so that $OC = OE$ or $\delta_{Ok_2}\delta_{Ok_1} = \delta_{Ok_1}\delta_{Ok_2}$.

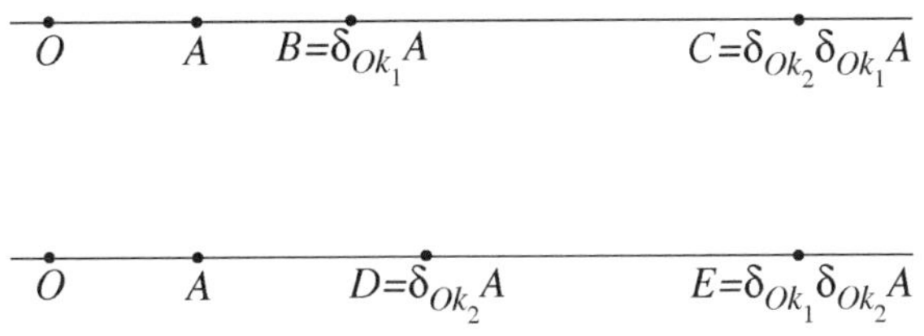

$$\overset{\bullet}{O} \quad \overset{\bullet}{A} \quad \overset{\bullet}{B=\delta_{Ok_1}A} \quad\quad \overset{\bullet}{C=\delta_{Ok_2}\delta_{Ok_1}A}$$

$$\overset{\bullet}{O} \quad \overset{\bullet}{A} \quad \overset{\bullet}{D=\delta_{Ok_2}A} \quad\quad \overset{\bullet}{E=\delta_{Ok_1}\delta_{Ok_2}A}$$

Figure 2.93

7. For the first dilation we have

$$x' = k_1 x + (1-k_1)x_1$$
$$y' = k_1 y + (1-k_1)y_1$$

and for the product

$$x'' = k_2 x' + (1-k_2) x_2$$
$$y'' = k_2 y' + (1-k_2) y_2$$

Substituting for x' and similarly for y'

$$x'' = k_1 k_2 x + k_2 (1-k_1) x_1 + (1-k_2) x_2$$
$$y'' = k_1 k_2 y + k_2 (1-k_1) y_1 + (1-k_2) y_2$$

For the equivalent center of dilation for this product we write

$$(1-k_1 k_2) x_P = k_2 (1-k_1) x_1 + (1-k_2) x_2$$
$$(1-k_1 k_2) y_P = k_2 (1-k_1) y_1 + (1-k_2) y_2$$

Solve for x_P, y_P

$$x_P = \frac{k_2 (1-k_1) x_1 + (1-k_2) x_2}{(1-k_1 k_2)} \quad \text{and} \quad y_P = \frac{k_2 (1-k_1) y_1 + (1-k_2) y_2}{(1-k_1 k_2)}$$

Section 2.6

1. For any point A construct $A\widehat{A}$ parallel to the direction of affinity, construct AO perpendicular to the direction of affinity (Figure 2.95). We know that $\angle OA\widehat{A} = 90°$, and that A' is the image of A under a perspective affinity. Also A, $\widehat{A}$, and A' are collinear, forming $\triangle OAA'$. O and $\widehat{A}$ are each their own images under the perspective affinity. $\angle OAA' \neq \angle OA'A$, therefore, angles are not preserved under a perspective affinity.

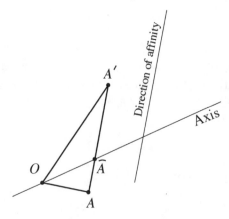

Figure 2.95

Answers to Selected Exercises

3. $x = \dfrac{x'}{k_x}$ and $y = y'$

5. The inverse of a strain is another strain (see Exercise 2, above). The product of two strains in the same direction is another strain in that direction, say $\varepsilon_{x,2}\varepsilon_{x,1} = \varepsilon_{x,total}$. Let the strain coefficients be $k_{x,1}$ and $k_{y,1}$, then the equations for the product $\varepsilon_{x,2}\varepsilon_{x,1}$ are $x' = k_{x,1}k_{x,2}x$ and $y' = y$. So that now we may say that strains in the same direction form a subgroup.

7. Given a parallel projection from points on l to points on m, where n is in the direction of the projection; find the perpendicular bisector b of the angle formed by l and m (Figure 2.96). Then an equivalent perspective affinity is defined as follows: b is the axis of affinity, n is the direction of affinity, and the scale factor $k = 1$ (since $A\hat{A}/\hat{A}A' = 1$).

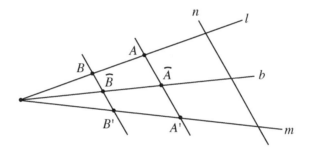

Figure 2.96

9. $x' = kx$, $y' = y/k$

Chapter 3 Answers

Section 3.1

1. It is not closed with respect to multiplication by real numbers, k, since if $k < 0$, then the vector $k\mathbf{p}$ does not belong to the first quadrant.

3. a. $k = -2$
   b. $k = -1$
   c. $k = 1$

5. Linearly dependent, since $2\mathbf{r} + \mathbf{s} - \mathbf{t} = 0$

7. a. $|\mathbf{a}| = 5$; 0.6, 0.8
   b. $|\mathbf{b}| = 2$; 0, -1
   c. $|\mathbf{c}| = \sqrt{34}$; $-\dfrac{3}{\sqrt{34}}, -\dfrac{5}{\sqrt{34}}, 0$
   d. $|\mathbf{d}| = \sqrt{26}$; $\dfrac{1}{\sqrt{26}}, \dfrac{4}{\sqrt{26}}, -\dfrac{3}{\sqrt{26}}$
   e. $|\mathbf{e}| = \sqrt{x^2 + y^2 + z^2}$; $\dfrac{x}{|\mathbf{e}|}, \dfrac{y}{|\mathbf{e}|}, \dfrac{z}{|\mathbf{e}|}$

9. a. 18.435°
   b. 40.601°
   c. 90°
   d. 58.249°
   e. 88.556°

11. Let $\mathbf{p} = \left(-\dfrac{1}{3}, \dfrac{2}{3}, \dfrac{2}{3}\right)$, $\mathbf{q} = \left(\dfrac{2}{3}, -\dfrac{1}{3}, \dfrac{2}{3}\right)$, $\mathbf{r} = \left(-\dfrac{2}{3}, -\dfrac{2}{3}, \dfrac{1}{3}\right)$
    $\mathbf{p} \bullet \mathbf{q} = 0$, $\mathbf{p} \bullet \mathbf{r} = 0$, and $\mathbf{q} \bullet \mathbf{r} = 0$.

13. $|\mathbf{a} + \mathbf{b}|^2 = (\mathbf{a} + \mathbf{b}) \bullet (\mathbf{a} + \mathbf{b})$
    $= |\mathbf{a}|^2 + |\mathbf{b}|^2 + 2(\mathbf{a} \bullet \mathbf{b}) \leq (|\mathbf{a}| + |\mathbf{b}|)^2$
    $= |\mathbf{a}|^2 + |\mathbf{b}|^2 + 2(\mathbf{a} \bullet \mathbf{b}) \leq |\mathbf{a}|^2 + |\mathbf{b}|^2 + 2|\mathbf{a}||\mathbf{b}|$

    iff $\mathbf{a} \bullet \mathbf{b} \leq |\mathbf{a}||\mathbf{b}|$
    but $\mathbf{a} \bullet \mathbf{b} \leq |\mathbf{a}||\mathbf{b}|\cos\theta$
    if $\theta > 0°$, then $\mathbf{a} \bullet \mathbf{b} \leq |\mathbf{a}||\mathbf{b}|$, Q.E.D.

15. $\mathbf{r} = \mathbf{r}_o + u(\mathbf{r} - \mathbf{r}_o)$, where u is a real number.

Section 3.2

1. a. $\begin{vmatrix} \sqrt{2}/2 & \sqrt{2}/2 \\ -\sqrt{2}/2 & \sqrt{2}/2 \end{vmatrix} = 1$; therefore the matrix is proper and orthogonal.

   b. $\begin{vmatrix} 1 & 1/2 \\ 2 & 0 \end{vmatrix} = -1$; therefore the matrix is improper and orthogonal.

Answers to Selected Exercises

c. $\begin{vmatrix} 3 & 1 \\ 5 & 2 \end{vmatrix} = 1$; therefore the matrix is proper and orthogonal.

d. $\begin{vmatrix} -\sqrt{3}/2 & 1/2 \\ 1/2 & \sqrt{3}/2 \end{vmatrix} = -1$; the matrix is improper and orthogonal

3. $\begin{bmatrix} 0 & 1 & 0 \\ 0 & 0 & -1 \\ 1 & 0 & 0 \end{bmatrix}$

5. $\mathbf{AB} = \begin{bmatrix} -\frac{\sqrt{3}}{2}+\frac{1}{4} & \frac{1}{2}+\frac{\sqrt{3}}{4} \\ -\sqrt{3} & 1 \end{bmatrix}$, $\begin{vmatrix} -\frac{\sqrt{3}}{2}+\frac{1}{4} & \frac{1}{2}+\frac{\sqrt{3}}{4} \\ -\sqrt{3} & 1 \end{vmatrix} = 1$

Section 3.3

1. a. 6, $\begin{bmatrix} k \\ -2k \end{bmatrix}$ and -1, $\begin{bmatrix} 2k \\ 3k \end{bmatrix}$ c. 1, $\begin{bmatrix} k \\ -3k \end{bmatrix}$ and 2, $\begin{bmatrix} 0 \\ k \end{bmatrix}$

   b. 7, $\begin{bmatrix} k \\ 2k \end{bmatrix}$ and -5, $\begin{bmatrix} 5k \\ -2k \end{bmatrix}$ d. 3, $\begin{bmatrix} k \\ k \end{bmatrix}$

3. a. Similar   b. Not similar

Section 3.4

1. $n = 2$, $\mathbf{r} = r^1 \mathbf{e}_1 + r^2 \mathbf{e}_2$
   $n = 3$, $\mathbf{r} = r^1 \mathbf{e}_1 + r^2 \mathbf{e}_2 + r^3 \mathbf{e}_3$
   $n = 4$, $\mathbf{r} = r^1 \mathbf{e}_1 + r^2 \mathbf{e}_2 + r^3 \mathbf{e}_3 + r^4 \mathbf{e}_4$

3. Use the rules of transformation of R_b^a and S^{cd} to show that $R_b^a S^{cd} = a_b^i b_j^a b_k^c b_l^d R_i^j S^{kl}$

5. $T_{ijk}^i = a_i^k a_j^l b_m^i T_{kl}^m$, $a_i^k b_m^i = \delta_m^k$, and $a_j^l \delta_m^k = a_j^l T_l$, $T_j = a_j^l T_l$. Therefore T_{ij}^i is a covariant vector.

Chapter 4 Answers

Section 4.1

1. In slope-intercept form, the translation transformation equation of a straight line is $y' = mx' + (b - mx_T + y_T)$, where m and b are the slope and intercept, respectively, of the original line. For a line passing through the origin it must be true that $y' = mx'$. This means that the intercept must equal zero (i.e., $b - mx_T + y_T = 0$). One way this can happen is if $\Delta_x = 0$, $\Delta_y = b$; another is if $\Delta_x = b/m$, $\Delta_y = 0$.

3. The simplified equation is $Ax' + By' + Cz' = 0$. This is the result if $A\Delta_x + B\Delta_y + C\Delta_z = D$. There are any number of translations that produce the desired result, for example: $\Delta_x = D/A$, $\Delta_y = \Delta_z = 0$.

5. $d = +\sqrt{a^2 + b^2 + c^2}$

7. a. $x' + y' - 2 = 0$ c. $y' = 2x'^2 - 8x' + 7$
   b. $x'^2 + y'^2 - 4x' + 2y' = -1$

9. a. $\mathbf{t} = (5,0)$ d. $\mathbf{t} = (-1,0,6)$
   b. $\mathbf{t} = (-3,2)$ e. $\mathbf{t} = (a,b,c)$
   c. $\mathbf{t} = (0,0,0)$

Section 4.2

1. $\mathbf{p}'^T = \mathbf{p}^T \mathbf{R}^T$ or $\begin{bmatrix} x' & y' & z' \end{bmatrix} = \begin{bmatrix} x & y & z \end{bmatrix} \begin{bmatrix} r_{11} & r_{21} & r_{31} \\ r_{12} & r_{22} & r_{32} \\ r_{13} & r_{23} & r_{33} \end{bmatrix}$

3. a. $|\mathbf{R}_{30}| = 1$ c. $|\mathbf{R}_{90}| = 1$
   b. $|\mathbf{R}_{45}| = 1$ d. $|\mathbf{R}_{180}| = 1$

5. $x' = (x - x_c)\cos\theta - (y - y_c)\sin\theta + x_c$ and
   $y' = (x - x_c)\sin\theta - (y - y_c)\cos\theta + y_c$

Answers to Selected Exercises

7. a. $\alpha = \beta = 0$ and $\mathbf{R}_{zyx} = \mathbf{R}_z = \begin{bmatrix} \cos\gamma & -\sin\gamma & 0 \\ \sin\gamma & \cos\gamma & 0 \\ 0 & 0 & 1 \end{bmatrix}$

    b. $\alpha = \gamma = 0$ and $\mathbf{R}_{zyx} = \mathbf{R}_y = \begin{bmatrix} \cos\beta & 0 & \sin\beta \\ 0 & 1 & 0 \\ -\sin\beta & 0 & \cos\beta \end{bmatrix}$

    c. $\beta = \gamma = 0$ and $\mathbf{R}_{zyx} = \mathbf{R}_z = \begin{bmatrix} 1 & 0 & 0 \\ 0 & \cos\alpha & -\sin\alpha \\ 0 & \sin\alpha & \cos\alpha \end{bmatrix}$

9. $\mathbf{p}'_i = \mathbf{R}_y(-\omega)\mathbf{R}_z(\phi)\mathbf{R}_y(\omega)\mathbf{p}_i$

11. The angle of rotation: $\phi = 45°$

    a. $x' = y'$
    b. $x' = 0$
    c. $x' = -3/\sqrt{2}$
    d. $x' + y' = 3/\sqrt{2}$
    e. $x'^2 + y'^2 = 1$
    f. $-x'^2 + y'^2 = 2$
    g. $x'^2 + y'^2 - \sqrt{2}(x' + y') = 1$
    h. $(x' + y')^2 = \sqrt{2}(-x' + y')$

Section 4.3

1. The answer comes from analyzing the properties of matrix multiplication. Given any sequence of rotations expressed as homogeneous transformations, the net resultant is another rotation, and thus orthogonal. For example, given two rotations $\mathbf{R}_1$ and $\mathbf{R}_2$, we have

$$\left[\begin{array}{c|c} \mathbf{R}_2 & \begin{array}{c} 0 \\ 0 \\ 0 \end{array} \\ \hline 0 \; 0 \; 0 & 1 \end{array}\right] \left[\begin{array}{c|c} \mathbf{R}_1 & \begin{array}{c} 0 \\ 0 \\ 0 \end{array} \\ \hline 0 \; 0 \; 0 & 1 \end{array}\right] = \left[\begin{array}{c|c} \mathbf{R}_3 & \begin{array}{c} 0 \\ 0 \\ 0 \end{array} \\ \hline 0 \; 0 \; 0 & 1 \end{array}\right]$$

where $\mathbf{R}_3 = \mathbf{R}_2\mathbf{R}_1$, and $|\mathbf{R}_3| = 1$. In addition, given any combination of a rotation and translation, the net resultant is the rotation and some translation. For example,

$$\left[\begin{array}{ccc|c} & & & 0 \\ & \mathbf{R}_1 & & 0 \\ & & & 0 \\ \hline 0 & 0 & 0 & 1 \end{array}\right] \left[\begin{array}{ccc|c} 1 & 0 & 0 & t_x \\ 0 & 1 & 0 & t_y \\ 0 & 0 & 1 & t_z \\ \hline 0 & 0 & 0 & 1 \end{array}\right] = \left[\begin{array}{ccc|c} & & & t_{14} \\ & \mathbf{R}_1 & & t_{24} \\ & & & t_{34} \\ \hline 0 & 0 & 0 & 1 \end{array}\right]$$

where the t_{i4} are expressions containing r_{ij} and t_x, t_y, t_z terms. The presence of the zeros in the last row and column of the homogeneous transformation matrix describing the rotation $\mathbf{R}_1$, and the presence of 1's on the diagonal and 0's elsewhere in the 3×3 submatrix ensures and yields $\mathbf{R}_1$ in the 3×3 submatrix of the resultant homogeneous transformation matrix.

Chapter 5 Answers

Section 5.1

1. A demonstration using the synthetic approach is trivial. Analytically we might proceed as follows: Assume the point of inversion is coincident with the origin. Every straight line through the origin has an equation of the form $y = ax$. The transformation equations for a central inversion are $x' = -x$, $y' = -y$. The inverse equations are simply $x = -x'$ and $y = -y'$. Substitute into the equation for a straight line $-y' = -ax'$, or $y' = ax'$. The line is fixed, although not point-wise.

3. No. This transformation is orientation reversing. Look at the face defined by vertices 1, 2, 3. In the original we traverse these vertices in a clockwise direction if done in the order 1, 2, 3. In the image, traversing the vertices 1', 2', 3', in that order, proceeds counterclockwise (Figure 5.71).

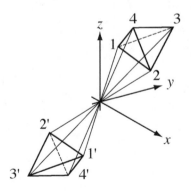

Figure 5.71

Answers to Selected Exercises

Section 5.2

1. $\mathbf{p}' = \begin{bmatrix} -1 & 0 & 2d \\ 0 & 1 & 0 \\ 0 & 0 & 1 \end{bmatrix} \mathbf{p}$ and $\begin{cases} x' = -x + 2d \\ y' = y \end{cases}$

3. $\begin{bmatrix} 0 & 1 & 0 \\ 1 & 0 & 0 \\ 0 & 0 & 1 \end{bmatrix} \begin{bmatrix} 0 & -1 & 0 \\ -1 & 0 & 0 \\ 0 & 0 & 1 \end{bmatrix} = \begin{bmatrix} -1 & 0 & 0 \\ 0 & -1 & 0 \\ 0 & 0 & 1 \end{bmatrix}$, and $\det \begin{bmatrix} -1 & 0 & 0 \\ 0 & -1 & 0 \\ 0 & 0 & 1 \end{bmatrix} = +1$

   therefore it is a rotation of 180° about the origin.

Section 5.3

1. $\mathbf{H} = \mathbf{T}_c \mathbf{R}_\phi \mathbf{R}_\psi \mathbf{R}_f \mathbf{R}_{-\psi} \mathbf{R}_{-\phi} \mathbf{T}_{-c}$. Translate along the z axis $c = D/C$.

   $\phi = \tan^{-1} n_x / n_y, \quad \psi = \sin^{-1} n_z$

   $\mathbf{R}_f = \begin{bmatrix} 1 & 0 & 0 & 0 \\ 0 & -1 & 0 & 0 \\ 0 & 0 & 1 & 0 \\ 0 & 0 & 0 & 1 \end{bmatrix}$

Section 5.5

1. The center of the circle is $(2,-3)$. The lines $x = 2$ and $y = -3$ both pass through the center, and therefore they are lines of symmetry.

3. All diameters.

5. All lines.

7. The x axis.

9. It is symmetric about the origin, but not about the x or y axis or either of the lines $x = y$ and $x = -y$.

11. It is symmetric about the origin and both axes, but not about the lines $x = y$ and $x = -y$.

13. It is symmetric about the origin and the lines $x = y$ and $x = -y$.

15. It is symmetric about the line $x = y$.

Section 5.7

1. a. F_2^1, b. F_1^2, c. F_2^2, d. F_1, e. F_1^3, f. F_2, g. F_1^1

Section 5.9

1. Each four-fold axis passes through a vertex and the center of the octahedron. Each three-fold axis passes through the center of a face and the center. Each two-fold axis passes through the center of an edge and the center.

Chapter 6 Answers

Section 6.1

1. Vectors $\mathbf{p}$ and $a\mathbf{p}$ are parallel, and $k(a\mathbf{p})$ is parallel to them both.

3. The angle θ between the two vectors $\mathbf{p}$ and $\mathbf{q}$ before dilation is

$$\theta = \cos^{-1} \frac{\mathbf{p} \bullet \mathbf{q}}{|\mathbf{p}||\mathbf{q}|}, \text{ and after dilation is}$$

$$\theta' = \cos^{-1} \frac{\mathbf{p}' \bullet \mathbf{q}'}{|\mathbf{p}'||\mathbf{q}'|} = \cos^{-1} \frac{k\mathbf{p} \bullet k\mathbf{q}}{|k\mathbf{p}||k\mathbf{q}|} = \cos^{-1} \frac{k^2(\mathbf{p} \bullet \mathbf{q})}{k^2|\mathbf{p}||\mathbf{q}|} = \cos^{-1} \frac{(\mathbf{p} \bullet \mathbf{q})}{|\mathbf{p}||\mathbf{q}|}$$

or $\theta' = \theta$.

5. $x_c = \dfrac{c_1}{1-k}, \quad y_c = \dfrac{c_2}{1-k}$

Section 6.2

1. $\mathbf{H} = \mathbf{T}_b \mathbf{R}_\phi \mathbf{D} \mathbf{R}_\phi^{-1} \mathbf{T}_b^{-1}$, where $\phi = \tan^{-1} m$.

Section 6.3

1. $x' = x, \quad y' = k_x x + y$

3. $\dfrac{l'}{l} = \sqrt{(\cos\phi + k_y \sin\phi)^2 + \sin^2 \phi}$

Answers to Selected Exercises

Section 6.4

1. Let $b_1 = ka_1$, $b_2 = ka_2$, $b_3 = ka_3$, then by substitution

$$\begin{vmatrix} a_1 & a_2 \\ ka_1 & ka_2 \end{vmatrix} = ka_1a_2 - ka_1a_2 = 0$$

and similarly for the remaining two determinants.

3. a. $h_1 - 3h_3 = 0$ d. $2h_1 + 6h_2 - 3h_3 = 0$
   b. $h_2 - 5h_3 = 0$ e. $h_1 - h_2 - h_3 = 0$
   c. $h_1 + h_2 = 0$

Section 6.5

1. In Figure 6.57, we see that:

$$\gamma + 2\alpha = 180° \qquad \theta + 2\beta = 180°$$
$$\gamma + \theta = 180° \text{ and } 2(\alpha + \beta) = 180°$$
$$2\alpha = \theta \qquad \alpha + \beta = 90°$$

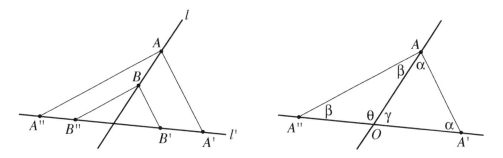

Figure 6.57

Section 6.6

1. In general, no. However, the distances are equal when the center of projection O is exactly midway between the lines.

Chapter 7 Answers

Section 7.2

1. Construct an auxiliary point n times as far from the origin as the given point, and then construct its inverse. This point will be $(1/n) \times$ the distance of the required point from the origin. Find the required point by multiplying the last distance by n.

3. a. $(70/13, 40/13)$
   b. $(10, 0)$
   c. $(-5/4, 0)$
   d. $(6, 2)$
   e. $(215/58, -175/58)$
   f. $(5, 0)$

INDEX

Abelian group, 37
Affine geometry, 82:
    affinities, 56, 82
    construction, 85
    deformations, 83
    distance in, 90
    equiareal transformations, 89
    group, 92
    in the plane, 84
    parallel projection, 94
    perspective, 86
    shear, 87
    strain, 88
    transformations, 93
Affine group, 92
Affinities, 82
Angle-preserving transforms, 53
Area-preserving mappings, 301
Associative property, 34, 37
Associativity, 34

Basis vectors, 114:
    change of, 118
    coordinate systems, 115, 116, 120
    objectivity of a vector, 125
    orthogonal, 126
    reciprocal, 124

Cayley, Arthur, 39
Cayley table, 39
Cells, unit, 243

Centered affine group, 46
Centerless uniform expansion, 258
Center of symmetry, 202
Central inversion, 194
Central projection, 97, 293
Characteristic equation, 129
Closure, 36
Collineation, 52
Commutative property, 35
Commutativity, 35, 37
Composite motion, 184
Computer graphic projections, 305:
    oblique, 308
    orthographic, 307
    perspective, 308
    picture plane cs, 396
Conformal mappings, 302
Conformal transformations, 302
Congruent figures, 53
Conical projection, 284, 304
Conics, 284
Connectivity, 22, 23
Contraction, 49, 52, 76, 256, 257, 261
Contravariant vectors, 138, 140, 141, 144
Coordinate systems (cs), 13:
    affine transformation of, 93
    basis vectors and, 105, 114, 119-126, 137, 139
    computer graphic display, 305

curvilinear, 14, 315
oblique, 13
picture plane, 306
rectangular, 107, 111
rotation of, 160, 165
transformation of, 24
Covariant vectors, 137-140
Crystallographic restriction, 242
Curves, parametric, 316
Curvilinear cs, 14, 315, 320
Cyclic group, 39, 227, 228
Cylindrical projection, 305

Deformations, 323:
   affine, 83
   bivariate, 324
   curvilinear, 324
   sweep with deformation, 325
   tapering, 326
   trivariate, 325
   twisting, 327
Descartes, René, 7
Dihedral group, 228
Dilation, 78:
   anisotropic, 261, 265, 266
   centerless uniform expansion, 258
   fixing the origin, 257, 261
   in space, 80, 266
   in the plane, 78, 262
   isotropic, 256, 259
   product of two, 79, 80, 260
Direction cosines, 107
Display projection, 305
Distance in affine geometry, 90
Distance, invariance of, 24
Distance-preserving transformations, 53
Domain, 26

Eigenvalues, 128:
   characteristic equation, 129
   symmetric transformations, 134
Eigenvectors, 128:
   diagonalization of matrices, 134
   equivalent rotations, and, 177
   geometric interpretation, 128
   similarity transformations, 133
Einstein convention, 44
Ellipse, 16
Equiareal transformation, 89, 270
Equivalent rotations, 172
Euclidean space, 12
Euclid's *Elements*, 6
Euler angle rotations, 167
Euler, Leonhard, 172
Euler's theorem, 177
Expansion, 258

Finite symmetry groups, 232
Fractals, 1
Frieze groups, 235
Functions, 26:
   associative property, 34
   commutative property, 35
   composition of, 33
   domain, 26
   identity property, 35
   input set, 28
   into mappings 28
   inverse, 36
   involution,
   machine, 27
   mappings, 26
   one-to-one functions, 30
   onto mappings, 28, 32
   range, 26

Geodesic mappings, 302

Geometry, 1:
- abstract, 18, 20
- affine, 21
- analytic, 18, 19
- classification of, 21
- congruent figures, 53, 82
- coordinate-dependent, 16
- coordinate-free, 15
- deductive, 19
- history of, 5
- incongruent figures, 21
- logical system, as a, 6
- metric, 18, 20
- projective, 22
- similar figures, 21
- symmetry, 192, 220
- synthetic, 18, 19
- topology, 22
- visual, 18
- ways of thinking about, 18

Geometric objects, 10

Glide reflection, 66

Groups, 36:
- abelian, 37
- affine, 92
- associative, 37
- Cayley table, 39
- closure, 37
- commutative, 37
- cyclic, 227, 228
- dihedral, 228
- even or odd numbers, group of, 38
- finite symmetry, 232
- frieze, 235
- identity, 37
- integers, group of, 37
- inverse, 37
- isometries, 72
- ornamental, 234
- rational numbers, group of, 37
- rotation, 178
- similarities, 81
- symmetry, 225
- translation, 150
- wallpaper groups, 236

Halfturn, 58:
- product of three, 60
- product of two, 59

Hilbert, David, 1

Homogeneous coordinates, 279:
- ideal points, 277, 279, 280
- parallel lines, 280

Homogeneous transformation matrix, 185

Homeomorphism, 100

Homothety, 76, 256

Hyperbola, 17

Ideal points, 277, 279, 280

Identity property, 35, 37

Improper rotations, 162

Into mappings, 28

Invariance, 20:
- geometric, 20, 51

Inverse, 36, 37:
- circle, of a, 314

Inversion, 60:
- center of symmetry, 202
- central, 194
- fixing an arbitrary point, 196
- fixing the origin, 60, 194
- by geometric construction, 312
- in a circle, 311, 313
- in a point, 58, 67, 194
- in the origin, 67
- product of three, 199
- product of two, 197

Involution, 36, 40, 56

Intuition, limits of, 4

Isometries, 54:
- even, odd, 55
- group of, 72

in space, 67, 71
rotation and translation, 72

Julia, Gaston, 2

Kinematics, 189
Klein, Felix, 8, 9, 32, 276
Kock, Helge von, 2
Kock snowflake, 2
Kronecker delta, 108

Lattice, 237, 243
Linear dependence, 104
Linear equations, 44, 309
Linear transformations, 44, 256:
    singular and nonsingular, 49
Linear vector spaces, 101, 102
Line(s) of symmetry, 210
Line-preserving transformations, 52
Lobachevsky, N., 8
Lorentz, H. A., 25
Lorentz transformation, 24

Magnitude, vector, 107
Mandelbrot, Benoit, 2
Mappings, 26, 300
Map projections, 300:
    area-preserving mappings, 301
    conformal mappings, 302
    continuous, 302
    cylindrical, 305
    geodesic mappings, 302
    projection onto tangent plane, 303
    projection onto a cone, 304
Maps, 300
Matrices, 127, 134
    diagonalization of, 134
    homogeneous transformation, 185
    linear transformations, and, 47

reflection, 203, 211, 218
symmetric, 134
Menger, Karl, 2
Mercator projection, 305
Metric tensor, 141, 142
Motion, composite, 144, 184, 187

Nonlinear equations, 309
Nonlinear transformations, 2, 309:
    curvilinear coordinate systems, 315
    deformations, 323
    inversion in a circle, 311, 313
    tapering, 326
    twisting, 188, 327

Oblique coordinate system, 13
Oblique projection, 308
Oblique reflection, 210
Octahedral symmetry, 233, 252, 253, 255
One-to-one functions, 30
Onto mappings, 28
Orientation, 52, 54, 192
Orientation-preserving transformations, 54
Ornamental groups, 234
Orthogonal basis vectors, 126
Orthographic projection, 307

Parabola, 17
Parallelogram law of addition, 103-106
Parallel projection, 94, 286:
    nonorthogonal, 289
    of a line, 288
    of a plane, 290
    of points in space, 291
Parametric variable, 316, 322
Peano, Giuseppe, 1

INDEX

Perspective affinity, 85
Perspective projection, 293, 308
Picture-plane coordinate system, 306
Plane of symmetry, 222
Polygonal symmetry, 246
Polyhedral symmetry, 252, 254
Product of transformations, 33, 34
Projection theorems of trigonometry, 273
Projective geometry, 96, 271:
    central projection, 97, 98, 293, 295, 298
    collineations of the plane, 274
    conics, 284
    cylindrical, 305
    display projection, 305
    equations of, 96, 276, 282, 283
    Felix Klein on, 276
    homogeneous coordinates, 279
    in space, 285
    maps, 300
    nonorthogonal, 289
    oblique, 291, 308
    orthogonal, of a line, 288
    orthographic, 307
    parallel, 94, 286, 290, 291
    perspective, 293, 308
    picture-plane cs 306
    points onto the line $y = x$, 273
    points onto the x axis, 272
    projective plane, 277, 281
    projectivities, 96
    stereographic, 302
    telescopic, 299
    vector, 274
Projective plane, 277, 283:
    special points on, 281
Projectivities, 96

Quaternions, 180
    modulus, 181
    multiplication, 180
    product of two, 180
    rotation in space, 180
    vectors, and, 180

Range, 26
Real numbers, the set of, 29
Reciprocal basis vectors, 124
Reciprocal radii, 310
Reflection, 192:
    fixing a line, 203, 204, 205, 211
    fixing an arbitrary plane, 213, 216
    glide reflection, 66
    in a line, 57, 61, 203
    in intersecting lines, 63
    in intersecting planes, 70, 208, 217
    in parallel planes, 212, 217
    in principle axes, 211
    in space, 211
    in the plane, 56, 62, 68, 203
    in the x axis, 67
    in the $z = 0$ plane, 68
    in three parallel lines, 208
    in two parallel lines, 207
    in two perpendicular planes, 70
    matrices, 203, 211, 218
    noncommutative products, 64
    oblique, 210
    orientation-reversing, 57, 68
    product of three or more, 65
    product of two, 62, 69
    summary of matrices, 218
Regular tilings, 247
Riemann, G. F. B., 8
Rigid-body motion, 144:
    composite, 184
    rotation, 151
    translation, 144
Roll, pitch, and yaw, 167
Rotation(s), 151:
    about an arbitrary axis, 170

about an arbitrary point, 157
about principal axes, 160
about the origin, 152
angles about principle axes, 161
coordinate system, 160, 165
eigenvectors and, 177
equivalent, 172, 177
Euler angle, 167
groups, 178
line segment, 156
product of, 65
quaternions, 180
roll, pitch, and yaw, 168
successive, 155, 158, 162

Scalar product, 108
Schläfli symbol, 247
Screw transformation, 187
Shear, 87, 267:
    equiareal transformations, 88, 270
    fixing the x axis, 87, 268
    in space, 269
    rotation as the product of, 268
Sierpinski, Waclaw, 2
Similarity(ies), 75:
    dilation and isometry, 76
    group of, 81
    similar triangles, 75
    transformation, 133
Solids, parametric, 321
Space, 11
Strain, 88
Surfaces, parametric, 319
Sweeps, 187, 325
Sweep with deformation, 325
Symmetry, 192, 220:
    analysis of, 222
    center(s) of, 202, 222
    crystallographic restriction, 242
    curve, of a, 224
    cyclic groups, 227, 228, 232
    dihedral groups, 228, 230
    equilateral triangle, of a, 220
    finite groups, 232
    frieze groups, 226, 235
    groups, 225, 253
    lattice, 237
    line(s) of, 210, 222
    ornamental groups, 234
    planes of, 222
    plane lattices, 237, 243
    polygonal, 246
    polyhedral, 252, 254
    reflection and halfturn, 233
    regular tilings, 247
    rotational, 253
    square, of a, 228
    symmetrical figures, 223
    tiling, 246
    wallpaper groups, 236

Tapering, 326
Tensors, 136:
    contravariant vectors, 137
    covariant vectors, 137
    metric tensor, 141, 142
    notation, 140
    orthonormal cs, 139
Tetrahedral symmetry, 233
Tiling, 246:
    congruent, 248
    duals, 248
    regular, 247
    semiregular, 249, 251
    three regular tilings theory, 248
    vertex varieties, 249
Topological transformations, 99
Topology, 22:
    connectivity and order, 22, 23
Transformations, 20, 26, 256:
    affine, 82, 93
    angle-preserving, 53
    area-preserving, 301

central inversion, 194
collineation, 52
composite motion, 144, 184, 187
deformation, 323
dilation, 256
distance-preserving, 53
equiareal, 89, 270
glided reflection, 66
halfturn, 58
homogeneous coordinates, 279
homogeneous trans matrix, 185
identity, 41
inversion, 311
isometries, 54
kinematics, 189
linear, 44, 256
line-preserving, 52
nonlinear, 2, 309
orientation-preserving, 54
orientation-reversing, 55
parametric variable, 316, 322
product of, 33, 34
projection, central, 97, 293
projective, 96
reflections, 192
rotation, 151
screw, 187
shear, 267
similarities, 75
strain, 88
sweep, 187, 325
symmetric, 192
tapering, 326
theory of, 26
topological, 99
translation, 144
twisting, 188, 327
Translation, 144:
    coordinate system, 148
    group, 150
    invariants under, 150
    line, of a, 146
    succession of, 148
    two point determination of, 145
    unequal over two points, 148
    vectors, and, 146
Twist transformation, 188, 327

Unit cells, 243
Unit vector, 107

Vector spaces, 101:
    linear, 101
    linear dependence,
Vector(s), 105:
    angle between two, 108
    basis, 114
    components, 106
    contravariant, 138, 140, 141, 144
    covariant, 137-140
    direction cosines, 107
    geometry, 109, 111
    linear combinations, 103
    linear dependence, 104
    magnitude, 107
    objectivity of, 125
    parallelogram law, 106
    scalar product, 108
    unit, 107
    vector product, 109
    vector projection, 274
    vector properties, 110
Visualization, limits of, 4

Wallpaper groups, 236
What is geometry?, 1

DATE DUE

SCI QA 601 .M85 2007

Mortenson, Michael E., 1939-

Geometric transformations for 3D modeling